FARMERS "MAKING GOOD"

PARKS AND HERITAGE SERIES

ISSN 1494-0426

The Parks and Heritage series focuses on topics related to national parks and historic sites in North America. Both historical and contemporary, these books raise our awareness about the many facets of national parks, including the warden service, religion, ethnohistory, and environmental studies.

No. 1 · Where the Mountains Meet the Prairies: A History of Waterton Country Graham A. MacDonald

No. 2 · Guardians of the Wild: A History of the Warden Service of Canada's National Parks Robert J. Burns with Mike Schintz

No. 3 · The Road to the Rapids: Nineteenth-Century Church and Society at St. Andrew's Parish, Red River Robert J. Coutts

No. 4 · Chilkoot: An Adventure in Ecotourism Allan Ingelson, Michael Mahony, and Robert Scace · Copublished with University of Alaska Press

No. 5 · Muskox Land: Ellesmere Island in the Age of Contact Lyle Dick

No. 6 · Wolf Mountains: A History of Wolves along the Great Divide Karen R. Jones

No. 7 · Protected Areas and the Regional Planning Imperative in North America Edited by Gordon Nelson, J.C. Day, Lucy M. Sportza, James Loucky, and Carlos Vasquez · Copublished with Michigan State University Press

No. 8 · The Bar U and Canadian Ranching History · Simon Evans

No. 9 · Heritage Covenants and Preservation: The Calgary Civic Trust Edited by Michael McMordie, Frits Pannekoek, E. Anne English, Kimberly E. Haskell, Sally Jennings

No. 10 · The Lens of Time: A Repeat Photography of Landscape Change in the Canadian Rockies Cliff White and E.J. (Ted) Hart

No. 11 · Farmers "Making Good": The Development of Abernethy District, Saskatchewan, 1880–1920, Second Edition Lyle Dick

Lyle Dick

FARMERS "MAKING GOOD"

The Development of Abernethy District,
Saskatchewan, 1880–1920

Second Edition

UNIVERSITY OF
CALGARY
PRESS

© 2008 Lyle Dick

University of Calgary Press
2500 University Drive NW
Calgary, Alberta
Canada T2N 1N4
www.uofcpress.com

First edition published by Canadian Parks Service, Environment Canada, in 1989.

No part of this publication may be reproduced, stored in a retrieval system or transmitted, in any form or by any means, without the prior written consent of the publisher or a license from The Canadian Copyright Licensing Agency (Access Copyright). For an Access Copyright license, visit www.accesscopyright.ca or call toll free 1-800-893-5777.

LIBRARY AND ARCHIVES CANADA CATALOGUING IN PUBLICATION

Dick, Lyle
    Farmers "making good": the development of Abernethy District, Saskatchewan, 1880–1920 / Lyle Dick. — 2nd ed.

(Parks and heritage series, ISSN 1494-0426 ; 11)
Includes bibliographical references and index.
ISBN 978-1-55238-241-7

    1. Abernethy Region (Sask.)—Colonization. 2. Agricultural colonies—Saskatchewan—Abernethy Region. 3. Frontier and pioneer life—Saskatchewan—Abernethy Region. I. Title. II. Series.

FC3545.A26D43 2008   971.24'4   C2008-901322-0

The University of Calgary Press acknowledges the support of the Alberta Foundation for the Arts for our publications. We acknowledge the financial support of the Government of Canada through the Book Publishing Industry Development Program (BPIDP) for our publishing activities. We acknowledge the financial support of the Canada Council for the Arts for our publishing program.

This book has been published with the aid of a grant from Parks Canada.

Cover design, page design and typesetting by Melina Cusano

Front cover: "With the look of a winning team, a threshing crew took a break to pose in front of their harvesting marchinery." This scene was photographed in the Lorlie District, Saskatchewan, only about 10-12 miles from Abernethy, in 1911. Saskatchewan Archives Board, Regina, Photo no. R-B 8853-3.

To RTF

CONTENTS

Acknowledgments	IX
Preface	XIII
Abbreviations	XVI
Introduction to the Second Edition	XVII
1. The Settlement of the Abernethy District	1
2. Estimates of Homesteading Costs in the Abernethy District in the Settlement Era	47
3. Economic Development of the Abernethy District, 1880–1920	67
4. Work and Daily Life at the Motherwell Farm	101
5. Abernethy's Social and Economic Structure	133
6. Social Relationships of the Settlement Era	147
7. Abernethy's Social Creed	171
8. Agrarian Unrest in the Central Qu'Appelle Region	191
9. Conclusions	223
Appendix A. Research Design for the Quantitative Analysis of Abernethy Settlement History, by David Greenwood	233
Notes	237
Bibliography	265
Index	285

ACKNOWLEDGMENTS

I am indebted to many persons for their assistance in the research and preparation of this book. Special thanks go to David Greenwood, who provided the research design for the statistical analysis of land acquisition in Chapter 1, and to Professor Douglas Sprague, who provided guidance in the development of quantitative methodology. Recording of quantitative data from the Homestead Files and associated Fortran coding for statistical tabulations and analysis was expertly performed by Sarah Carter. The Fortran data coding was performed by Sarah Carter. I would also like to thank Michael Camp of the Statistical Analysis System (S.A.S.) Institute at Raleigh, N.C., for permission use the S.A.S. computer package for this project.

Lloyd Rodwell of the Saskatchewan Archives Board in Saskatoon was extremely helpful in devising a scheme to permit the pulling of more than 800 homestead files and in making many suggestions about source material. Archival assistance was also provided by Doug Bocking and D'Arcy Hande in Saskatoon; Trevor Powell, Ruth Dyck Wilson, and Ed Morgan in Regina; Elizabeth Blight and Barry Hyman at the Provincial Archives of Manitoba; and Brian Corbett and Terry Cook at the National Archives of Canada. Many of the non-archival primary materials used in this study were obtained at the Manitoba Legislative Library.

Outside commentary was generously provided by Professors James Richtik, Gerald Friesen, Robert Ankli, Michael Percy and Judy Wiesinger. Professor Irene Spry provided a critique of an earlier version of the second chapter of this study published in the Fall 1981 issue of *Prairie Forum*; Spry's critique appears in the Spring 1982 issue. Illustrations, maps and graphs were designed and drafted by Brent Richard, Darlene Stewart, Kathy Lausman and myself.

I would also like to acknowledge the contribution of Dr. Frits Pannekoek, former Chief of Historical Research, Prairie Region, Canadian Parks Service, who supervised this study in its formative stages.

Finally, I would like to thank the many members of the Motherwell family and residents of the Abernethy district who have contributed such a great deal of information on its early history: Laura Jensen, Pat Motherwell, Laura Murray, Gertrude Barnsley, the late Jack Bittner, Marie Bittner, the late Walter Brock, Allan Burton, H.C. Burton,

Howard Dinnin, the late Dan and Olive Gallant, Bert Garratt, the late Nina Gow, Dick and Elizabeth Large, Margretta Evens Lindsay, the late Alma Mackenzie, Major McFadyen, Elizabeth Morris, Annie Morrison, George Morrison, Ben Noble, Mr. and Mrs. Rich Penny, Harry Smith, the late Ralph Stueck, Nelson Stueck, Mr. and Mrs. Hugh Stueck, and Louis Wendell.

Acknowledgments, Second Edition

I am indebted to Alice Gavin for her superb redesign of the maps and other graphics for this edition. Special thanks also go to Ron Frohwerk for encouraging me to pursue the republication of this book. He provided invaluable feedback during the process of revisions and also expert advice on visual presentation. My continued research in still images of the settlement era and the social creed of the early settlers was ably assisted by Trina Gillis, Trevor Powell, Catherine Holmes, Paula Rein, and Tim Novak at the Saskatchewan Archives Board in Regina, and Nadine Charabin at the Saskatchewan Archives Board in Saskatoon. Over the intervening years, I also benefited from various discussions on rural and prairie settlement history with Sarah Carter, Cecilia Danysk, Betsy Jameson, Valerie Korinek, Royden Loewen, Marvin McInnis, Ian MacPherson, Ruth Sandwell, the late Irene Spry, and John Thompson. Thanks as always go to the staff of the University of Calgary Press for their high level of professionalism throughout, especially the editors Peter Enman and John King. Also appreciated were the comments of two anonymous assessors, who made a number of useful suggestions regarding the more recent historiography of the rural prairies.

"Just a word of how we have made good.... I always kept one Ideal in view, and have worked steadily to it from a little boy of 13 without a cent in Toronto. I have steadily climbed till today 40 years after I have $10,000 of stock and 9 quarter-sections of land"

– John Teece

Source: Saskatchewan Archives Board Regina, Department of Agriculture Statistics Branch. File re: General Publicity, 1914. Letter of John Teece to J. Cromie, 27 December 1913.

Make good b. To prove to be capable or efficient; also to succeed; to justify by successes a course of action or expectation.

Source: *Webster's New International Dictionary of the English Language.* 2nd Edition. G. & C. Merriam, publishers, Northampton, Mass., 1950, p. 1485.

PREFACE

A neighbour of W.R. Motherwell, John Teece was the largest landowner in Saskatchewan's Abernethy district before the First World War. He viewed his success as the product of his honesty, plain living, astute farm management, and initiative. In Teece's words, "what I have done any hardworking man can do if he means to." Teece's Horatio Alger recipe for success has continued in popular perceptions of Saskatchewan rural life. A biography of former Saskatchewan premier James Gardiner, who farmed six miles from Abernethy at Lemberg, Saskatchewan, is entitled *None of It Came Easy: The Story of James Garfield Gardiner*.[1] The local history of a nearby community bears the alliterative title *Grit and Growth: The Story of Grenfell*.[2] Another biography of a Saskatchewan farm leader is called *Stout Hearts Stand Tall*.[3] A constantly recurring theme in the folklore of Anglo-Canadian settlement on the prairies is the early homesteaders' triumph through adversity, of "making good" against all odds.

Do these accounts present a balanced picture of the nature of early economic development at Abernethy and other Anglo-Canadian prairie communities? History tends to be written of and by the successful; they are the ones who have lasted long enough and acquired the resources to write about their own experience. As well, the documents historians use may reflect a very biased account of the past, and so it is with the documentary record of Abernethy history. Personal testimonials and reminiscences are often suffused with emotion or subjective perception. Just as the early settlers believed they had to make good the expectations they held for themselves, so current popular historians often feel obliged to justify the actions of their ancestors.

In the regard they follow an alternative definition of "making good," that is, "to justify by successes a course of action or expectation." The concept that success is its own vindication is reminiscent of the Calvinist paradigm in which rewards are naturally granted to the Elect, whose thrift, hard work, intelligence, and other qualities justify their election. As is discussed in Chapter 7, Calvinistic concepts loomed large in Abernethy's early history, particularly in the pivotal role assumed by the pillars of the Presbyterian Church, especially W.R. and Catherine Motherwell, in community affairs.

While not disputing that these qualities may have played a role in the fulfillment of a settler's promise, it is pertinent to ask if other factors were at play. Land is a finite resource and speculation in it is endemic in pioneer economies. It would be useful to determine the role of timing of settlement in a farmer's eventual success or failure. Did early arrival give some settlers a discernible advantage in terms of land selection? Similarly, did the prevailing free homestead system grant all settlers equal access to cheap land? Were there important differences in terms of access to political power between different settlement groups? We clearly need a more comprehensive look at the questions of who were the winners and losers in the prairies' early settlement era, and why.

This study of Abernethy's early social and economic history is one of a series written in support of the interpretive and restoration program at the W.R. Motherwell Homestead National Historic Site of Canada near Abernethy, Saskatchewan. The site was first recognized in 1966 when the Historic Sites and Monuments Board of Canada recommended that W.R. Motherwell be designated a person of national historic significance and that his farmstead should be preserved. This recommendation was subsequently approved by the Minister responsible for Parks Canada on behalf of the federal government. As the co-founder of the Territorial Grain Growers' Association at Indian Head in 1901, Motherwell had been identified as a prominent figure in prairie farmers' struggle to promote their economic interests. His subsequent achievements as Saskatchewan's first Minister of Agriculture in the province's formative period and as federal Minister of Agriculture after 1922 were also recognized at that time.

In the course of initial planning for the Motherwell Homestead National Historic Site of Canada, four historical themes were selected for the interpretive program: W.R. Motherwell, the Ontarian settler; Motherwell and the development of scientific agriculture in western Canada; his role in the agrarian unrest; and his political career. It was considered that his political career was already well served by two M.A. theses.[4] To address the theme of scientific agriculture, the Historical Research Section with the former Prairie and Northern Region of Parks Canada commissioned a survey of the scientific agriculture movement by David Spector entitled *Agriculture on the Prairies*, which was subsequently published in Canadian Parks Service's *History and Archaeology* series.[5] The present study has been written to provide a more comprehensive basis for interpreting the two remaining themes, of Motherwell as a representative of Anglo-Canadian settlement on the prairies, and his role as an agrarian activist.

Inasmuch as political activity is inextricably connected to socio-economic forces, the approach here has been to trace early Abernethy settlement, development of the agricultural economy, and community social structures. The study of socio-economic structures provides an interpretive framework but does not address the important questions of how people behave, interact, and conduct their daily affairs. Therefore this study also addresses the issues of social creed, relationships, and work processes in defining the character of Abernethy society before the First World War. While not claiming Abernethy to be representative of all Anglo-Canadian settlements in the west, this history endeavours to identify local phenomena with potentially broader implications.

If Abernethy's history is a story of making good, it is paradoxical that this district was a centre of the early farmers' agitation around 1900. The existing literature of the agrarian unrest leaves unanswered many questions that must be resolved if we are to view this phenomenon from a modern perspective. What socio-economic groups sponsored the agitation? Were they representative of the overall rural population on the prairies? In terms of individual contributions, was W.R. Motherwell's role the pivotal one or might it rather be viewed as a catalyst in a much broader movement? In the larger scheme of things, did the founding of the Territorial Grain Growers' Association really herald the beginning of the farmers' era in Canadian politics, as has been so often been asserted?

ABBREVIATIONS

DLB	Dominion Lands Branch
LAC	Library and Archives Canada
AM	Archives of Manitoba
SABR	Saskatchewan Archives Board, Regina
SABS	Saskatchewan Archives Board, Saskatoon
SC	Statutes of Canada, Ottawa
WNSC	Western and Northern Service Centre, Parks Canada

Introduction
to the Second Edition

First published in 1989, *Farmers "Making Good"* is a microhistorical study of the Abernethy District, a farming community in eastern Saskatchewan settled during the late nineteenth century. Specifically commissioned to provide a basis for interpreting the historical themes of the Motherwell Homestead National Historic Site of Canada, this book was also written to make a more general contribution to the historical understanding of prairie agricultural settlement during the region's formative era.

The settlement of the Canadian prairies is one of the great themes of Canadian history and has periodically attracted the attention of scholars across the country and internationally. Associated with such nation-building events as the acquisition by Canada of the former Hudson's Bay Company Territory of Rupert's Land, the building of the Canadian Pacific Railroad, the Métis resistances of 1870 and 1885, and massive immigration that developed the prairies into a major settled region, prairie settlement was integral to the forging of Canada as a transcontinental country. Throughout the period of 1870 to 1930, the primary focus of settlement policies was agricultural development. In 1931, at the end of the era, 1,468,147 of the 2,353,529 residents of the prairie region enumerated by the Census of Canada, or more than 62 per cent of the population, were classified as rural.[1] The overwhelming majority of these people were farmers, farm labourers, persons working in villages or towns providing services to the farming population, and members of their families. While the relative numbers and percentages of persons engaged in agriculture has dropped dramatically since the Second World War, a large proportion of the population of the prairies continues to be descended from families formerly or currently engaged in agriculture or related economic activities. Agricultural settlement, then, goes to the root of prairie identity.

Nevertheless, our understanding of the agricultural settlement process and its implications for people within and outside the region has changed substantially over the years and has been the subject of periodic debate. A survey of prairie historiography since the end of the settlement era around 1930 reveals a succession of changing concerns of historians in different historical eras. In the 1930s, the Canadian Frontiers of Settlement series was launched as a major interdisciplinary effort to place the society of this recently settled rural prairie region in historical perspective. Broadly intended to aid in the reconstruction of rural society ravaged by drought and depression, the books in this series addressed a range of issues in agricultural economics, sociology, and history and featured the work of some of Canada's leading academics of the era. In historiographical terms, it implicitly constituted a critique of the nation-building approach of such earlier series as The Makers of Canada or the books of Canada and its Provinces. At the same time, the Frontiers of Settlement volumes were marked by an underlying spirit of accommodation to the nation's political-economic structures. Its authors seem to have been principally concerned with studies oriented to facilitating adaptations to the dominant ideology and practice of rural modernization.[2]

The anomalous structures of the National Policy received a harsher, and perhaps more penetrating analysis in the books of the Social Credit in Canada series. Several classic texts, including works in sociology by Jean Burnet, social psychology by John A. Irving, political economy by C.B. Macpherson, and economic history by V.C. Fowke, among others, explored the often destructive impacts of the existing political economic system on the prairies' rural populations. In focussing on general or national issues, the Social Credit books sometimes presented a monolithic picture of the region's rural population that suppressed local differences in its composition. For example, types of farming other than wheat monoculture were largely ignored, and while C.B. Macpherson identified the critical role of social class, he largely avoided such subtleties as socio-economic differences within the class of farm proprietors, as well as ethnocultural variations. While such differences were brought out to a degree in Jean Burnet's *Next Year Country*, her judgment of Hanna, Alberta in the 1940s as a failed experiment in settlement reinforced a prevailing stereotype of prairie farmers as failures.[3]

The Social Credit series did represent a significant advance in both scholarship and the conceptualization of prairie rural history as a subfield of study. Yet the interdisciplinary insights of both the Frontiers of Settlement and Social Credit in Canada series had little impact on the post–Second World War historiography of the rural prairies. For a number of reasons, perhaps including complacency occasioned by thirty years of relative farming prosperity after 1945, agricultural history reverted to organicist, often romanticized presentations of rural life. In both amateur and professional writing, rural history was typically presented from the victorious perspective of the dominant Anglo-Canadian settlement group.

Elsewhere, however, the field of agricultural history was taking a different turn. By the 1960s and 1970s, the American agricultural historians Allan G. Bogue and Robert Swierenga were calling for a 'new rural history.' In his seminal article "Social Theory and the Pioneer," Bogue advocated a wide-sweeping application of models from diverse fields such as sociology and social psychology to the historical study of rural areas. Swierenga, who looked to the French Annales scholar Marc Bloch and the American historical geographer James C. Malin for his models, promoted a shift to detailed interdisciplinary studies to establish a comprehensive basis for challenging and revising long-standing generalizations about the rural past. By this time, other models relevant to rural historiography included: Merle Curti's early intensive study of a Wisconsin county published in the 1950s; Robert Swierenga's work on the new rural history; studies by John Gjerde and Robert Ostergren on cultural transfer at transplanted Scandinavian communities in Minnesota; and Michael Conzen's historical geographical study of a Wisconsin community, which provided the Abernethy study with a strong methodology that could be adapted to the statistical study of land selection in other settlement communities.[4] The remarkable body of scholarly work on agriculture in the American grassland regions carried out in the decades following the Second World War was summarized by Bogue in a bibliographic essay published in 1981. In it, Bogue referenced detailed studies on agricultural history dealing with the region's natural environments, demographics, ethnicity, land tenure, mobility, agricultural organization, economics, technology, and farming techniques.[5]

By the 1980s, in addition to the impressive body of work in the United States, other models for the study of agricultural history were available from the great French *Annales* tradition of social and economic history. They included: Marc Bloch's *French Rural History*; Emmanuel Le Roy Ladurie's *Montaillou*; and especially Fernand

Braudel's *The Mediterranean and the Mediterranean World in the Age of Philip II*.[6] These monographs explored the interrelationships between politics, economics, society, culture, world views, and the natural environment in rural societies, underscoring both their complexity and the potential for integrating these diverse dimensions of the human condition into historical work. While no Canadian social or environmental historical studies of comparable depth had yet been produced for the prairies, studies of continuing relevance to prairie rural history included the classic works in economic history and political economy produced by Vernon Fowke and C.B. Macpherson, as well as an insightful ethnographic study of ranchers and farmers in the Maple Creek area of Saskatchewan, published by anthropologist John W. Bennett in 1969.[7]

In western Canada, a number of prairie versions of the "new rural history" appeared by the late 1980s and early 1990s. Prepared before the advent of microhistory as a sub-discipline, these books could nevertheless be placed under the rubric of microhistorical studies, as they were animated by concerns similar to those expressed in the writings of Carlo Ginzburg and Carlo Levi, the leading exponents of this method of historical research and representation.[8] Entailing the intensive study of small units of history, microhistory seeks insights in local and personal experience that would not be apparent to practitioners applying a broader brush to the study of social or cultural history.[9] Microhistorians have also been inclined to be skeptical of overarching historical syntheses as potentially oversimplified or reductionist.[10] Alternatively, practitioners of this sub-field have argued that a microcosm, or a small component of a larger whole, can carry within it the potential to illuminate the macrocosm if its examples can be shown to be representative of more general historical experience. For western Canadian settlement history, works of this period displaying the characteristics of microhistories included: Paul Voisey's *Vulcan*; Lyle Dick's *Farmers "Making Good"*; Royden Loewen's *Family, Church, and Market*; and Ken Sylvester's *The Limits of Rural Capitalism: Family, Culture, and Markets in Montcalm, Manitoba, 1870–1940*.[11] Among other aspects, these studies employed systematic research methods, especially quantification, and focussed on detailed treatments of such topics as the impact of land policies and markets on prairie agriculture, the economic development of local communities, and the study of social relationships, family, and kinship ties in newly settled rural areas of the prairie provinces.

As mentioned, a specific objective of *Farmers "Making Good"* was to provide an enhanced understanding of the society formed by Anglo-Canadian newcomers to the prairie region during the formative early years of the National Policy. By the term "National Policy," I am not referring specifically to the election platform of John A. Macdonald's Liberal Conservative party in 1878 but rather to the collectivity of federal policies pursued by various federal governments in the period 1867 to 1896 to encourage immigration and the settlement of Canada, especially in the prairie provinces. For this historical study, the research methodology was deliberately systematic and scientific. Following initial archival and library research, a report on historiography was prepared to take into account then-current trends in North American rural history and to identify major unresolved issues in the western settlement experience.[12] Using historiographical issues as a guide, initial heuristic frameworks were developed alongside methodologies for testing hypotheses. These methodologies were then applied to the collection, recording, and analysis of data, entailing the assemblage of more than 800 homestead files from two districts of southeastern Saskatchewan, followed by statistical methods to chart the process of land disposal in these areas. After first identifying the settlers' rationale for selecting particular tracts of land and tracking the initial disposal, the study then turned to tabulation and analysis of homestead cancellations and land turnover rates by decade after entry in order to chart both short-term and long-term trends of persistence for settlers in two adjacent farm communities settled in different decades in the late nineteenth century.

Beyond the *Annales* school, the quantitative methodologies of *Farmers "Making Good"* reflected the influence of quantitative history or cliometrics, a subfield of history oriented to applying economic theory and econometric models to the study of the past, which reached its zenith in Canada in the 1980s. At that time, several talented practitioners, such as Kenneth Norrie, Michael Percy, and Frank Lewis were engaged in applying statistical methods to help explain the prairie wheat boom and the extent of its impact on the growth of the Canadian economy.[13] While the interpretive preoccupations of the cliometric specialists largely fell outside the scope of enquiry for the Abernethy study, their methodologies offered fruitful examples of how quantitative approaches might be applied to the study of aspects of prairie settlement history. To the degree possible, I also followed the advice of John Herd Thompson, a long-term colleague in Canadian rural history, who advised that whenever social historians can count, they should count. In other words, whenever available data lends itself to quantification, historians should take advantage of such opportunities to try to establish the absolute and relative numbers pertinent to the reconstruction of patterns of social history. Therefore, for *Farmers "Making Good,"* my colleague Sarah Carter and I compiled quantitative data on the Abernethy community's population in a wide range of environmental and social categories, including the soil quality for their homesteads, the numbers of homesteaders settling on lands of varying quality; demographic data, including the numbers and proportions of settlers who were married or single in the year of their homestead entries, the number of children per family, the age profile of this population, the size and form of the settlers' initial dwellings, the relative numbers belonging to different religious denominations; economic indicators such as breakdowns by size of farm, accumulated capital over time, farm income for different groups of farmers, economic differences between districts; and other measurable indices. Such data and distributions enabled the development of fairly precise generalizations regarding several key questions of prairie settlement historiography.

As every practising historian knows, not all aspects of past life can be quantified, as much of the surviving data is incomplete or does not lend itself to counting. Therefore, the Abernethy study also entailed the collection of a variety of qualitative data on work and daily life, social structure, social relations, and the creed of the community, aimed at addressing a series of historiographical questions: to what degree did the settlement societies of the prairies reproduce the cultural inheritance brought by the settlers from their regions of origin to western Canada? Alternatively, to what extent was the natural environment of the prairies a factor in shaping the emerging settlement communities in the west? Were prairie settlement societies largely the product of market forces, or, alternatively, did non-market inputs, such as family and kinship relationships, play a significant role in the development of these communities? What was the early demographic and social composition of rural communities in the settlement era, and how did the character of these populations change over time? What does the compilation and analysis of data on land disposal and ownership over time reveal about the persistence and/or success of farmers and their families? Specifically, what kinds of settlers or social groups persisted in rural areas and what were the characteristics of individuals who left these communities or moved on to other pursuits? Was the settlement process selective, and if so, whom did it select over the long term and why? Further, regarding farmers who persisted on their farms for extended periods, the book posed the basic question: were they persisting in prosperity or poverty?

For all homesteaders in three townships in the Abernethy District, largely comprising Anglo-Canadian and British settlers arriving in the 1880s, the book traced the entire settlement process, from initial land selection, through assemblage of additional farm lands, development of the local agricultural economy, changes in the settlers' material culture and the emerging farm landscape, development of social structures, social relationships, and associated world views. For comparative purposes, all homestead files for three townships in the nearby Neudorf District, which received its German-speaking settlers from continental Europe in the 1890s, were also examined. Further, the Abernethy study correlated the settlers' political activity in the agrarian movement to the emerging socio-economic differences in rural Saskatchewan society. It thereby took an integrated approach to the study of settlement through tracing the historical interplay of economy, society, and politics in a single farm community, which happened to be the centre of the early agrarian movement on the prairies in the early twentieth century.

Pursuing an intensive analysis entailing the empirical testing of various hypotheses enabled *Farmers "Making Good"* to challenge several prevailing notions in prairie historiography. One popular notion critiqued in the book was the myth of the self-made man, that settlers were the makers of their own destiny and that their successes derived primarily from hard work, perseverance, and personal ingenuity. Without discounting the importance of such personal attributes, *Farmers "Making Good"* asked if other factors were at play and concluded that certain early-arriving settlers enjoyed advantages in land disposal that were not available to subsequent settlement groups. By arriving first and claiming the best available lands for agricultural production, these settlers were positioned to compete far more effectively than subsequent settlement groups in the emerging agricultural economy of the prairies.

That these early arrivals continued to enjoy the fruits of such advantages long after the homesteading era was demonstrated in the book's compilation of statistics on persistence rates for two different groups of settlers – the Anglo-Canadian settlers who claimed the best lands by virtue of arriving at the optimal time in the 1880s, and the German-speaking settlers at nearby Neudorf, who settled in the 1890s on secondary land and in a period in which settlers qualified for fewer tracts of free-grant land. The fact that such differences were reflected in reduced economic opportunities for these subsequent arrivals over the long term led to the conclusion that, in addition to the personal attributes required for success in the emerging prairie economy, luck also played a significant role. In this case, the lucky settlers were the settlers who arrived early, claimed the best lands, and qualified for additional quarter-sections of free-grant land, such as pre-emptions and second homesteads. They were well positioned to reap significant capital gains as the price of land increased. As David Jones has elaborated eloquently in his now re-issued classic *Empire of Dust*, luck, or the lack of it, also played an important role in terms of environmental impacts on prairie settlement history. Jones showed how homesteaders who had been encouraged to settle in the dry belt of the Palliser Triangle in southwestern Saskatchewan and southeastern Alberta saw their hopes and dreams literally dry up as drought descended on these areas within only a few years after their arrival.

Part of the received wisdom of prairie history has been the notion that the shared experience of settlement gave rise to an egalitarian cooperative society, expressed in the rise of the Co-operative Commonwealth Federation to power in Saskatchewan in 1944. Left unanswered by this myth was the question as to why sharp political divisions nevertheless developed between right and left, Protestants and Catholics, and established settlers and new arrivals, throughout Saskatchewan's first half-century, and found fertile ground in the same prairie soil that nurtured the rise of the C.C.F. In this regard, while drawing on Macpherson's insights, *Farmers "Making Good"* refined his analysis of apparent contradictions in the farmers' movement by showing that the varying political positions in the farmers' movement reflected different political orientations connected to socio-economic differences between both individuals and settlement groups.

The activists associated with the Territorial Grain Growers' Association, the prairies' first successful agrarian organization, were largely drawn from the ranks of the more prosperous settlers. Positioned by the size of their holdings and extent of grain production to compete more effectively than smaller producers, this group was content to focus on legislative amendments to the Manitoba Grain Act involving a reapportionment of grain cars from elevators to individual producers. Successful in achieving these limited objectives, this group turned away from activism after this attainment. As was noted in this book, the early grain handling legislation and its application had the perhaps unintended effect of reducing the number of cars available to recent settlers or smaller producers, who were unable to produce enough grain to fill a grain car and therefore obliged to sell "on the street." In the process, the price of street grain was effectively lowered, reducing the poorer farmers' proportionate share of overall receipts from grain sales and contributing further to inequalities among the rural farming populations of the prairies.

Following the publication of *Farmers "Making Good,"* the recent discovery of the minute book of the Abernethy chapter of the Territorial Grain Growers' Association reinforced this book's arguments, based on more general sources, that the agrarian unrest reflected socio-economic differences within the farmers' movement. This local minute book shows that the officers and prime movers of key motions in the local association, several of whom were also active in the larger umbrella association, were among Abernethy's most prominent and prosperous farmers.[14] By contrast, E.A. Partridge and his brother Henry O. Partridge, both prominent in the more radical Grain Growers' Grain Company, employed only one male labourer each and farmed smaller holdings than their counterparts in the TGGA. Perhaps more telling than accumulated wealth or the employment of farm labour were the ideological assumptions embedded in the material culture of the TGGA farm leaders, especially the tendency of such individuals as W.R. Motherwell, Elmer Shaw, Peter Dayman, and W.H. Ismond, to build large masonry residences with a distinctly formal organization of space, suggesting shared hierarchical assumptions held by their respective owners.[15]

Farmers "Making Good" was one of a handful of detailed community studies that attempted to determine the extent to which long-standing assumptions derived from general works actually applied to local communities when placed under the historical microscope. These studies focussed on the experience of different ethnocultural settlement groups, whether Mennonites (Loewen), Franco-Manitobans (Sylvester), Anglo-Americans (Voisey), or Anglo-Canadians (Dick). Anglo-Canadians were also deliberately included within the classification of "ethnic groups" in Randy Widdis's book on Anglo-Canadian migration to the prairies and other regions of North America. As a dominant group – economically, socially, and politically – the Anglo-Canadians' distinctive cultural characteristics have often been invisible to scholars, perhaps owing to the tendency of members of this group, including its historians, not to view it as an ethnocultural community.[16] In an enduring cultural mindset, members of this group historically presumed to constitute themselves as the normative majority, set against the presumed 'ethnic' or 'racial' differences of all other cultural groups.

In one way or another, each of these settlement studies has been concerned with issues of culture and environment, although their authors have drawn differing conclusions regarding the relative importance of these factors in shaping the emerging prairie society. Studies such as Dick's and Loewen's have emphasized cultural transplantation, while Voisey stressed the role of the environment and the frontier experience in changing the culture of the newcomers in the west. While Voisey, Loewen, and Sylvester might disagree as to whether cultural values were transplanted or developed in the context of settlement, these studies appeared to share implicit assumptions regarding the nature of these communities, that they were cohesive and even organic collectivities. Whatever pressures were exerted by the larger economy or society of western Canada, these authors have generally concluded that the communities being studied retained their distinctive cultural identities throughout the settlement era and after.

Farmers 'Making Good' departed from other studies in that it identified major social and economic differences within prairie agricultural societies, as well as differences between communities. Unlike such works as Seymour Martin Lipset's *Agrarian Socialism* or John Bennett's *Northern Plainsmen*, *Farmers "Making Good"* approached rural prairie society, not as an organic entity, but rather, from the standpoint of classical political economy, as a house divided.[17] The book discerned that rural prairie society of the settlement era was divided into three socio-economic classes: an absentee business class in

Winnipeg and central Canadian metropolitan centres, an intermediate class of farm proprietors, and a proto-working class of farm labourers. The book also argued that economic imperatives were a major motivating force for most settlers, in the sense that the market-oriented society established by the National Policy generally demanded accommodations to commercial agriculture by most persisting settlers. In some respects *Farmers 'Making Good'* shared affinities with the concerns explored in Jeremy Adelman's comparative monograph *Frontier Development*, which similarly identified the ownership of land as a key determinant of economic success and persistence of settlers in Canada and Argentina.[18]

This was not to suggest that all settlers were oriented to surplus production or the profit motive, or in the same degree. As several historians of settlement have usefully pointed out, non-market imperatives, including establishment of a lifestyle for one's family or extended network, were also powerful motivating factors, at least for some individuals and settling groups. Nevertheless, the book argued that if we are to understand why inequalities characterized much of rural prairie life during and after the settlement era and since, the compilation and analysis of economic indicators will be essential to understanding cross-generational differences of opportunity and life chances for members of different rural communities. However localized, the socio-economic analysis of *Farmers 'Making Good'* established relevant socio-economic benchmarks for the settlement era that will need to be followed up by future work in this field if we are to better understand both the subsequent character of rural societies in the prairie provinces and the evolving social structure of the region as a whole.

Since the heyday of the new rural history, more recent settlement-era studies have turned their attention to other issues unanticipated in the earlier works. An aspect that has attracted considerable attention is the role of the international boundary as a factor in the historical evolution of communities in the west. The new borderlands studies have attempted to identify differences and similarities in settlement societies on the edges of the international boundary through the exploration of particular themes as they pertain to the experience of settlers both north and south of the forty-ninth parallel. These works include Beth LaDow's *The Medicine Line*, Sheila McManus's *The Line Which Separates*, and Stirling Evans' *The Borderlands of the American and*

Canadian Wests, a collection of essays on the borderlands of western North America.[19] In a related vein, a recent book by Warren Elofson on Alberta's ranching frontier has identified the important influence of immigrating American cowboys on ranching practice in that province in the late nineteenth and twentieth centuries.[20] Contrasting with their microhistorical counterparts, these studies treat larger geographical areas – for LaDow, the Saskatchewan – Montana borderlands, for McManus, the Alberta-Montana frontier, while still others cover the borderlands across the entire prairie region. Further, these works necessarily employ different methodologies in the collection, interpretation, and synthesis of historical data. They are crafted to address large-scale themes, including the interactions of culture and nature on both sides of the border, interconnections between gender, race and the international boundary, cross-border movements of First Nations, Métis groups, and itinerant labourers in the settlement era.

By increasing the scale of reference to a larger regional frame, and integrating an impressive range of thematic concerns, the borderlands studies have expanded our concepts of settlement history. For example, the role of women and categories of gender were examined in a promising new collection of essays assembled by Sarah Carter and other scholars.[21] For the Montana-Saskatchewan borderlands LaDow described the development of nurturing female networks, contrasting with a male ethos of individualism, which imposed constraints on men's interactions in keeping with prevailing notions of masculinity. She also well illustrated the negative consequences attending the unfettered pursuit of market-driven agendas of progress, which wrought havoc on the human and ecological balance in borderlands areas prior to the entrenchment of the international boundary, and its impacts on the structuring of social and economic relationships in the region. In her study, Sheila McManus convincingly argued that categories of race and gender constituted filters essential to entrenching Euro-Canadian and American notions of possessing and controlling the land.[22] These scholars have added significantly to our understanding of the complex process of ideologically constructing human relationships in the region, with long-term consequences for prairie residents belonging to various social and cultural categories, whether Canadian or American, Aboriginal or non-Aboriginal, male or female.

The larger regional studies also display certain limitations arising from extensive approaches driven by thematic concerns, as opposed to intensive documentation and synthesis undertaken "from the ground up." In casting such a wide net, the thematic works sometimes lack the comprehensive web of interconnections between individuals and their social environments that can generally only be reconstructed through microhistorical "thick description," that is, through the intensive study of a limited number of people interacting with one another in a specific place over a defined period.[23] As well, certain historiographical questions do not readily lend themselves to a regional level of analysis. Specifically, how did various settler elites establish their privileged status within their adopted communities? What were their relative financial and social assets of these actors vis-à-vis other settlement groups at the outset and how might these have changed over time? How did notions of gender and race feed into the socio-economic position, social relationships, and overarching worldviews of specific settlement groups? How did social class, ideology, and gender concepts relate to the settlers' political activities, whether formal or grassroots politics? As such intersections can generally be comprehensively addressed only through reconstructing a complex set of interrelationships, made visible through the intensive research of smaller groups in dynamic interaction, we are unavoidably drawn back to microhistory as a preferred method of testing the applicability of various interpretive rubrics to particular settlement communities.[24]

This is not to suggest that microhistorians have yet addressed all significant dimensions of the social history of the rural West. For example, to date prairie settlement historians have generally avoided the serious examination of human sexuality, choosing instead to subsume conjugal relationships under the safe rubric of the "family," while avoiding overt references to sexual relations of any type. The pervasive assumption of universality of the patriarchal nuclear family of two opposite-sex spouses with children in current settlement historiography has distorted our understanding of the actual diversity of both historical and current-day domestic relationships on the prairies.[25] Imbued with "heteronormative" assumptions, that is, operating from a presumption of normative heterosexuality, most settlement studies have resisted acknowledging even the possibility of alternative forms of sexuality, such as same-sex or bisexual relationships and/or identities, in the settlement era or since. A welcome exception to this tendency is Sarah Carter's trenchant critique of the federal administration of marriage and divorce among prairie First Nations, which showed that the federal government assiduously channelled conjugal relationships of Aboriginal peoples into Western models as part of Canada's imperial project of nation building in the West.[26] Generally, western Canadian settlement historiography continues to lag at least a generation behind the other social science disciplines in the study of sexuality on the prairies. Only recently have a small number of studies dealing with sexuality begun to appear in the region, and much remains to be done to lift the veil of silence which has heretofore shrouded this important aspect of human experience.[27]

Another aspect of settlement experience warranting greater attention in future studies is the natural environment, a major factor influencing human history and in the case of the prairie region, a factor establishing the physical limits to and opportunities for settlement, farming, and ranching in this region. From the era of Walter Prescott Webb in the 1930s and James Malin in the 1940s and 1950s, American settlement studies stressed the influence of the prairie environment and the importance of adaptation to it.[28] Recently, American anthropologists John Bennett and Seena Kohl applied an environmental thesis to western settlement in their book *Settling the Canadian-American West*.[29] This book had the merit of incorporating the voices of men and women whose families settled on the American Great Plains and the Canadian Prairies. A drawback is that the authors did not fully place these voices into contexts of time, place, or cultural or socio-economic group. Barry Potyondi's book *In Palliser's Triangle: Living in the Grasslands, 1850–1930*,[30] was another early effort to explain prairie settlement in terms of environmental factors. David Jones' treatment of the destructive impact of drought on nascent communities in the dry belt of southeastern Alberta remains an exemplary study of the interplay of culture and the environment in settlement history.[31] We still await a comprehensive cross-regional account of the interplay of humans and their environments in settling the prairies, especially one that might show both differences and similarities in cultural-environmental dynamics as played out in varying circumstances in the respective geographical areas of this vast region.

Beyond the natural environment, a further environmental dimension that is critical to understanding the settlement history of the prairie provinces is the cultural landscape and built environment. In this regard, published scholarship has also lagged behind other aspects of enquiry. At a highly theoretical level, Rod Bantjes's *Improved Earth* sought to identify the ideological underpinnings of the geometrical division of prairie space in the settlement and subsequent eras.[32] His conclusions regarding the cultural and economic impacts of the grid survey and the associated privatization of prairie space on the prairies' populations, were well founded, anticipated to a degree by my earlier discussion of some of these issues in a monograph on prairie settlement patterns in 1987.[33] As pointed out by Bantjes, the prairie settlement monograph, and other specialized studies of ethnocultural variations in settlement, the rectilinear grid was not universally adopted in the settlement period, and in various districts its hegemony was established only after one or more generations of resistance. Other explorations of alternatives to the standard grid and associated individual dispersed settlement have been carried out by historical geographers, including Carl Tracie's book on Doukhobor settlements, earlier works on Mormon and Ukrainian settlement by John Lehr, and John Warkentin's seminal dissertation on Mennonite settlement patterns.[34]

As discussed by Lyle Dick and Jean-Claude LeBoeuf, and elsewhere by Ian Clarke, even within the confines of the grid survey, farmers took radically different approaches to both the layouts of farmsteads and the spatial configurations of farmhouse interiors.[35] For example, the formal, class-oriented organization of the homes of prosperous farmers and hierarchical organization of their farmsteads stood in sharp contrast to the smaller, more informal vernacular approaches of poorer settlers and their families to their domestic and farmstead environments.[36] Sarah Carter's materials history of the W.R. Motherwell home suggested that even in established Anglo-Canadian communities, such rigid hierarchies in farmhouse layouts were beginning to break down by the First World War.[37] In North America, beginning with the work of the eminent material culture specialist Henry Glassie, a number of interesting studies addressing the form and significance of farm houses and other tangible artefacts of rural life have been published in both Canada and the United States.[38] In western Canada in the 1980s, other notable contributions to our understanding of rural material culture were published as monographs by the Historic Sites Service of Alberta Culture and the Historic Resources Branch of the Province of Manitoba.[39] To date, these detailed studies of folk or vernacular material culture have not yet been extensively integrated into the work of settlement historians. While material history studies have yet to gain wide acceptance in Canadian prairie settlement studies, they are suggestive of promising approaches that could usefully inform research in the years to come.

Future studies of the region's settlement history will need to address a wide range of questions, but will need to do so in an integrated way if we are to better understand prairie society as it was, is, or might be. I suggest that the answers will probably come through further microhistorical studies of a larger cross-section of rural communities, coupled with comparisons between and among these studies. For these investigations, quantification will play a necessary role, so that measurable findings might readily be compared to the results of similar investigations of communities in other areas. Microhistorians of settlement will need to become more historiographical. That is to say, in both research design and execution, they will need to respond more directly to the concerns, data, and interpretations presented in other microhistories of rural communities if they wish their studies to contribute useful building blocks for regional syntheses. Microhistories of rural societies will need to continue to be concerned with issues of land disposal, demography, economic change and socio-economic differentiation, as well as the roles of gender, sexuality, and the environment as they played out in the settlement and development of rural communities. Historians will also need to keep an eye on the subsequent character of these communities and how their diverging historical experiences contributed to the varied and uneven development of prairie societies as they exist today. A sound approach to the historical study of prairie agricultural communities would be to acknowledge that all major identified factors – political, institutional, economic, social, ideological, gender-related, and environmental – are relevant to the future practice of settlement history. Whatever rubric is pursued, future practitioners might well be obliged to demonstrate not only that a particular thesis is plausible but also the extent to which its supporting evidence and interpretive conclusions complement or refute alternative versions. Approached in this way, microhistorical studies could go far to further illuminate our understanding of the settlement process and by extension much of the character of Canadian prairie society, past, present, and future.

1

The Settlement of the Abernethy District

THE GEOGRAPHICAL SETTING

The Abernethy district is an area of gently undulating prairie to the east of the Qu'Appelle Lakes in southeastern Saskatchewan. The district is located on the second prairie level in the Aspen Parkland belt, also known as the boreal forest and grassland transition zone, which extends in an arc from the escarpment of the Manitoba Lowland in eastern and south-central Manitoba through central Saskatchewan to north-central Alberta, and south along the escarpment of the Missouri Coteau to the east of the Rocky Mountains. To the south, the park belt is bordered by the "Palliser Triangle," the semi-arid region of mixed low to medium prairie stretching from the southwestern corner of Manitoba, across southern Saskatchewan, and including much of southern Alberta. The parkland belt thereby forms a transition between the northern forests and the Great Plains regions and ecosystems. The natural vegetation of the parkland is predominately made up of fescue prairie (principally, rough fescue grass) with intermittent stands of trees that give this region a "park-like" appearance. This zone has a sub-humid climate, receiving an annual precipitation of between 35 and 50 centimetres, of which 23 to 35 centimetres fall during the five months of the growing season. Predominantly black soils in this region are generally high in organic matter and its areas of open prairie are generally suitable for grain cultivation. However, the proximity of districts such as Abernethy to the Palliser Triangle ensures that these areas are not beyond the reach of drought in a dry year. With elevations on the second prairie level higher than the first level to the east, the parkland belt in Saskatchewan also possesses a shorter growing season and is vulnerable to early frosts in the late summer. Further, the extreme variability of the climate, characterized by major oscillations in annual, seasonal, and daily temperatures, poses particular challenges for agriculture in this region.[1]

The district is currently defined by the boundaries of the Rural Municipality of Abernethy, an area of local jurisdiction extending approximately eighteen miles west from Lake Katepwa and twenty-two miles north from the Qu'Appelle River (Fig. 1). The lakes and river form a continuous waterway in the Qu'Appelle River Valley that extends from the lakes north of Regina to its confluence with the Assiniboine River east of the Manitoba boundary. In the Abernethy area the other major feature is a tributary of the Qu'Appelle, called Pheasant Creek, and its coulee, that meanders through the district in a generally northeast to southwest direction before joining the river. While the valley floor can be cultivated in certain places, its walls and the steep slopes of Pheasant Creek and other tributary ravines and spurs cannot be cultivated and are suitable only for grazing animals. The Pheasant Hills, an upland area of undulating terrain dotted with sloughs, prairie, and wood bluffs, border the district to the east. To the north lie the "File Hills," an area of alternating prairie and woodland possessing a number of small lakes (Fig. 2). These wooded lands are of somewhat limited productivity. In addition to the loss of much of the land to sloughs and trees, their clay-loam soils are generally of secondary quality and the undulating character of the land is sometimes an impediment to effective cultivation.

On the "Pheasant Plains" around Abernethy, however, are found high quality clay soils and some of the richest farmlands in the province. The predominant topography of gently undulating to undulating prairie in the Abernethy district provides a relatively level surface, suitable for croplands. Its undulations and a general gradual slope towards the major drainage basin of the Qu'Appelle River Valley allow for effective drainage of excess precipitation in the spring. As well, the predominantly heavy clay soils of this area are both rich in nutrients and effective in retaining soil moisture to foster the germination of grains and other plants. Further, the area north of the Qu'Appelle

River receives more rainfall than its counterpart districts just to the south of the valley. At the time the district was systematically settled in the 1880s, the lands around Abernethy held the potential to become among the richest farmlands in the District of Assiniboia. Not surprisingly, when southern Saskatchewan was first settled systematically a hundred years ago, the townships in this area were among the first to be claimed once opened to homestead entry.

When W.R. Motherwell came west to claim his homestead on the Pheasant Plains in 1882, he was participating in a great national experiment. Only twelve years earlier, the Canadian Confederation had assumed control of the Hudson's Bay Company's western territory of Rupert's Land. In the first decade after its acquisition, thousands of immigrants, principally from Ontario, settled on farms in Manitoba. Yet, as the Canadian economy entered a serious depression in the 1870s, large numbers of Ontarians passed over the western prairie opportunities and emigrated to such eastern states as New York, and the Great Plains states of the United States.[2] Notwithstanding preexisting Aboriginal populations and the first wave of central Canadian settlers, the vast expanse of prairie west of the Manitoba boundary remained sparsely populated. In 1878, John A. Macdonald's governing Conservative Party commenced its most concerted efforts to address the problem of settling the prairies. With the inauguration of the National Policy of immigration, settlement, tariff barriers, and railroad promotion, the Conservative government hoped to realize the dream of a settled West that had so far eluded them. The resultant survey and railroad construction paved the way for the systematic settlement of the North-West Territories.

THE FREE HOMESTEAD SYSTEM AND SURVEY

Since the Dominion government's land grant system provided the framework for the subdivision and disposal of most lands in the Abernethy area, this system was investigated in some detail. Essentially, the Dominion lands administration was the outgrowth of national policies formulated by Macdonald's Conservatives at the time of Confederation. These policies were rooted in several causes, but centred on their desire to maintain British North America as a political entity separate from the American republic to the south. After the American Civil War, United States expansionism, articulated under the banner of Manifest Destiny, seemed to threaten the territorial integrity of the Hudson's Bay Company's territory of Rupert's Land, particularly the southern prairie regions. Faced with serious economic problems arising from surplus population in rural areas of the United Province of Canada, Macdonald and his colleagues saw the potential to siphon this population to the western Canadian prairie. Moreover, the settlement of the prairies held the promise of creating an agricultural hinterland for eastern financial, manufacturing, and food processing industries, while promoting transportation development. As the clarion calls of American annexationists grew louder in the late 1860s, the need for rapid Canadian settlement became apparent. The original public lands policies were, therefore, drafted in a climate of urgency and haste.

The prevailing system of land subdivision on the prairies was established by order-in-council on 25 April 1871. By the *Dominion Lands Act* of 1872, the western Canadian prairie lands were to be divided into a grid of townships six miles square. Each township was to be composed of 36 sections of 640 acres, to be divided in turn into four quarter-sections of 160 acres. Surveying of townships extended from two base lines: the principal meridian, which ran through Fort Garry, and an east-west axis on the American-Canadian boundary at the forty-ninth parallel. Townships were numbered north from the international boundary, in ranges measured west from the principal meridian. Additional meridians were added as the survey moved westward. Land in the Abernethy study area, for example, lies west of the 2nd Meridian.

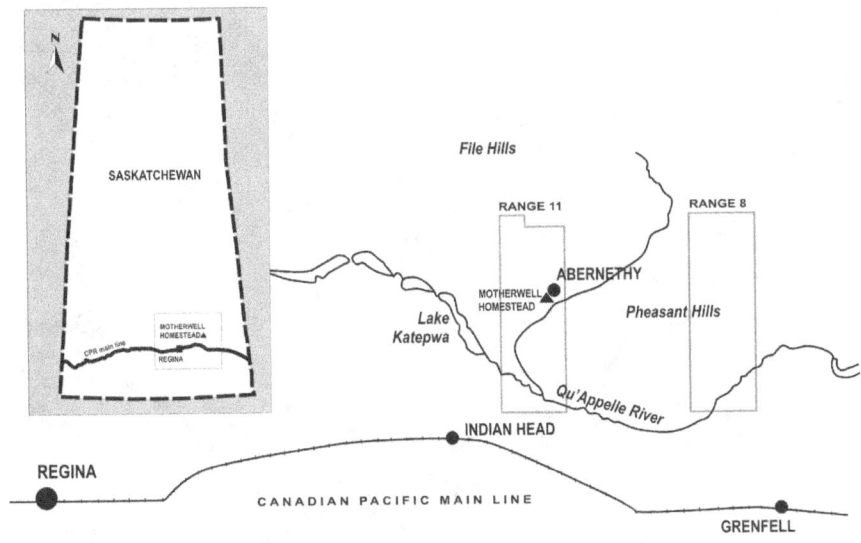

Fig. 1. Location of sample townships in southeastern Saskatchewan. The left block containing the village of Abernethy represents the three townships in Range 11 west of the 2nd Meridian selected for quantitative analysis of Anglo-Canadian settlement. The right block includes three townships of predominantly Eastern European German settlement in Range 8 around the town of Neudorf.

The townships in the vicinity of Abernethy were surveyed in 1881 and 1882. Dominion land surveyors, appointed in eastern Canada, brought their surveying parties west in the spring and worked until snowfall. At Abernethy the major survey party was led by C.F. Miles, whose diary remains the best source of information on the pre-settlement geographic character of the land. Dominion land surveyors were required to report on the vegetative cover, presence of marsh and water, and soil quality of individual quarter-sections rated on a scale of one to four. Miles reported that the townships now encompassed by the Rural Municipality of Abernethy consisted "of the finest land that has come under my observation."[3] Two years later another surveyor, J. Bourgeois, commented on Township 20, Range 11, West of the 2nd Meridian, in which the Motherwell homestead is situated:

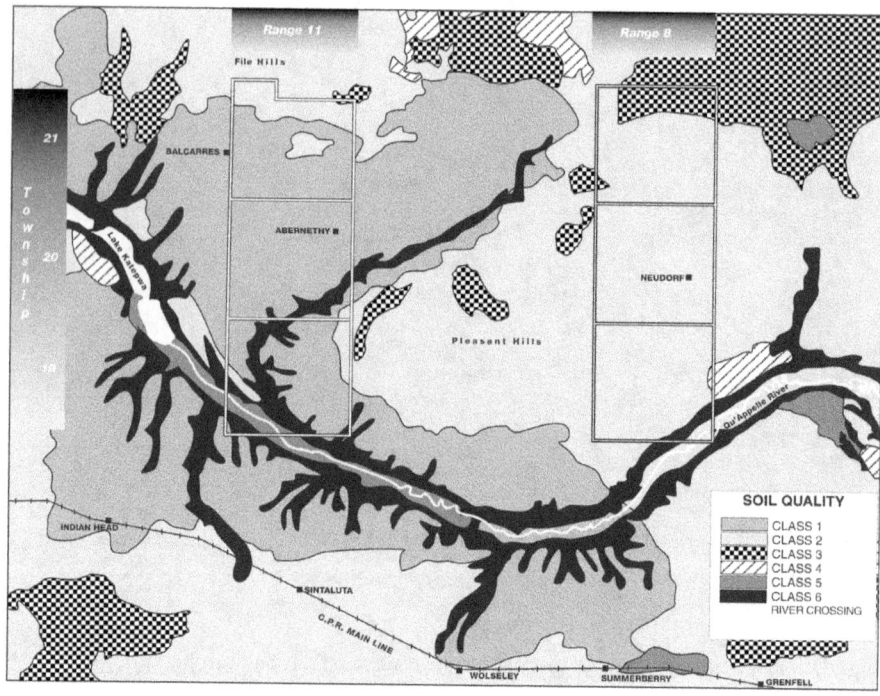

Fig. 2. Soil quality in the Abernethy and Neudorf districts. Map adapted from ARDA Canada Land Inventory, Melville Map Sheet Area 62L [1967], reproduced with the permission of the Surveys and mapping branch, Department of Energy, Mines and Resources.

Soil is of the very best quality. Nearly all the even-numbered sections are occupied by settlers who have made extensive improvements. Pheasant Creek traverses the south-east portion of the township, entering in Section 4 and leaving from Section 13. The land is undulating on level prairie, and soil a good clay loam. The only timber in the township consists of a few bluffs of small poplar and willow in the western tier of sections, there are numerous ravines running towards Pheasant Creek and a chain of small marshes traverses the northern part of the township.[4]

Regulations governing the disposal of public lands in Manitoba and the North-West Territories were outlined comprehensively by the original order-in-council.[5] Under its terms, a "free" homestead system was inaugurated, patterned on the 1862 Homestead Act of the United States. The new regulations permitted all males 21 years or older and persons who were heads of families to make entry for a homestead upon payment of a $10 entry fee; fulfillment of certain duties entitled the entrant to a patent for this land. The duties included residence and cultivation on the homestead during a three-year "proving-up" period, during which the entrant was expected to prove that he was a *bona fide* settler. Shortly after the issuance of the order-in-council the Secretary of State recommended the creation of a new branch within the Department of the Interior, known as the Dominion Lands Branch, to be charged with the administration of public lands in the West.

Concurrently, the Dominion Parliament passed the first *Canadian Pacific Railway Act* in 1872, which granted a charter to Sir Hugh Allan and associates to build a transcontinental rail line.[6] A central feature of the Act was the provision of a block of 50 million acres to the company to subsidize rail construction. The block was to be allocated from a belt 40 miles wide – twenty on each side of the railway. Within this belt all odd-numbered sections were to be granted to the company; even-numbered sections were to be retained by the government for public lands disposal. There were two exceptions to this pattern. Two sections in each township were set aside for the support and maintenance of educational facilities, and 1 3/4 sections were claimed by the Hudson's Bay Company. In every fifth township, this grant was raised to two full sections (Fig. 3).[7]

After the CPR company failed to meet its initial commitments and was, with the Conservative government, implicated in the Pacific Scandal of 1873, Alexander Mackenzie's new Liberal administration abandoned the initial railway statute. The Liberals pursued a policy incorporating both private and public railway construction. The return to power of Macdonald's Conservatives in 1878 occasioned the unveiling of its National Policy of immigration and settlement, protective tariffs, and renewal of the CPR concept. New railway legislation in 1881 to incorporate the Canadian Pacific Railway Company provided a new syndicate with generous federal assistance, including a grant of $25 million and 25 million acres of suitable agricultural land. Included in the package of inducements were extensive tax exemptions and a

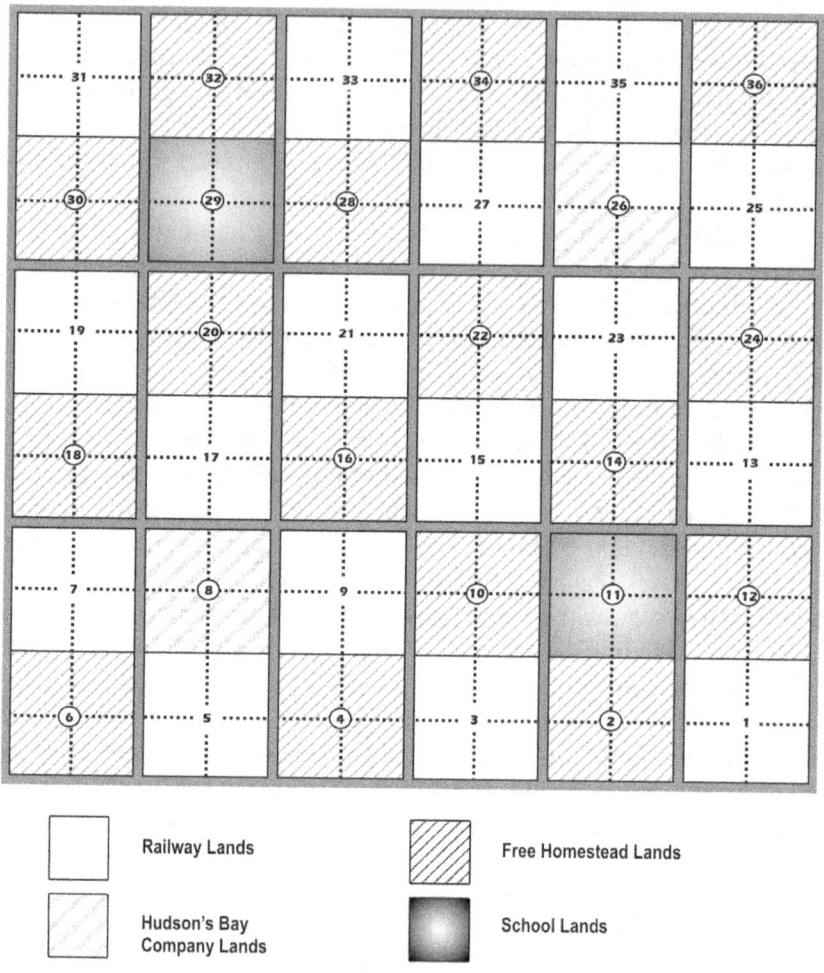

Fig. 3. Plan of a typical township, as outlined in the Dominion Lands Act (after 1879).

FARMERS "MAKING GOOD"

monopoly provision that guaranteed no competitive rail lines would be approved for the West within twenty years after incorporation.[8]

Meanwhile, Dominion lands regulations underwent frequent changes. Many of these changes were influenced by land disposal policies in the United States, which was competing for many of the same settlers.[9] Following the American lead, the Dominion government created the pre-emption privilege in 1874. In the 1880s this provision permitted the homesteader to reserve and purchase an additional quarter-section of Dominion land adjacent to his homestead. The government hoped that this additional inducement would not only increase settlement on the Canadian prairies, but also help to avert the outflow of Canadians to the United States.

When immigration to western Canada increased in 1878, the Macdonald Conservative government reduced the size of homestead and pre-emption allotments from 160 acres to 80 acres each.[10] Homestead entries tailed off sharply after the new regulations took effect, and the government quickly reversed this policy two months after its inception. Yet, within a decade after its introduction, it was apparent that the Dominion Lands Policy had not lived up to expectations. While Manitoba's population increased from about 19,000 to 60,000 in this period, the western agrarian sector represented only a fraction of the anticipated numbers.

During the following five years, the Conservative government enacted numerous changes to the Dominion Lands Act and Regulations in an attempt to increase immigration to western Canada. An important new provision permitted settlers to make second homestead entries. The intent of this provision was to induce experienced homesteaders to sell their improved homesteads in order to take up new land on the margins of settlement. In practice this goal was seldom realized. Since the regulations did not prohibit settlers from making second homestead entries in the vicinity of their original quarter-sections, many simply used it to add to their existing holdings.

New regulations also provided greater flexibility in the fulfillment of homestead duties. In addition to the original terms providing for three years' cultivation and residence, the 1884 amendments created two new systems from which the settler could choose:

i) Residence for two years and nine months within two miles of one's homestead, followed by residence in a habitable house on the homestead for three months at any time prior to the application for patent. Cultivation duties under this system included the breaking of 10 acres in the first year, 15 in the second, and 15 in the third and the cropping of 10 acres in the second year and 25 in the third.

ii) A five-year system permitting the settler to reside anywhere for the first two years, although he was obliged to begin cultivation within six months of entry. Cultivation duties included the breaking of five acres in the first year, the cropping of five acres and breaking of 10 acres in the second year. After two years the settler had to maintain residence and cultivation on his homestead for at least six months in each of the three succeeding years.[11]

While about one-third of the homesteaders arrived in 1882, Abernethy received the majority of its settlers in the period following these amendments to the land regulations in 1884. The more liberal residency requirements of 1884, coupled with the concurrent provisions for pre-emption purchase and second homestead entry, made the 1880s one of the most favourable periods in which to homestead in the entire settlement era.

INDIAN TREATY NO. 4

Euro-Canadian settlement at Abernethy marked only the last layer in a series of occupations by successive cultures in this region. For thousands of years, Plains First Nations cultures had occupied southern Saskatchewan. At the time of the Hudson's Bay Company's transfer of the Territory of Rupert's Land to Britain and thence to Canadian government in 1870, the predominant Aboriginal culture in the Qu'Appelle area was the Plains Cree, who had moved out on the open prairie by the seventeenth century.[12] Like most other plains Aboriginal cultures, the Cree pursued a hunting and gathering economy based largely on the buffalo.[13] With the extension of the European-Aboriginal trade to the prairies in the late eighteenth century, the Qu'Appelle region became an important source of provisioning and of the commercial trade in buffalo robes and pemmican.

While the Dominion Lands Act and survey were designed to prepare the western lands for systematic agricultural settlement, the government was also obliged to deal with its existing occupants, the Aboriginal inhabitants. In 1874 Lieutenant-Governor Alexander Morris negotiated Treaty No. 4 with representatives of the various Indian bands at Fort Qu'Appelle and later at Fort Ellice, providing for the surrender of the area now encompassing southeastern Saskatchewan.[14] A reading of Morris's correspondence with his Ottawa superiors and others reveals that the treaty signing was a well-orchestrated event. Employing scarlet-coated Mounted Police to impress his Indian counterparts, Morris was intent on obtaining their acquiescence with the smallest possible commitment and expenditure.[15] The staged character of the event could not mask significant differences of opinion between the federal government and First Nations representatives as to the nature of the Treaty, what the government was actually offering, and what the First Nations were agreeing to.

Under the written terms of Treaty No. 4, the Cree were to receive settlements in land, cash, assorted gifts, and agricultural and educational assistance. The land reserve was to consist of one square mile for each family of five, or lands in that proportion for larger and smaller families. Each chief was to receive $25 in cash, a coat every three years, a Queen's silver medal, and a "suitable flag"; and each head man, or councillor, $15 and a coat every three years. In addition, each Cree band was granted $750 worth of powder, ball, shot, and twine. To promote self-sufficient agricultural development among the Cree, the treaty provided rudimentary implements of cultivation – two hoes, a spade, a scythe, and an axe for each family, as well as seed wheat, barley, oats, and potatoes to sow the lands they had prepared for crop. One plough and two harrows were allocated to every ten families; each band was to receive a yoke of oxen, one bull, four cows, carpenter's tools, handsaws, augers, two saws, files, and a grindstone. Finally, the Government of Canada agreed to maintain a school on each reserve.[16]

The surveying of reserves followed the signing of the treaty. By 1880, three contiguous reserves had been surveyed to the north of Abernethy: Star Blanket's Reserve, twenty square miles; Little Black Bear's Reserve, 45 square miles, and Peepeekisis's Reserve, 45 square miles. The following year an additional tract was surveyed and the Okanese Reserve was created. Some of the Cree bands were slow to move on to their reserves. J.L. Tobias has described the attempts by the major Plains Cree chiefs to forestall their forced resettlement pending their efforts to negotiate better terms and for the right to occupy contiguous treaty lands. One of the principal holdouts was Piapot, whose band was eventually forced to settle in the Qu'Appelle Region in 1885.[17]

In *Bounty and Benevolence*, a recent study of Treaty No. 4, Arthur R. Ray, Jim Miller, and Frank Tough have noted that, following the negotiation of the Treaty, the First Nations signatories to the Treaty encountered various problems in its implementation. At an early date, the Cree leaders expressed their strongly-held views that the federal government was not honouring its obligations in relation to issues of subsistence specified in the Treaty. Further, at a meeting with Governor General Lorne at Fort Qu'Appelle in 1881, First Nations spokespersons advised Lord Lorne that the terms of the Treaty needed to be extended to provide the necessary means of subsistence. Ray, Miller, and Tough have written that the concerns of the Cree and other First Nations extended well beyond matters of basic subsistence to the enabling of means of livelihood to replace their former economies based in hunting and associated lifeways. The Cree also subsequently discovered that the government's text of the Treaty was more restrictive in terms of hunting, trapping, and fishing rights than the words that were used by Lieutenant-Governor Morris in the actual negotiations. These were just a few of the problematic issues surrounding Treaty No. 4 that have been brought to light in *Bounty and Benevolence*, a thoughtful exploration to which readers are referred for a more detailed account.[18]

On the reserves, a policy of promoting agricultural self-sufficiency through small-scale production was pursued, although the Dominion government's policy of keeping the Cree separate from the new society and discouraging their entrepreneurship prevented them from assuming a more important role in the settlement of the region as a whole. While the Canadian government stated its intention to assist the Indians in adapting to the new order, the reserve lands assigned to them were of secondary soil quality and the simple agricultural tools provided to them were suitable only for cultivating very small holdings for basic subsistence rather than market-oriented agriculture.[19] Reports of Indian agents and other federal officials suggested that, from their perspective, the File Hills Cree were responsive to government-sponsored agricultural instruction on their four reserves, which their leaders had secured for them through the treaty process. As historian Sarah Carter has extensively shown, however, the Cree wanted more than basic self-sufficiency, which would place them in a marginal position in the new society; rather, they wanted to pursue commercial agriculture alongside the arriving immigrant farmers. Instead, the federal government initiated policies designed to undermine the First Nations farmers' capacity to compete in the marketplace, which the Cree persistently protested and sought to revise. Meanwhile, the government's policies of acculturation and assimilation of First Nations, carried out through a combination of government and missionary activities, met similar resistance. The persistence of First Nations cultural practices, such as the Sun Dance, after resettlement on these reserves suggested continuing opposition to the assimilationist policies of the government, extending well into the settlement era.[20]

THE QU'APPELLE MÉTIS

The land claims of anther Aboriginal group – the Métis – remained to be considered. In the mid-1860s and throughout the 1870s, Métis settlers from Manitoba had taken up land along the banks of the Qu'Appelle River and the shores of the Qu'Appelle Lakes to the south and west of Abernethy. Some of these early settlers led a semi-migratory existence, as they alternated between the seasonal activities of agriculture, fishing, and the buffalo hunt. Others had been assimilated, for all practical purposes, into First Nations lifeways and economies. A third group possessed a predominantly agricultural economy and remained on the land most of the year. Within this group, considerable differences existed in terms of the extent of land cultivated, value, and number of buildings. A reading of the homestead files for these settlers indicates that several were engaged in fairly extensive farming operations, particularly in the raising of livestock. For example, in 1891 one Métis farmer reported the ownership of 70 head of cattle and 50 horses.[21] Other Métis settlers possessed dwellings and outbuildings that they assessed at up to $2,500 in value. The point to be made here is that many Métis were engaged in agriculture on a fairly sophisticated level. Yet a large proportion of the Qu'Appelle Métis left their lands in the first few years after the influx of Euro-Canadian settlement. Did they leave because of they could not, or would not, make the necessary adjustments to the new social and economic order? Or were they, in subtle or other ways, forced out? To a significant degree the answers to these questions lie in the original disposition of Métis farmlands in the early 1880s.

Land claims of the Qu'Appelle Métis were addressed in two orders-in-council, passed in 1885. The first, dated 30 March, gave Métis settlers who had predated the survey the option of surrendering title to their waterfront acreages in favour of homestead and pre-emption entries on other Dominion lands as yet unclaimed. Alternatively, the government would pay scrip in the amount of one dollar per acre to the claimants. A second order-in-council, dated 18 April, amended the first by adding provisions permitting the Métis who were in *"bona fide* possession, by virtue of residence and cultivation,"* of water frontages, to purchase these lands for one dollar per acre. In no instance were these lands to exceed forty acres, and claimants were given two years in which to make the payment. Settlers who opted to purchase waterfront acreages were deemed eligible to select one quarter-section from lands open to homestead and pre-emption entry. Moreover, those Métis heads of families residing in the North-West Territories before 15 July 1870 were granted certificates entitling them to 160 acres of land in lieu of $160 of scrip.[22]

From the point of view of the Dominion government, the decision to grant scrip represented an expedient solution to the problem of extinguishing the Métis land claims. It did not, however, take into consideration the vulnerability of a people unschooled in the technicalities involved in land registration. Earlier, many Métis of Manitoba had been dispossessed through a combination of political manipulation, prejudicial judicial decisions, and extensive defrauding of Métis settlers by speculators after 1870.[23] Alternatives to the adopted policy had been proposed. For example, the North-West Territorial Council and the Roman Catholic clergy had recommended that a scrip system not be adopted. Under their proposed system, land would be granted to the Métis but title would be retained by the Crown during the first five to ten years. Dominion authorities ignored these suggestions.

When the North-West Halfbreed Commission sat at Fort Qu'Appelle to dispose of the claims in 1885, the majority of Métis holders of waterfront acreages chose to accept money rather than land. The Commission, accordingly, issued scrip in amounts ranging between $20.58 and $152 to 14 Métis who had settled on the banks of the Qu'Appelle River and the lakes of the valley.[24] In retrospect, the choice of money scrip was unfortunate. For a small amount of cash, the Métis had relinquished ownership of lands that would soon appreciate to many times the value of the cash settlement. Yet their decision should be viewed in the specific context of the time in which it was made.

Speaking on the subject of Indian and Métis concerns in 1886, T.W. Jackson of Fort Qu'Appelle discussed the reason for the Métis acceptance of money scrip. Jackson, a member of the Territorial Assembly and a legal advisor to the Métis, stated that just prior to the sitting of the North-West Halfbreed Commission he and others had urged the Métis to accept land rather than money. At this time, 70 per cent of the Métis had indicated their preference for land, but before they had formally accepted, news arrived of the outbreak of hostilities at Duck Lake, the first confrontation of the North-West Rebellion. Rumours spread to the effect that "Riel was to prove victorious, that the halfbreeds had better not take land; that they should take scrip, buy what they could and the land would ultimately belong to them."[25] For this reason, Jackson related, they had accepted the money.

Insofar as it concerned only the even-numbered Dominion Lands sections, the issue of scrip represented only part of the general question of land disposal. Since most lands bordering the Qu'Appelle Valley and its lakes fell within the 40-mile CPR belt, the odd-numbered sections had been issued to the railway as part of its land grant. Further complications arose when the CPR sold 150,000 acres of its land in the central Qu'Appelle region to the Ontario and Qu'Appelle Land Company. Many of these lands were already occupied by Métis settlers.

In 1882, 45 Métis settlers in the Qu'Appelle Valley sent a petition to Edgar Dewdney, Lieutenant-Governor of the North-West Territories, in which they alerted him to apparent attempts by the Ontario and Qu'Appelle Land Company to dispossess them of their lands. The dispute stemmed from surveys that

> have discovered some of us to be on Railway Land now owned by the Ontario and Qu'Appelle Land Company – whose agent has informed us that we must either buy the land from them or move off – in fact we are informed that they have sold some of the land at present occupied by a bona fide settler....[26]

The petition also stated that the Métis refused to comply with the land company's request, claimed patents to their lands, and requested Dewdney's intervention on their behalf. Dewdney forwarded the Métis petition to Sir John A. Macdonald, Prime Minister and Minister of the Interior, on 29 August 1882. Receiving no reply he wrote again on 19 March 1883 impressing the urgency of the settlement of their claims. He noted that many of the Métis settlers were living on the same section, and

> as land became valuable a scramble was made by land speculators to obtain the right titles and interest of those settled in the most favoured locations. The sooner the claims of these Half Breeds are determined the better as a number of them are 'bona fide' settlers and deserve consideration.[27]

Finally on 6 July 1883, Dewdney received a reply to his letter of the previous August from John R. Hall, Acting Secretary of the Department of the Interior. Hall stated that the correspondence regarding Métis land claims had been referred to Commissioner James Walsh for investigation.[28] Yet, despite these assurances, no action was taken, and on 8 December 1883, T.W. Jackson, a Fort Qu'Appelle lawyer, wrote on behalf of the Métis:

> If their own grievances upon which the Government have been repeatedly petitioned and memorialized, were brought personally to your notice some immediate action would be taken. Under any circumstances there would not be very many to deal with and the settlement of their claims need

not be a troublesome one, but there are half breeds in the territories who have never received anything from Government and who it has been admitted are entitled to some consideration.[29]

Hall replied to Jackson's letter on 13 March 1884. He stated that the Minister had been very anxious to have the question of land claims settled and instructed Walsh to conduct an investigation. "Owing to an unusual pressure of business," however, Walsh had been unable as yet to visit the Métis. Hall stated that Walsh had again been required to investigate the matter "at the earliest possible opportunity."

At this point the file ends. It is not known what action was eventually taken by the Department, but within a year the majority of Métis settlers had disappeared from their lands. The possibility exists that they had assigned their quarter-sections to the Ontario and Qu'Appelle Land Company in exchange for money or some other consideration. It is also possible that the Métis were divested of their lands by fraudulent means.

The second scenario finds some substantiation in a separate Department of the Interior file dealing with the land claims of three settlers vis-à-vis the Ontario and Qu'Appelle Land Company lands.[30] Quit claim deeds for two of the three are still extant. By the first, a Métis settler, Albert Fisher, surrendered all claim to his quarter-section to the Ontario and Qu'Appelle Land Company for one dollar. A perusal of this document reveals the signature of only one witness, one R.J. Dodd; Albert Fisher's signature is denoted by an "X." The other quit claim deed provided for a payment of $300 to Stephen H. Caswell. Caswell's own signature on the document shows he was somewhat literate; his Anglo-Saxon surname also implies that he may have been capable of reading the document. In Fisher's case, however, it is difficult to escape the conclusion that some fraud was at play. There is no way of knowing whether Fisher actually signed the release or received the dollar, but having obtained the release, the Ontario and Qu'Appelle Land Company stood in position to sell the quarter-section for a price of up to $1120.

Beyond the difficulties encountered with land companies and speculators, many Métis were hampered by the grid survey system, which made no allowances for the existing pattern of cultivation and land use. This arbitrary method of land subdivision frequently placed two or more Métis settlers on the same quarter-section or divided a Métis holding in two. In these cases settlers who had been dispossessed by the survey were usually minimally compensated for the improvements they had made on their lands and were granted an opportunity to make entry for other available homestead quarters. But since Métis society was based on a system of economic and social interdependencies, the dispersion of the formerly closely knit community at Qu'Appelle may have contributed to a deterioration in their economic position.

By 1886 the economic effects of this dislocation were evident. In June of that year one Norbert Welsh wrote to Lieutenant-Governor Dewdney on behalf of his fellow Qu'Appelle Métis. Claiming that his compatriots were "striving against adversity, and are not able to provide themselves with the necessaries of life,"[31] Welsh pleaded for assistance in finding work in freighting or other jobs. Dewdney passed the request on to the Department of the Interior, but directed his clerk to add that:

> His Honor has learnt that, although one or two instances of individual distress may exist, there is no general destitution prevailing in the District, and Mr. Welsh himself is in fairly comfortable circumstances.[32]

At the request of the Department of the Interior, Dewdney commissioned W.A. Clarke of Fort Qu'Appelle to investigate the circumstances of the Qu'Appelle and File Hills Métis. Clarke subsequently reported that, "as a rule, we found the people happy, healthy and contented, although in poor circumstances, and not very bright circumstances for the future."[33] Noting that many of the Métis owned working horses and were prepared to undertake freighting work for the government, Clarke recommended that they be given this work. He added, however, that they "should be compelled to earn what they get, otherwise it will cause discontent amongst the whites."

Clarke's detailed commentary on individual Métis confirms that while the majority were considered to be in good health – "good health" was not defined – they faced a bleak economic future. Those who were employed were engaged in gathering and selling wood, hunting, or fishing on the Qu'Appelle Lakes. Each of these primary products was a fast-diminishing resource. With the expansion of settlement and the exhaustion of supplies of wood, game and fish, agriculture represented the only realistic alternative lo economic stagnation. But market-oriented agriculture required capital to finance the purchase of implements, buildings, and livestock. Most of the Métis did not possess the requisite financial resources to participate fully in the new agriculture.

A 1906 map of land tenure in the Indian Head and Abernethy districts shows few Métis still engaged in agriculture.[34] Some continued to live in log houses in the valley on land that was now owned by Euro-Canadian settlers with whom they occasionally found work. The Fayants, for example, continued to live on valley lands now possessed by J.A.R. Blackwood, who employed them in harvest and other operations.[35] The Métis also acquired a reputation as fencers and performed contract fencing for various new settlers.[36] Others continued to eke out a living from fishing on the Qu'Appelle Lakes, and the small settlement at Lebret continued to be the centre of Qu'Appelle Métis life. Generally speaking, however, their employment prospects remained poor, as they were effectively pushed to the sidelines by the newly dominant society of Euro-Canadian and European newcomers.

SETTLEMENT ON THE PHEASANT PLAINS

The Pheasant Plains district received its first permanent Euro-Canadian and British settlers in 1882 and 1883. In this early period, the "free" homestead was the dominant form of land disposal. To the southeast, near the future community of Rosewood, a handful of North West Mounted Police officers from the Fort Qu'Appelle detachment had taken homesteads as military scrip in 1881.[37] But in 1882 large-scale settlement began as Ontarian and British immigrants headed west by ox cart from the then western terminus of the CPR at Brandon.[38] On reaching the Dominion Land Office at Fort Qu'Appelle, these newcomers discovered that all the suitable surveyed government land south of the Qu'Appelle River had been already claimed. Accordingly, they selected homesteads and pre-emptions on the fertile plains bordering Pheasant Creek to the north of the valley.

The process and timing of settlement is a question that has preoccupied historians and historical geographers of the American Midwest and Great Plains states, as well as of the Canadian prairie provinces.[39] Insofar as land selection may indicate a predisposition to a particular type of farming or of economic and social relationships among farmers, its study is central to an understanding of the settlement experience. The most sophisticated attempts to analyze land acquisition in settlement have employed quantitative methodologies. Foremost among these is Michael Conzen's *Frontier Farming in an Urban Shadow*.[40] Conzen's method was to identify the principal factors bearing on land selection, and to code these as dependent variables in a multiple regression equation in which the year of sale was the independent variable.

A similar approach was adopted for the Abernethy settlement study, entailing the development of a regression model of land selection in which the year of homestead entry was the independent variable. To measure the relative significance of the most probable factors bearing on land selection, this model incorporated dependent variables identified in the contemporary settlement literature and more recent historical studies. These variables included: soil quality; accessibility to wood and water; proximity to grain handling facilities, and proximity to supply centres. The multiple regression model based on these variables showed that approximately 60 per cent of the variation among years of entry was explained by variables based on distance from the railway (for further detail see Appendix A). Each of the other variables appeared to explain only minor portions of the remainder of the variability. These results gave credence to the hypothesis that proximity to the railway was the dominant factor in the settlers' initial choice of homestead lands.

The desirability of most available quarter-sections around Abernethy was demonstrated by the land rush when Townships 19 and 20 were opened for entry in March and July 1883.[41] In Township 19 in particular, nearly all quarters were claimed in a matter of days, a fact that may have affected the degree of variability that could be explained by the regression model. The apparent lack of significance attributable in the model to the presence of wooded areas may have related to the fact that most homestead parcels in close proximity to the coulee and river banks had been reserved prior to formal entry. A perusal of the 1883 surveyor's township plan showed the presence of squatters, or pre-entrants, on most of the quarters. As was the case in

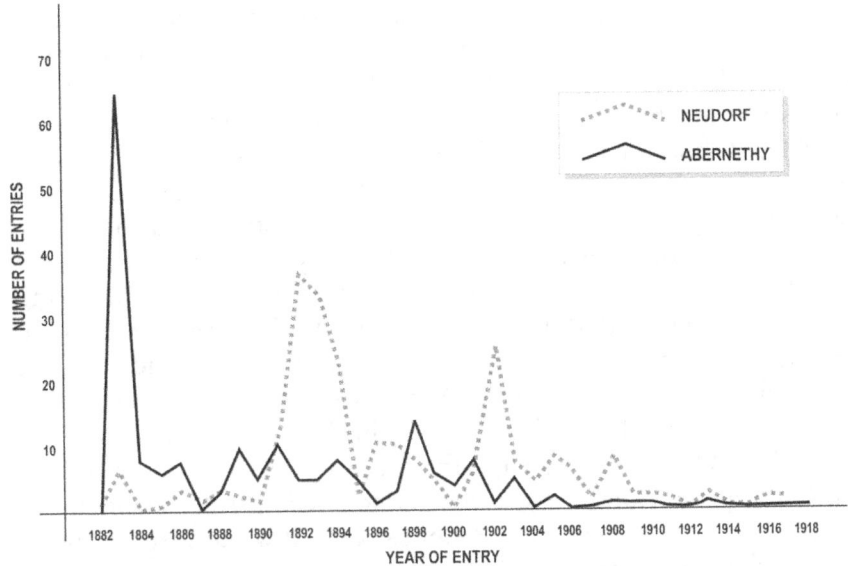

Fig. 4. Distribution of homestead entries by year of entry, Abernethy and Neudorf districts, 1882–1918.

Manitoba, however, the heavily wooded areas were passed by in favour of the open prairie that was at least accessible to woods. It would appear that from the outset, the Abernethy settlers were intending to specialize in grain farming and selected lands that were most readily cleared for this purpose.

In the Abernethy district a preponderance of settlers arrived during the land rush of 1882–83. In 1883 alone, 66 settlers made entry for homesteads and pre-emptions in Townships 19, 20, and 21, in Range 11, West of the 2nd Meridian. This early group represented 43 per cent of the total number of homesteaders in these townships. Such a large influx can be explained by the initial opening of these townships for homestead entry, the building of the CPR, and the settlers' expectation of early rail service through the Pheasant Plains (Fig. 4).

What is less easy to explain is the timing of later settlement. Assuming that prospective eastern Canadian settlers had access to information on the political and economic conditions in western Canada, it might be expected that the chronology of immigration into the area would reflect their knowledge of those conditions. After the initial rush, homestead entries in the Abernethy district plummeted to eight in 1884 and six in 1885. By the former year delays in local rail construction had already become evident, and it is reasonable to conclude that this fact had some impact in curtailing immigration. With regard to 1885, the settlement historian André Lalonde has argued that settlement was impeded by the outbreak of the North-West Rebellion that spring.[42] While the Indian and Métis inhabitants of the Qu'Appelle area did not join the resistance, numerous instances of vandalism and theft were reported in the area, and First Nations movements off the reserves created a climate of apprehension within the Euro-Canadian communities.[43] Chief Star Blanket led the men of the File Hills reserves to the Qu'Appelle Valley, where they conducted warlike manoeuvres.[44] Meanwhile, members of Chief Piapot's band also travelled off their reserve and were reported as far west as Swift Current.[45] At Qu'Appelle, the local immigration agent wrote in his annual report that a number of immigrants who were on their way to the Qu'Appelle district changed their destinations as a result of the Rebellion.[46]

In 1886 entries rallied slightly, but dropped to zero in 1887. The complete tailing off in immigration might well be attributable to the drought and near total devastation of crops in 1886. At this point many settlers had suffered three successive crop failures. The seriousness of the economic situation is illustrated by a tabulation of homestead attempts in this period, which showed that cancellations far outstripped entries that year.[47]

Agency	Homestead		Pre-Emption	
	Entries	Cancelled	Entries	Cancelled
Qu'Appelle	149	255	60	190
Souris	160	265	56	265
Winnipeg	87	104	17	28
Dufferin	13	77	1	55

The figures might also illustrate the rooting out of land speculators or non–bona fide settlers. Such a large number of cancellations, however, must surely indicate great hardship among Qu'Appelle area farmers.

Indeed, farming conditions deteriorated so greatly during the decade that a large number of settlers abandoned the North-West Territories for greener pastures in the United States. An indication of the extent of the out-migration in the late eighties is provided by census statistics. In 1885, the 5,147 farm-operators in the District of Assiniboia occupied 1,641,752 acres, of which 160,133 acres were cultivated. In 1891, despite continuing immigration, the net number of occupiers had risen only slightly to 5,694, while occupied acreage had dropped to 1,599,156 acres and improved acreage to 151,699.[48]

In consequence of the outflow, settlement remained scattered, and the development of schools and social amenities was further inhibited. Economically, farmers experienced a chronic shortage of labour; psychologically, life on the frontier continued to be a lonely experience for many. With the failure of rail service to appear in the area, Abernethy was yet another isolated community, a distant adjunct of the communities of Indian Head and Sintaluta on the CPR main line to the south and Fort Qu'Appelle to the west.

LAND SPECULATION

Little direct evidence exists that could identify the extent of land speculation in the Abernethy district in the settlement period. Abernethy was not provided with local rail service before 1904, and therefore it was probably spared the rather blatant speculation in homestead lands that took place in the areas situated closer to the CPR main line. Speculation was a common feature of settlement life in both the American and the Canadian Wests, as many settlers seized the opportunity to turn their "free" homesteads into quick profits by sitting on these lands and making minimal improvements while the price of land went up. Stories abounded of borrowed shacks and cattle that settlers fraudulently moved onto their homestead lands in anticipation of the homestead inspector's arrival. After he had come and gone, the shanties and cows were quickly moved to the next homestead.

In the Qu'Appelle district particularly flagrant examples of speculation occurred, as the Dominion Lands official William Pearce observed in the period:

> Along the line of the Canadian Pacific Railway, in the Qu'Appelle District, there is one great trouble regarding homesteading which will have to be met next year. Many parties have gone ahead of the townships being open for entry, and after the survey was begun did a little breaking, varying generally from one-half to four or five acres, and erected a small shanty – or what is stated in affidavits as such – which in many cases is not more than a few poles. Sometimes one person will make the above amount of improvements on several quarter sections. This 'squatting,' so called, is done by a ring of speculators. At each station of the Canadian Pacific Railway, in the vicinity of the lands so improved, there is one of the ring or his agents placed. He meets the settler looking for land, informs him that all the land in the vicinity is taken up but for a consideration varying from $100 to $1,000, he will buy out the party holding a claim, and then entry by the purchaser can be made for it. I have personally witnessed the above operation, and would suggest that some steps be taken to checkmate the operators. The Dominion Lands Act, in such cases appears rather vague....[49]

Most speculative activities in the area north of the Qu'Appelle River were confined to petty holdings by individuals. In 1884 Homestead Inspector Rufus Stevenson sent an enumerated list of entrants on quarter-sections that had been allotted to the Touchwood-Qu'Appelle Land and Colonization Company, about thirty miles north and west of Abernethy.[50] He reported a high incidence of absenteeism among these settlers, some of whom were working at Fort Qu'Appelle or other locations throughout western Canada. Few had built habitations and what little breaking they had accomplished had been hired out. A follow-up report a year later showed 30 to 99 settlers were still absent from their homesteads. Stevenson voiced his doubts that these settlers would become bona fide settlers and recommended that a strict interpretation of the homestead law be adhered to, that is, that these entries be cancelled.⁵¹ In 1885, the *Qu'Appelle Vidette* reported that 10 of 35 homestead entrants in a township just north of Balcarres had not visited their lands in the two years since the entry, and that their lands were subject to cancellation.⁵²

THE GREAT NORTH WEST CENTRAL RAILWAY

Beyond specific considerations such as soil quality and proximity to wooded areas, settlers chose lands north of the Qu'Appelle Valley because they anticipated the early arrival of rail service to their locality. A reading of Department of the Interior files confirms that in 1882 the Souris and Rocky Mountain Railway Company received a charter from the Dominion government to build a line from Melbourne, Manitoba, to the Rocky Mountains via Fort Ellice and Battleford.⁵³ For its efforts the company was to be permitted to purchase land up to 6,400 acres per mile of completed track, at the rate of $1.06 per acre, which it could then sell to immigrants. In 1884 the company made a slight start in building the railroad, but construction was soon bogged down in disputes with its labourers, who claimed that they had never been paid for their work. Almost immediately the Department of the Interior was flooded with petitions from groups of settlers along the proposed route of the Souris and Rocky Mountain line.⁵⁴ These petitions stated that the settlers had selected lands in these districts only on the assumption that rail service would soon be provided. Since their lands were a prohibitive distance from the CPR main line, they claimed that the failure to build the railway would force most of them to abandon their lands.⁵⁵

In 1885, the Dominion government issued, by order-in-council, a charter to the North West Central Railway Company to assume the charter of the Souris and Rocky Mountain Railway. The company petitioned the government to convert the land purchase to a free grant, subject only to a charge of ten cents per acre to cover the cost of survey. Their request was approved by order-in-council on 29 July 1885.⁵⁶ The route to be followed remained essentially the same as the one outlined in the earlier charter, although it was to terminate at Battleford, rather than the Rocky Mountains. Maps outlining the original contemplated route show that the railway was intended to run straight west from Fort Ellice and curve gradually northward after passing through Pheasant Forks, the centre of the Primitive Methodist Colonization Company to the north and east of Abernethy. According to the terms of the 29 July order-in-council, the company was bound to build the first 50 miles by 31 December 1885 and to reach Battleford no later than 31 December 1889 (Fig. 5).

Signs of future financial difficulties with the railway were not long in appearing. On 30 July 1886, Senator Francis Clemow, on behalf of the syndicate, declined to accept the company's charter on grounds that its approved bond issue was restricted by the Dominion government to $20,000 per mile of track. Clemow maintained that the company, by then known as the Great North West Central Railway Company, required a minimum of $25,000 per mile to finance rail construction. The Macdonald government approved Clemow's request on behalf of the company by order-in-council on 3 August 1886.[57] Further extensions of the initial construction deadline were then passed by three orders-in-council in November 1887, August 1888, and July 1889.[58] At that point the Great North West Central Railway Company had been incorporated for more than three years but it had failed to build even the first 50 miles of track. While the reasons remain unclear, the principal failing of the syndicate appears to have been its inability to raise sufficient money to proceed with the work. By September 1889, the company had spent less than $80,000 on construction.[59]

On 16 September 1889, the five original principals of the GNWC Railway concluded an agreement with financier J.A. Codd and a group of European capitalists to change the corporate membership of the company and to provide a new injection of capital to permit the building of the rail line. Under the terms of this agreement:

> i) Bonds to the amount of £1,500,000 would be sold to a European financial group, provided that the first 50 miles had been built and opened to traffic, and that all debts under Clause 27 of the railroad's charter had been paid.
>
> ii) Arrangements would be made for a private syndicate to pay £200,000 upon the final inspection of the first 50 miles.

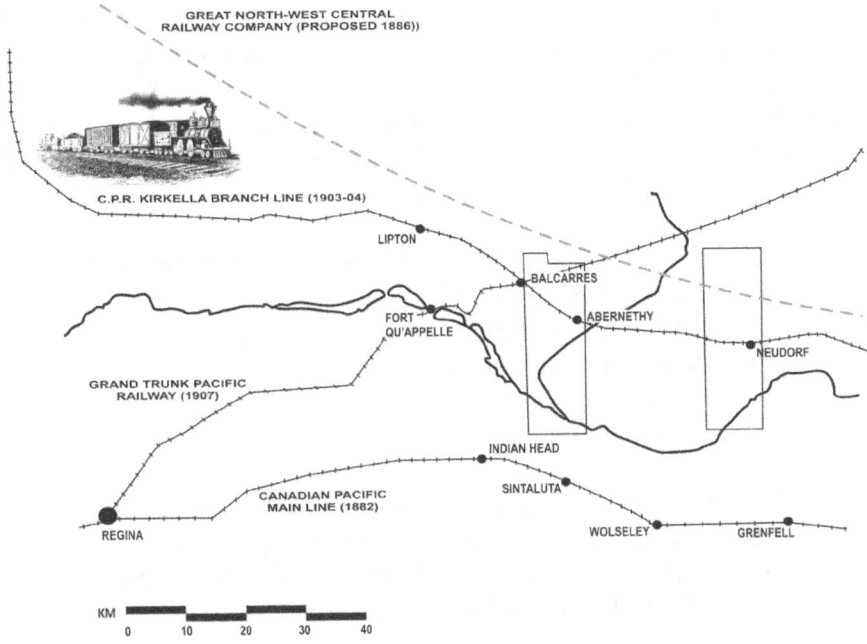

Fig. 5. Rail linkage in the Abernethy district, 1882–1920. The dotted line indicates the proposed route of the Great North West Central Railway, adapted from the map accompanying a memo to Privy Council, 23 November 1886. Library and Archives Canada, Department of Interior Records, RG 15, Vol. 247, File no. 251149, part 3.

All the original incorporators thereby transferred their shares to Codd and his associates, while the former shareholders, represented by A. Charlebois, became subcontractors to build the track for the new owners. Only two years later, however, the subcontractors brought suit against the GNWC Railway Company for non-payment. The Chancery Court of Ontario subsequently ruled that the company had not deposited bonds as required by its charter and that all owed monies ($622,226) and interest were then due and payable.

The history of the GNWC Railway was further complicated by charges of corruption. During the 1886 House of Commons debate on the company's contract extension, it was revealed that a Member of

Parliament had received a gratuity of $386,000 worth of Souris and Rocky Mountain stock. D.W. Woodsworth, a Nova Scotia MP, charged that James Beaty (West Toronto) had been the recipient. Woodsworth stated that he had introduced the original Great North West Central Railway Bill in 1884 at Beaty's behest. Beaty had a subsequent bill drawn up to amend this charter. Woodsworth related:

> I looked at the Bill and found that all the guards, all the checks, ensuring payment to the workmen upon the road – the old Souris and Rocky Mountain Railway, of which this was a revival – had been left out.

Woodsworth continued:

> There was not an honest attempt to build one foot of this road. There was not an honest attempt to put a theodolite on the road but merely to hawk (the charter) of a road ... I say this was a charter selling and nothing else.[60]

Whether Woodsworth's charges were completely true is difficult to determine, but Beaty's involvement in the GNWC Railway seemed to be a particularly blatant example of conflict of interest in the awarding of railway contracts. While the GNWC and other railway companies continued to be the subject of hot debate in the Commons, railroad construction remained a chimera for the settlers north of the Qu'Appelle River.

In 1898 the charter of the Great North West Central Railway was assumed by the Canadian Pacific Railway Company with all unearned land rights. Yet further delays occurred. In 1902, W.R. Motherwell and his fellow Abernethy settlers petitioned Prime Minister Wilfrid Laurier:

> The settlers north of the Qu'Appelle Valley have been for the past 15 years petitioning, praying, beseeching and imploring with your Government and the Government that preceded yours, for the carrying out of the charter obligations entered into by the N.W. Central Railway with the Parliament of Canada, just 20 years ago. We have season after season placed reliable statistics before you setting forth

the number of settlers, acreage under cultivation and total grain grown, and the immediate necessity of Railway facilities to carry these products to the markets of the world. You have also been shown that no matter how rich a country's natural resources may be, that it is impossible for farmers to haul wheat from 20 to 35 miles across the Qu'Appelle whose banks are 300 to 400 feet high, and leave anything like a reasonable result of the tiller's toil in his hands at the end of the year....[61]

Interminable delays in railroad construction could hardly fail to have serious economic consequences for the farming population north of the Qu'Appelle River. One way to determine the impact of accessible rail service on the eventual success or failure of homesteaders is to tabulate homestead cancellation rates for townships of varying remoteness to the railway. In the Abernethy district, Township 19, Range 11, which possessed lands ranging between six and fourteen miles from grain-loading facilities on the CPR, recorded forty-one patents to forty-two cancellations, for a cancellation rate of slightly more than 50 per cent. Township 20, ranging between fourteen and twenty-two miles from the railway, recorded forty patents to fifty-two cancellations, for a 57 per cent rate of cancellation. Township 21, however, ranging between twenty-two and thirty straight-line miles, recorded only thirteen patents to thirty-nine cancellations, for a cancellation rate of 75 per cent. Another measure of the economic difficulties experiences by settlers is the time lag between entry and patent. In Townships 19 and 20, homestead entrants in the 1880s took an average of approximately six years each to qualify for their patents. In Township 21, on the other hand, entrants occupied their lands for an average of nearly twelve years before they applied for patents. The inference that the lack of rail service may have been largely responsible for the district's lagging development is supported by a 1905 article in the Abernethy *Abernethan*. The article asserted that in the previous decade, early settlers in the Balcarres area northwest of Abernethy had dropped out owing to the lack of rail service.[62] Balcarres is located in Township 21, where the highest rates of homestead cancellation were identified in the Abernethy study tabulation.

Thus, settlement in the Abernethy area was initially encouraged and ultimately inhibited by the failed Great North West Central Railway.

While it is difficult to sort out all of the details of the railway's complicated history, the Dominion government must bear a large degree of responsibility for granting extensions to the company in the face of overwhelming evidence of its inability to carry out the charter. Particularly damning were revelations of financial involvement by members of the House of Commons and Senate. Individuals such as James Beaty and Senator Clemow evidently reaped a large return from the proceeds of the original land grant, while many settlers in the area north of the Qu'Appelle Valley were forced to the brink of bankruptcy.

GERMAN SETTLEMENT AT NEUDORF

In 1889 German-speaking immigrants from central and eastern Europe began to settle in the Pheasant Hills District fifteen miles to the east of Abernethy. They were preceded by a number of Ontarian and British immigrants who had taken up lands in this area in the early eighties. Most of these early settlers had dropped out by the end of the decade, and the area was relatively unpopulated at the time of the German influx. While the German settlement was not formally organized under the auspices of a colonization company, it quickly came to assume a monolithic character. German settlement at Neudorf began with a trickle of newcomers in 1889, followed by a flood of immigration in the early nineties, when more than half of the settlers in the district arrived. By 1895 settlement tailed off, possibly related to spreading word of three successive years of depressed wheat prices. In the late nineties, the flow of settlement recovered and reached another peak in 1902, just two years before the building of the Kirkella branch line of the CPR through the district. With this last surge of immigration, lands in the townships to the north of Neudorf were claimed and the initial settlement process was essentially complete.

The Germans' reasons for settling *en bloc* in the wooded and marshy lands of the Pheasant Hills are not definitively known, but a subsequent article in the Abernethy *Abernethan* is suggestive. The author observed that while the land in the area of German settlement was more broken up than land on the open prairie, it was more conducive to stock raising.

> After the experience of drought in the 1880s the advisability of having stock to fall back on when the wheat crop failed became apparent and large numbers in the early 1890s began to settle in the district, principally peoples of German nationality.[63]

Whether or not the prior experience of drought had induced the Germans to settle together in the Pheasant Hills, it remained for each settler to select his own quarter-section. The application of the land acquisition model to land selection in the Neudorf area showed a high correlation between the year of entry and distance from the railway. Since the Germans arrived in the area via the CPR and then travelled north to claim their homesteads, it is reasonable to suppose that the direction of population flow influenced these results. Other variables explained only part of the remaining variability, and one might infer that many settlers did not employ a sophisticated selection process. It seems probable that language barriers impeded communication with Dominion Lands officials respecting the character of different lands. Correspondence with the Department of the Interior reveals that some of these settlers made entry for their homesteads sight unseen.[64] Some discovered after the fact that they were five or six miles from water.[65] Others, arriving in midwinter, claimed lands that were found to be covered with stones, scrub, and marsh after the snow cover melted in the spring.[66] Another factor not included in the model was the tendency of settlers to homestead in close proximity to friends and relatives. In Township 19, south of Neudorf, thirty of seventy-eight entrants chose lands within two miles of settlers possessing identical surnames. Clearly many of these name matches were immediate relations. This figure would be considerably augmented if friends were added, since a number of Germans wrote to the Department of the Interior to request homesteads adjacent to those of specific individuals. The proximity of friends and relatives was not only socially desirable, but fulfilled essential economic functions in the initial period when pooled labour and the sharing of resources could constitute the difference between success and failure.

A tabulation of homestead cancellation rates for Neudorf area homesteaders shows a comparatively high rate of early economic survival among German settlers vis-à-vis their Anglo-Canadian counterparts. In three townships around Neudorf of distances from the railway comparable to the Abernethy study townships, settlers recorded only twenty-five cancellations to sixty-four successful applications, for a cancellation rate of 28 per cent. Overall this failure rate was only slightly smaller than that of the Abernethy area, but the ethnocultural breakdown reveals striking differences. Seventeen of twenty-seven, or 63 per cent, of Anglo-Canadian or British settlers cancelled, but only ten of sixty-two, or 16 per cent, of the German settlers failed to secure their patents. In other words, the cancellation rate of German settlers was half that of Anglo-Canadian settlers in the Abernethy area and only one-fourth as great as the rate registered by Anglo-Canadians in their own district of Neudorf.

In accounting for the Germans' apparent success in "proving-up" their homesteads and the Anglo-Canadian settlers' failure, certain demographic features of the two populations may be noted. First, clear differences existed in terms of marital status. Eighty-five per cent of the German settlers in the Neudorf area were married, compared with fewer than 60 per cent of the Anglo-Canadians. The much larger proportion of bachelors at Abernethy would imply a more mobile population. Second, the German settlers were older than their Abernethy neighbours. At the year of entry the average age of non-British nationals in the Neudorf area was thirty-four, compared with twenty-nine for the Abernethy settlers. Even more striking was the age distribution which showed that more than one-third of the Germans were forty years or older, while only one-sixth of the Anglo-Canadian settlers belonged to this age group (Fig. 6). Third, German families tended to be larger, averaging almost three children per settler, as opposed to 1.3 per Anglo-Canadian settler (Fig. 7). Larger families provided much of the necessary person-power for labour-intensive mixed farming operations. Labour demands were particularly heavy during the winter, when the German settlers frequently kept their children home from school to help on the farm.[67] In the summer, they sometimes also hired their sons out to work for Anglo-Canadian farmers for five dollars a month and their board.[68] In addition, the German settlers' wives participated in heavy farm labour, as they performed such tasks as breaking, cultivating, and hauling.[69] While lacking the financial

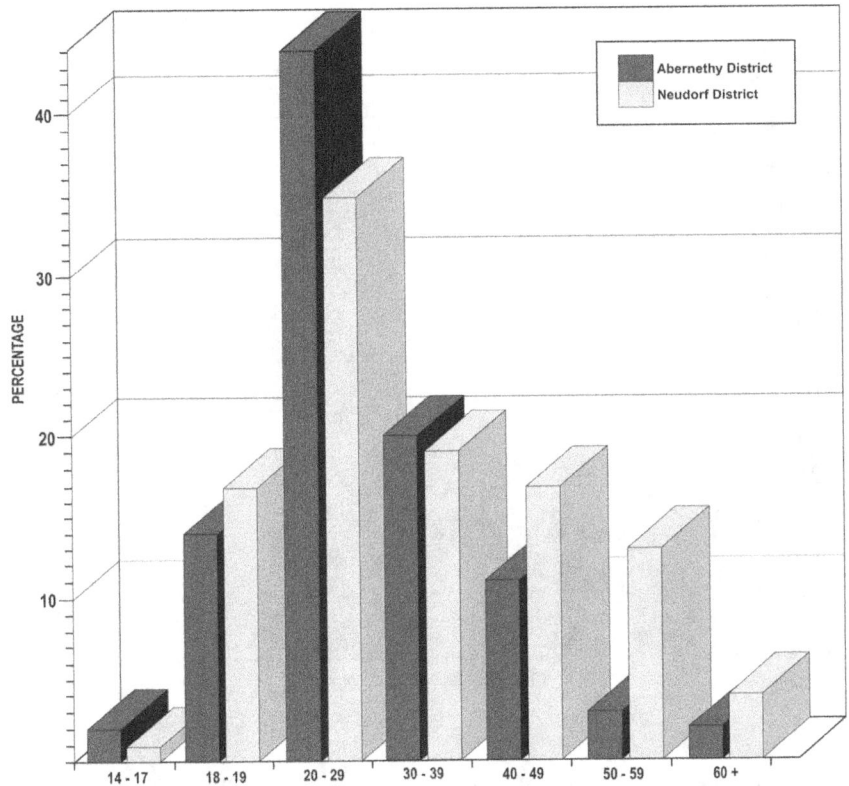

Fig. 6. Age distribution of Abernethy and Neudorf district homesteaders at the time of entry.

resources of their Ontarian counterparts, the Germans possessed an apparent advantage in terms of available labour.

Yet, the mere achieving of a patent does not constitute a measure of the long-term performance of the homesteader. At best, it indicates that the homesteader had complied with the requirements of the Dominion Lands Act by breaking and cropping a minimum acreage and by building a habitable dwelling and then living in it. To gain a better measure of longer-term success or failure, and following the lead of settlement studies performed in the United States, farm turnover and persistence rates were calculated at five-year intervals for one township in each of the two study areas. The data on farm ownership were obtained

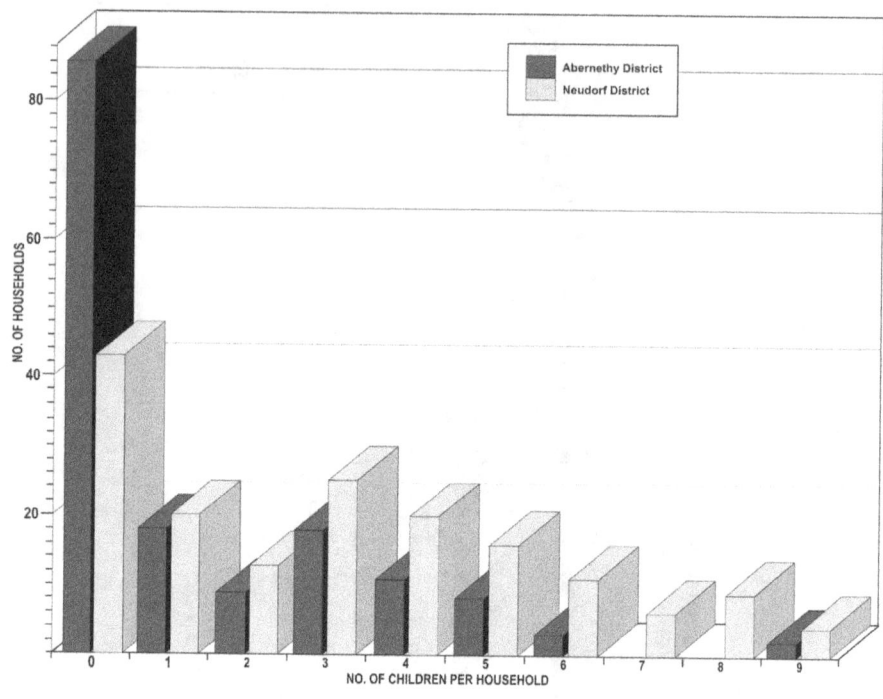

Fig. 7. Distribution of Abernethy and Neudorf district children per family at the time of application for patent.

from a comprehensive series of land titles searches for these townships over a period of 40 years following the initial date of patent.

The results of the persistence test indicate a reversal of the pattern established by the homestead cancellations, which showed the initial failure rate to be twice as high among the British Ontarians as it was among the Germans. After the issuing of patent, however, almost 30 per cent of the Neudorf homesteaders dropped out before the next five years had passed, most of these within the first year. Nearly 50 per cent sold out prior to ten years. In the Abernethy township, on the other hand, more than 70 per cent of the settlers were still present ten years after patent. After 20 years, 40 per cent of the Ontarians remained, a figure nearly double that of the Germans (Fig. 8).

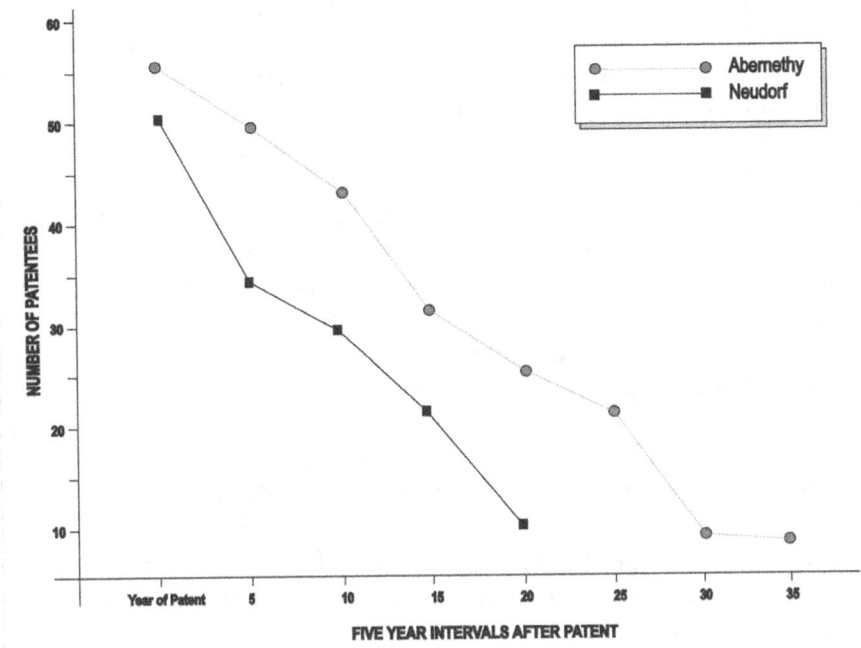

Fig. 8. Persistence of homesteaders in the Abernethy and Neudorf districts. Rates were calculated for 106 homesteaders in the Abernethy area township of Township 20, Range 11 and the Neudorf-area township of Township 19, Range 8, west of the 2nd Meridian. Owing to its later settlement, no titles were searched for Neudorf homesteaders beyond 20 years after patent.

To some extent the lower persistence of the Germans reflects the denser settlement of the Neudorf townships. German colonization at Neudorf took place principally after 1889, the year in which the right of pre-emption was abolished, effective 1 January 1890. Most Dominion Lands sections in the area found a different homesteader on each quarter. In the early period clear patches of land interspersed between wooded areas had provided a few easily broken acres sufficient to sustain a minimal lifestyle. Viability in the emerging commercial agricultural economy, however, demanded a much larger cultivable land base. Costs involved in clearing wooded areas ran to several times those entailed by breaking open prairie, and thus much of the land in the Neudorf area remained unbroken until the 1920s. Additional lands could be purchased when the CPR, Hudson's Bay Company and school lands were opened for sale, but if every patentee remained, he could look forward, on the average, to a modest expansion to 320 acres. On lands of secondary quality, such acreage was probably insufficient to sustain a farm operation by the 1920s. As late as 1918, the average farm size in Township 19, Range 8, south of Neudorf, was only 340 acres, and one-fifth of the farmers possessed a single quarter-section of 160 acres. For many German settlers, the road to prosperity was slow and marred by frequent failure.

The rapid disappearance of so many German homesteaders after the issuance of patent suggests that these settlers stayed only long enough to earn an equity before moving on to more suitable lands. Their perseverance through the "proving-up" period in some cases actually attests to their initial financial difficulties rather than success in homesteading. At least one settler doggedly tried to eke out a living because he did not possess the $10 required for another entry. But, according to a local historian, many others "would have gladly returned to their native lands, had they the means to do so."[70]

In 1895, Senator W.D. Perley of Wolseley wrote to the Deputy Minister of the Interior to appeal on behalf of his German constituents to the north. His description of their situation illuminates many of the problems they were then facing.

> I am told by several of them that for most of the summer season they haul water from one to five miles as they may happen to be near one or the other of four lakes. The want of water is the great difficulty they have to contend against. Most of the land is bluffy and light soil and this season the wheat is all frozen, I am buying to feed 200 bushels at 17 cents from one man. They haul wood to town 18 to 24 miles for a bag of poor flour at $1.50 per load at least 3 days and nights.[71]

German settlers in the other colonies of the District of Assiniboia also experienced difficulties in finding water. In 1894, one Frederick Seibold of Balgonie wrote to the *Regina Leader* to state that he had been forced by want of water to abandon his farm. Claiming that he had sunk seventeen wells to the depth of seventy feet on various parts of his homestead, he stated that it was impossible to draw water a long distance and prosper. Despite repeated promises by the local member of the Territorial Assembly to send an auger, no action had been taken, and Seibold warned that all his neighbours would similarly be obliged to leave.[72]

Where did the Germans go after abandoning or selling their lands? A perusal of *Cummins Rural Directories* in the period shows that by the First World War, some German farmers with names identical to the original Neudorf settlers had occupied lands in townships immediately west and to the north of the original settlement.[73] Township 19 in Range 9, to the west, had originally received a group of Anglo-Canadian settlers in the early eighties. At Lemberg, Richard Acton remembered that the early settlers found their mouldboard ploughs to be inadequate for the effective tillage of the heavy textured soils on their lands.[74] This technological gap precipitated their early departure from their homesteading ventures. By 1900, however, the appearance of the disc plough permitted the successful cultivation of these lands, and the German-speaking settlers moved in. Some Germans also migrated to Grenfell to the south, and others to Manitoba.[75] In 1897, the Department of the Interior reported that as a result of economic difficulties Germans from various settlements in the District of Assiniboia had emigrated to Texas. A number of these later returned when they discovered conditions in Texas to be worse yet than those they had encountered in the North-West Territories.[76] In sharp contrast to official accounts of prosperity, the German migrations were indicative of widespread hardship among immigrant settlers.

THE "FREE HOMESTEAD" POLICY AND ABERNETHY SETTLEMENT

Given the central role that the Dominion Lands administration played in the disposal of lands in the Abernethy area, it could not fail to have a profound impact on the district's early development. Historians of western Canadian settlement have been concerned with assessing the positive and negative effects of Dominion Lands Policy. Evaluated solely in terms of its original purposes the Dominion Lands Policy was, as Chester Martin has suggested, an "imposing" success. There is no doubt that between 1870 and 1930, the western Canadian prairie was transformed into a settled agricultural region. That the Dominion government's land disposal system and immigration policy was to a significant degree responsible for this fact seems indisputable. Yet the means by which this settlement was achieved have been subject to frequent criticism. The grid survey system in particular has been cited as a bureaucratically expedient method of land subdivision that created great problems for the settlers it was intended to serve. Drafted in the abstract, it bore no relationship to the pre-settlement geographic character and land use patterns of the prairie. Similarly, the adoption of the 160-acre quarter-section as the basic homestead unit was a bureaucratically neat solution, but officials may not have given enough consideration to the question of whether such a parcel was economically viable as a farm. Some writers such as Chester Martin and Vernon Fowke have drawn attention to the human toll that resulted from Dominion settlement policies, and have suggested that this factor must be considered in any evaluation of their overall impact.

The Dominion Lands Policy was formed in essentially two ways. Parliamentary legislation, in the form of the Dominion Lands Act, related orders-in-council, and amendments, constituted the cornerstone of government settlement policy. Beyond the statutory changes, the Department of the Interior was obliged to interpret the law and draft regulations to carry out the "Purposes of the Dominion." Individual cases became precedents upon which administrative policy was based. In the course of applying the legislative provisions, the Department came under criticism from various quarters. Many homesteaders who had been unable to comply with the provisions of the Dominion Lands Act and its regulations complained of inflexible treatment by federal officials. Others claimed that the Department was too lenient in its approach and urged the rooting out of apparent speculators and fraudulent homesteaders. Throughout the 1880s and 1890s, the Dominion Lands Policy was in a constant state of flux as the authorities experimented with new legislation in hopes of encouraging the rapid settlement envisioned in the National Policy.

It should be reiterated at the outset that a major discrepancy existed in the amended Dominion Lands Act in terms of access to free-grant land between the Anglo-Canadian settlers at Abernethy and the German-speaking settlers at Neudorf. Most Abernethy area settlers had arrived in the early eighties. Under the liberal provisions of the act at the time, these settlers were entitled to make entry for a homestead, pre-emption, and second homestead. For some settlers the total cash outlay required to obtain three quarter-sections of prime farmland was only $420.

As a result of perceived widespread land speculation, in 1884 the Canadian Parliament passed amendments to the Dominion Lands Act providing for the abolition of both the second homestead and pre-emption privilege to take effect in 1886.[77] Under normal circumstances, settlers arriving in 1882 and 1883 would have received their patents by 1886, at which time they could have bought a pre-emption for $2.50 per acre and also made entry for a second homestead. However, recurrent crop failures prevented most settlers from proving up their homesteads within the specified three-year period. Settlers in twenty-eight districts of the North-West Territories, principally Anglo-Canadian areas, petitioned Parliament for extensions to the period of eligibility for pre-emptions and second homesteads.[78] Their petitions were supported by a resolution passed by the North-West Territorial Assembly:

> Whereas it has been proved that for the success of the settler, it is necessary for him to engage in both grain and stock raising, and for this purpose he requires not less than three hundred and twenty acres of land; therefore the Assembly prays that the right of pre-emption be extended beyond the first of January eighteen hundred and ninety.[79]

Through the parliamentary efforts of N.F. Davin, MP for West Assiniboia, and later of Edgar Dewdney, Minister of the Interior, the period of entitlement for early entrants was eventually extended retroactively to 1889 in both instances.[80] Those settlers who did not elect to purchase their pre-emptions were given the opportunity to convert their pre-emption entries to second homesteads. The effect of this provision was to permit a settler to obtain two contiguous quarter-sections for an outlay of $20. Such generous provisions were a windfall to the predominantly Anglo-Canadian settlers in the 1880s. They were also a testament to their dominant position in the emerging political structure of the District of Assiniboia.

By the time German settlers had arrived at Neudorf after 1889, however, the right of pre-emption and of second homestead entry had been abolished. Not only were they restricted to a single quarter-section of free-grant land, but they settled on marginal land, only a fraction of which was arable. Lacking a political voice, the Germans were consigned to an economic backwater for decades after their arrival.

A second major anomaly in land administration was the indiscriminate opening of all manner of lands for homestead entry, regardless of their suitability for agricultural settlement. To some extent this policy was the result of the haste with which the original survey was conducted. Dominion Land surveyors had neither the time nor the expertise to appraise accurately the potential of various lands for agricultural purposes. They were required to rate the soils of the surveyed townships on a scale of one to four, but since no sophisticated criteria were established, the evaluations often gave a misleading impression. A case in point is the Pheasant Hills area to the east of Abernethy. In Townships 19 and 20, in Range 8, West of the 2nd Meridian, surveyors reported that most were predominantly second class although some lands were first class, and a significant proportion were of tertiary quality. Yet, despite early failures on these lands by some Anglo-Canadian homesteaders in the early eighties, all even-numbered quarters in these townships were reopened to entry by German settlers in the early nineties.

The adverse effects of opening these lands to settlement were worsened by the Germans' ignorance of the provisions of the Dominion Lands Act. The unfortunate case of Karl Hopp, a German settler at Neudorf, gives credence to the charges of "human wastage" in settlement. In December 1898, Hopp wrote to request the right to abandon his homestead in favour of another entry.[81] Stating that he had tried in vain to cultivate his lands during the previous five years, Hopp recorded his meagre yields in each year:

Year	Bushels Sown	Reaped
1894	15	30
1895	25	60
1896	40	200
1897	45	11
1898	75	100
Total	200	401

If contemporary maximum wheat prices are tabulated, Hopp's net revenue from wheat sales, after the deduction of seed grain, was $106 or $21 per year. From this revenue he was obliged to pay implement charges, feed his livestock, and provide for his wife and six children. Obviously, Hopp must have supplemented his income, through such means as hiring his children out to work for other farmers or through the sale of cordwood and hay, but such paltry wheat revenues may be taken as ample verification of his claims of destitution.

Hopp's case was complicated by his ignorance of technicalities in the Dominion Lands regulations, possibly resulting from language barriers. In February 1898 he had applied for a patent for his homestead, and the Department had accordingly issued a certificate of recommendation. Prior to receiving the actual patent, however, he wrote back in December to state the impossibility of making a living on his farm and to request an abandonment of his claim to permit him to make another entry.

Since Hopp had already been issued a certificate of recommendation, however, the Department ruled that Hopp's request for an abandonment could not be entertained. In consolation, a Department officer wrote:

> In view of your statement that after five years of trial you have found it impossible to make a living for yourself and family upon the land, an application from you for permission to buy a quarter section at the minimum price of $1 per acre, subject to the ordinary homestead conditions, would receive consideration.[82]

According to the Department's terms of reference, its decision to permit Hopp to purchase another quarter for a dollar per acre seemed a lenient gesture. To Hopp, however, the opportunity to buy a homestead for a price exceeding his total grain income over the preceding five years must have seemed a cruel joke.

It might be noted that Hopp was slow to request an abandonment. But even if he had wished to leave his homestead earlier, he probably could not have done so. Additional homestead entries required an entry fee. Hopp's neighbour Jacob Popp, who found himself in a similar situation, wrote that he "would have taken up another farm if I had been able to pay $10." Popp could not exchange his homestead because his house "cost me that last bit of money I had."[83] In August 1899, a Departmental Homestead Inspector visited the Hopp homestead to corroborate Hopp's statements and discovered that he had died several months earlier. He described the land as follows:

> The land is of a very inferior quality for agricultural purpose as it is very rough and stoney also being full of small sloughs and willow shrubs. It is best adapted for grazing pasture.[84]

Nor was Hopp's experience of settling on unfit land an isolated case. Several other settlers in the Neudorf area wrote to the Department of the Interior with similar complaints. In addition to problems with scrub, marsh, and stones, a large number discovered that there was no water to be found for miles around. It would be unfair to blame the Dominion Lands administration for the lack of water on these lands as it is probable that language barriers inhibited some settlers from seeking early abandonments.

In studying a cross-section of homestead files in the two areas, one gains the impression that notwithstanding inequities in the legislation, the actual administration of the Dominion Lands Act was carried out in a reasonably fair manner. Within the guidelines of the Act, Interior Department officials were concerned principally with encouraging bona fide settlers to stay on the land. Often they were caught in the unenviable position of trying to adopt a flexible approach to meet the needs of legitimate settlers, while being forced to accommodate the demands of "claim jumpers" seeking to cancel the original settler's entry. The administrators could not ignore the legislative provisions, but they frequently bent the rules to give the homesteaders a chance to comply with the Act.

In terms of the settler's duties, for example, Department of Interior officials were obliged to insist that the mandatory period of residence on the homestead be completed. At Abernethy James Gaddis, who had resided two and one-half miles from his homestead quarter, requested a waiver of the two-mile limit to permit him to obtain a patent without having lived on the land. His request was refused.[85] On the other hand, the Department frequently granted informal extensions to settlers to permit them extra time in which to fulfill their responsibilities. In 1895, the Neudorf settler Jacob Adolf stated that he wished to go into service to earn money for the purchase of farm implements. A Department official wrote back to state that while the Commissioner could not grant a leave of absence,

> in consideration, however, of your agreement to have the five acres cropped during the coming season and a farther acreage broken, the Agent will be instructed not to accept any application for cancellation without reference to this office it being understood that you will go into permanent residence not later than the 15th of May 1896.[86]

In other words, the Department was granting the equivalent to a leave, without technically breaking its own regulations.

At Balcarres, George Balfour wrote to state his intention of leaving his homestead to earn money during the summer of 1889. Fearing attempts by his neighbours to jump his claim Balfour asked, "will it please our Hon. to grant me leave of abstinence [*sic*] for say six months from the first of April 1889?"[87] Whether or not Balfour's choice of words was a Freudian slip, the Department granted him his request.

In those cases in which a settler had unknowingly broken the regulations, the Department tended to take a lenient approach provided it was convinced of his bona fide intentions. In 1882 William McKay had settled on the west half of Section 4 in Township 20, Range 11, West of the 2nd Meridian. McKay undertook fairly extensive improvements; by 1885 he had placed five acres under crop and possessed $400 worth of buildings. That year, in apparent ignorance of the law, he entered into an agreement to sell his lands prior to obtaining a certificate of recommendation for patent. H.H. Smith, Dominion

Lands Commissioner, charged Homestead Inspector Arsenault with investigating the matter. Arsenault reported that McKay

> is a good, honest, innocent, harmless soul, not well posted in the land regulations, and I am confident he signed the agreement in question ignorant of the land law. Seems the assignee is represented as a man who will take advantage of his neighbour if he can and lead an innocent person into a trap....[88]

On the basis of this report, Smith recommended leniency. While McKay had forfeited his entry, Smith recommended that he be given the opportunity to purchase his half-section at $1.25 per acre and his action was approved by the Ottawa authorities.

On the other hand, it appears that settlers of different ethnocultural origins were sometimes disadvantaged by their unfamiliarity with the English language in dealings with the Department of Interior. The Department conducted all business and correspondence in English. Correspondence from German settlers at Neudorf and French-speaking Métis at Qu'Appelle suggests that many individuals were ignorant of the provisions of the Dominion Lands Act, and suffered as a result.

The contrasting settlement experience of the Abernethy and Neudorf districts was representative of developments in other Anglo-Canadian and eastern European immigrant colonies in western Canada. The Anglo-Canadians arrived first, and claimed the best lands that were most suited to wheat production. By virtue of occupying a dominant position in the political structure of the North-West Territories, they were able to lobby effectively to extend the period of eligibility for bonus quarter-sections of land (the pre-emption and second homestead privilege). Those farmers who could not afford to purchase their pre-emptions were given the additional option of converting their pre-emptions to second homesteads. As a result many Anglo-Ontarians obtained 320 acres of excellent farmland for only $20; those who retained their pre-emption entries obtained 480 acres of land for $420. In either instance, they possessed an enormous advantage over subsequent settlers.

Despite these advantages in purchasing land the Anglo-Canadian homesteaders often experienced considerable hardship that attended the lack of accessible rail linkage during the Abernethy district's first two decades. The Macdonald government's repeated granting of extensions to the Great North West Central Railway Company, in which several prominent Conservatives held a healthy interest, helped retard settlement north of the Qu'Appelle River. Ottawa's perceived unresponsiveness to the farmers' needs later contributed to the outbreak of agrarian radicalism in 1901.

The German-speaking settlers at Neudorf arrived mostly after 1889, the year in which the right of pre-emption was abolished. They were thus restricted to a single quarter-section of free-grant land. The value of their homestead quarters was further reduced by the generally marginal nature of lands in the Pheasant Hills district, which tended to be broken up by scrub and marsh and whose soils were of secondary quality. Other difficulties apparently stemmed from language barriers that hindered communication with the Department of Interior regarding the German immigrants' homesteads.

The appearance of Euro-Canadian and European settlement in the central Qu'Appelle region was accompanied by a diminution in the fortunes of the Aboriginal inhabitants who had previously occupied these lands. For First Nations people who had experienced the loss of the buffalo, their economic mainstay, the reserve system at least provided them with the essential means of subsistence. The Métis had fewer protections. While federal authorities passed various bills and orders-in-council providing for Métis land claims, most of these lands fell into the possession of Euro-Canadian settlers or land companies shortly after their disposal. Cree and Métis residents continued to live in the region, but thereafter on the margins of the new Euro-Canadian society that had asserted its dominance.

2

Estimates of Homesteading Costs in the Abernethy District in the Settlement Era

Claiming land was only the first stage in the settlement process. Since free-grant land was unimproved, Abernethy settlers confronted the task of building necessary structures and preparing their land for crop. To fully understand the community's subsequent economic development it is necessary to determine how much capital was actually required in "proving up" a homestead. One recent study estimated that average start-up costs of settlement on the prairies were about $1,000 in 1900.[1] Yet homesteading costs were not fixed for all settlers. They could be affected by a multiplicity of variables, including the settlers' cultural background, the type of farming practised, family size, and relationships with neighbours. Environmental factors, such as access to wood and water, could also have a significant impact on start-up costs. Detailed analysis is needed to differentiate the costs among different settlement groups.

For this chapter, homesteading costs were tabulated both for Abernethy settlers and for the German-speaking settlers at the neighbouring community of Neudorf. The German immigrants were included as a basis of comparison to help establish the minimum amount required to set up a homestead. All 461 Department of the Interior homestead files relating to six townships – three at Abernethy and three at Neudorf – were examined. Most of these files contain the Application for Patent, a form requiring the homesteader to state, among other things, the extent and value of his or her improvements. These improvements usually include the homesteader's dwelling and outbuildings, as well as the amount of acreage broken, cultivated, and fenced. Since the Application for Patent was filled out after the settler had performed his requisite duties under the Dominion Lands Act, the values recorded in the files do not reflect the initial expenditures so much as the accumulated improvements made in the first three or more years. For the Abernethy area, the mean year of application was 1894, for Neudorf, 1901. In addition to the quantitative data, qualitative sources such as pioneer diaries, early newspapers, colonization journals, and immigration pamphlets have been consulted to provide a larger context in which to treat the Abernethy-area findings.

In approaching the question of homesteading costs, it would be useful to first establish the extent of financial resources available to settlers. The absence of concrete data in this instance requires that an approximation of resources be reconstructed through individual case studies. Settlers arriving in the townships surrounding Abernethy between 1882 and 1905 formed part of the preponderant Ontarian migration to western Canada in the thirty years after Confederation. In his studies of Peel County after 1850, David Gagan has shown how the prevailing Ontarian inheritance system forced younger sons off the farm, many of whom joined the massive migration to the prairies.[2] Gagan's figures respecting the average debt per farm suggest that little surplus was available to the second and third sons of most Ontarian farmers. These men might have worked for a few years before taking up a homestead, but a demographic breakdown of Abernethy homesteaders shows that this group contained a large portion of very young men – 40 per cent were twenty-four years or younger and 23 per cent were twenty-one or younger. Given the then current scales of pay in Ontario,[3] $1,000 seems a very large amount for men with only two or three years in the work force to have saved.

Wages were higher in western Canada than in the East, although in the 1880s the maximum wage obtainable there for experienced farm hands was $35 per month, including board.[4] Threshing labourers were paid at a higher rate, but the threshing season lasted only two months, and the thresher was often obliged, if he could find work for the winter, to hire on at a very modest salary. Agricultural historian John Herd Thompson has argued that most harvest workers would have experienced great difficulty in raising even $600, which was the figure cited by Saskatchewan premier Walter Scott in 1906 as the minimum amount required to establish a homestead.[5]

Similar inferences may be drawn from the case studies of individual settlers. After graduating from the Ontarian Agricultural College, W.R. Motherwell, the third son of a Lanark County farmer, came west in the spring of 1881. He did not immediately file for a homestead, but worked for two summers and harvest seasons in Manitoba before settling in the Pheasant Plains area of what was then the District of Assiniboia in 1882. It is difficult to determine how much money Motherwell brought to his homesteading venture, but given the straitened circumstances of his parents, farmers on the thin soil of Lanark County near Perth, Ontario, the possibility seems remote that he received much family assistance. It is equally improbable that he saved large sums during two expensive years at college. Taking into consideration the current farm wages and the cost of living, the most plausible range of capital that Motherwell might have earned and saved in the two years following his graduation is $400 to $600.

Testimonials of other settlers in the Qu'Appelle region indicate the amount of liquid capital many newcomers brought. John Burton, aged twenty-one, homesteaded two miles north of the present village of Abernethy in 1882. A native of Bruce County, Ontario, Burton related that he had only $65 when he arrived.[6] Samuel Copithorn, who took up land northwest of Balcarres, emigrated from Toronto with a year's earnings of $200 in his pocket.[7] Both settlers eventually became successful farmers. Indeed, the pages of contemporary agricultural journals and immigration pamphlets are filled with dozens of similar testimonials vis-à-vis meagre starting capital.[8]

Local newspapers from the settlement period confirm that many settlers brought only a few hundred dollars to their farming ventures. In April 1889, the *Regina Leader* reported the arrival of fifty-three Canadians and twenty-two settlers of other nationalities. The total value of their effects was $6,000, and cash, $12,000, giving an average figure of $240 per settler as a starting capital.[9] In another report the *Leader* noted that fifty Ukrainian families at Batoche each possessed capital ranging between $40 and $1,000.[10] In 1885, German immigrants established the colony of New Toulcha north of Balgonie. Of thirteen farmers surveyed by the *Leader*, none had owned more than $250 at the outset. Each German homesteader reportedly shared ownership of a team of oxen, harness, plough, and wagon with another settler.[11] It could be misleading to treat these testimonials and journalistic accounts as representative of the majority, but there is ample evidence to support the view, commonly held in the period, that "energy, experience, judgment, and enterprise"[12] were probably more important to the establishing of a successful farm than a large initial investment.

If a large starting capital was beyond the reach of many settlers, it is useful to explore alternative forms of financing that might have been available. Under the terms of the Dominion Lands Act, a settler was prohibited from mortgaging homestead lands prior to the issuing of a patent.[13] While the minimum "proving-up" period was three years, the majority of settlers took even longer to fulfill their homestead duties.[14] The homesteader might negotiate a bank loan, but with little collateral it seems unlikely that banks would have advanced him more than a few hundred dollars.

Land purchased from corporate interests could be mortgaged, but required a down payment. Canadian Pacific Railway land could be acquired with an initial payment of one-sixth or one-tenth of the purchase price, followed by five equal annual instalments. In the period before 1900, CPR land commonly sold at prices ranging from $2.50 to $5 per acre,[15] depending on its distance from the railroad and grain handling facilities. In these instances, a settler was obliged to pay out at least $66.67 initially for a quarter-section purchased on time, but often this payment was a much larger sum. As the agrarian sector finally began to fill out after 1900, CPR land prices jumped dramatically. After 1900, therefore, purchased land tended to be prohibitively expensive for the newcomer of limited means.

By tabulating the individual costs involved one can determine the minimum expenditure for establishing a farm. The first and most essential expenditure encountered by the prairie settler was the cost of building a shelter. Ankli and Litt have estimated the cost of housing at $200 to $300 for most settlers. They based this conclusion on two sources: an article on early housing written from a series of pioneer questionnaires distributed by the Saskatchewan Archives Board in 1955 and James M. Minifie's book *Homesteader*,[16] in which Minifie cited the valuation of $400 on their home placed by his father on his homestead Application for Patent form. For the late nineteenth century, however, a quantitative investigation of other homestead applications suggests a lower figure. In the Abernethy district, 106 settlers, largely of Anglo-Canadian origin, assessed their houses at an average value of $253, but at Neudorf, 147 Germans valued their dwellings at an average of only $168.

The criteria employed in these valuations may have varied from one applicant to the next. It is not clear, for example, to what extent the applicants took into consideration the cost of materials, the labour cost or the overall market value. The paltry values recorded by a large proportion of the settlers suggest, however, that capital expenditures for housing were often very small indeed. In the two study areas, 46 per cent of the homesteaders valued their houses at $100 or less and 17 per cent recorded values of $50 or less. Although some settlers placed values of $1,000 or more on their houses, it should be remembered that these valuations were made at the time of application, several years after the entry. Even these relatively prosperous settlers frequently put up cheaper temporary structures at the outset. These structures served as shelters for a year or two until a more substantial residence could be built. The values found in the homestead files are therefore probably higher than the actual initial investment. In the Neudorf area, the log dwellings listed in the homestead files were preceded by even more primitive initial shelters. A local resident has described the building of these habitations:

> The first thing they would do was dig a hole in the ground and bank it up with sods, put a few poles and hay on top, make a big oven of clay in the middle of the house, no floor, and live in it until they built a log house.[17]

Such mean lodgings were common among the poorer European peasant immigrants, but some of the Anglo-Canadians built similar shelters. An English family in the Primitive Methodist Colony northeast of Abernethy "spent their first winter in a dugout in the east side of a hill. The roof was covered with green poles, dirt, and hay."[18]

Savings on housing costs stemmed principally from the availability of indigenous building materials. Sod was obtained simply by ploughing the prairie turf. Logs could be collected at little or no capital cost, particularly in the early period, when settlement tended to cling to the parkland belt and the woodland-prairie margin. In these areas the settler's actual investment was limited to the purchase of nails and tools, lumber for doors, door frames, windows and, occasionally, flooring and roofing. Poorer settlers typically opted for a sod roof and a dirt floor. One farmer in Manitoba in the early 1880s suggested that $60 to $75 was required for pine flooring and window and door frames.[19] Another, writing in 1895, estimated the cost of lumber finishing for all his farm buildings to have been $110.[20] A 1902 immigration pamphlet set the lumber cost of doors and windows for a log house at $50.[21] At Neudorf valuations of homestead dwellings ranged as low as $5, suggesting that a settler could build a shelter with an almost negligible investment as long as he was prepared to forego all amenities and comforts for the first year or two.

Most Abernethy and Neudorf settlers had the initial advantage of being able to choose lands near woods. A tabulation of data from the homestead files for six townships in these two districts indicates that the great majority of homesteaders in both areas built their first dwellings of logs. Of 257 homesteaders who reported the building materials for their dwellings, 161, or about 63 per cent, fashioned their homes of logs. Another 30 homesteaders, or 12 per cent of the total, built their units with a combination of logs and sod. Fifty-one, or 20 per cent, of the dwellings were wood frame, 13 were a combination of frame and sod, and another dwelling was a combination of frame and log construction. While many Abernethy residents eventually built their permanent houses of masonry, only one of the 257 respondents reported a masonry building for his first dwelling, in this case, fashioned of stone. Figure 9 is a view of W.R. Motherwell's first log house near Abernethy after being re-faced with clapboard siding. As the range of homestead lands expanded into the true prairie of the Palliser Triangle after 1900, many settlers were obliged to purchase more of their building materials. Even there, however, a habitable shanty could be put up with a relatively modest outlay. Willem de Gelder, a Dutch immigrant who homesteaded north of Morse, Saskatchewan in 1910, built a ten-foot by twelve-foot lumber and shiplap shanty for $100 including extra labour costs.[22]

This last example illustrates that the cost of shelter was also related to the size of family that was to be housed. Western Canadian settlement was characterized by a large proportion of bachelor homesteaders, and in the Abernethy area, more than one-third of the homestead applicants were unmarried at the time of application. Since some of the applicants had married in the interval between entry and application, an even larger proportion was single at the outset. Bachelors tended to select their lands adjacent to the homesteads of their friends

Fig. 9. W.R. Motherwell and his family in front of their log house, ca. 1890. The house was built in 1883 and later faced with clapboard siding. W.R. Motherwell Collection, R87-219, No. 302.

with whom they would share accommodation during the first year or two. In addition to providing needed company on the lonely frontier, these arrangements were of economic importance. Settlers could save money by sharing the expenses of food and shelter and by pooling their equipment and labour.

The Dominion Lands Act permitted a settler to postpone establishing a residence on his homestead provided that he lived within a radius of two miles from his quarter-section after the initial entry.[23] He was still obligated to build a "habitable" dwelling and live in it for three months prior to application, but his provision allowed him to live with neighbours for up to thirty-three months of the thirty-six-month "proving-up" period. That many took advantage of the clause is evident from a reading of individual homestead files. Shared living arrangements were common not only among young bachelors, but also among married homesteaders who had left their wives in Ontario for one or two years until they were in a position to bring them out to a finished home.

Estimates of Homesteading Costs

The settler also required a shelter for his livestock. In the Abernethy area study, only one of the homesteaders reported the existence of a barn, although roughly half of the Anglo-Canadians and 75 per cent of German settlers stated that they possessed stables. The average value reported for the Germans' stables was $75 compared with $139 for the Anglo-Canadians. As was the case with the house values, these arrangements for stables probably reflect the market value and are higher than the actual capital investment. Most respondents stated that their stables were made from logs. This obviously represented a great saving in material costs. Granaries and other outbuildings were also principally constructed of logs. Abernethy applicants valued their granaries and outbuildings at an average of $79; Neudorf homesteaders appraised their outbuildings at $58. Overall, if the structural improvements on homesteads in the study areas are broken down by quartile, a wide variation in assessed values is revealed (Table 1).

Beyond the structural costs, establishing a residence entailed expenditures for sundry household items, including a stove, furniture, bedding, and kitchenware. Estimates for these costs vary from one pioneer account to the next, a fact which may be indicative of the diversity of needs and resources of different shelters. With respect to stoves, it appears that the Ankli and Litt estimate of $40 is higher than the essential minimum. In 1891 the *Qu'Appelle Vidette* carried an advertisement for "cheap patent stoves" ranging in price from $16 to $80.[24] Isaac Cowie's pamphlet *The Edmonton Country*, published in 1901, reported that cooking stoves in that locality could be purchased for $23 to $26.50.[25] Georgina Binnie-Clark, who settled north of Fort Qu'Appelle in 1905, later wrote that she had purchased a second-hand stove for $15.[26]

Other necessary chattels could be purchased relatively cheaply. Cowie's pamphlet recorded the price of hardwood chairs to be from 55 cents to $1; tables, $3 or more; and bedsteads, from $4 up.[27] Alternatively, all of these articles could be handmade. Tables and chairs might be constructed with the available supply of timber, and beds were commonly made by sewing a tick and filling it with straw. Under the most frugal circumstances, settlers were still obliged to purchase lamps, bedding, kitchen utensils, and tools, such as axes, saws, and nails. Contemporary immigration pamphlets suggest that the costs of furnishing a log house could range between $20 and $75.[28] The minimum tools required by a settler building his own log house would seem to include a spade, crosscut saw, hammer, chisel, brace and bits, planes, auger, axe, and some nails. In 1902, these items cost between $11.40 and $21.70 at Edmonton.[29]

TABLE 1. Value of homestead structural improvements by quartile, Abernethy and Neudorf Districts, 1882–1917, in current dollars.*

	I	II	III	IV	
Value of Dwelling	$5–75	$75–150	$150–250	$250–1500	(n:303)
Value of Stable	5–30	30–50	50–100	100–725	(n:147)
Value of Granaries & Out-buildings	2–20	20–50	50–75	75–400	(n:81)
Value of Fencing	5–20	20–50	50–100	100–500	(n: 111)
TOTAL	$17–145	$145–300	$300–525	$525–3,125	

* In terms of constant dollars, house values increased overall after 1900. The mean valuation for pre-1900 dwellings was $201 (1900 dollars) compared with a mean of $258 between 1900 and 1914. If valuations are grouped by five-year intervals. however, the trend is less clear.

INTERVAL	No. OF OBSERVATIONS	MEAN VALUATION
1886–89	53	$209
1890–94	40	354
1895–99	106	139
1900–04	43	228
1905–09	40	296
1910–14	14	238

The low average recorded between 1895 and 1899 reflects the preponderance of poorer settlers at Neudorf in this period (89 of 106). Apart from this era, house values averaged $271 before 1900, and actually dropped slightly after the turn of the century.

No settler could avoid the necessary expenditures for provisions, although estimates of food costs diverge greatly from one contemporary account to the next. In 1882 John Macoun wrote that a settler with a family of five would spend about $250 on provisions during the first year.[30] A year later, however, a British settler in Manitoba estimated his cash requirements for groceries to be only $20.[31] Obviously the outlay for groceries varied with the size of family and the kind of lifestyle the settler pursued. Two accounts – an immigration pamphlet published in 1902[32] and a Dutch homesteader's tabulation written in 1910[33] – place the cost of provisions for a bachelor settler at about $90 to $120.

This capital requirement would tend to increase in the case of a larger family, although poorer peasant settlers could and did make do with less. An immigration pamphlet published in 1882 stated that the cost of provisions for one family of five Mennonites at a subsistence level had been $93.[34] Their diet consisted almost solely of flour, pork, and beans. Alternatively a settler could spend much more. In 1907, for example, Georgina Binnie-Clark spent $245 for groceries, flour, meat, repairs, and veterinary fees, although in her view the "degree of necessity" was a much smaller amount.[35]

Fencing was a requirement for farmers who raised livestock, but the most striking aspect of farm fencing in the Abernethy district in the homesteading period was its absence. Of 461 applicants, only 111 reported the presence of fencing on their homesteads. Among this group, there were wide variations in acreage fenced. The average extent was forty-four acres at Neudorf and sixty-three acres at Abernethy, but the majority enclosed thirty-five acres or less and more than 20 per cent fenced in areas of twenty acres or less. Settlers also reported widely diverging expenditures for fencing. One appraisal stated a cost of only nineteen cents per acre; others valued homestead fencing at up to $4 per acre. Overall, the mean cost for Abernethy settlers was about $1.25 per acre, a figure that is confirmed by most estimates in the period.[36] Given that most settlers began farming with no more livestock than a yoke of oxen or team of horses, a sufficient initial expenditure for fencing in the 1880s and 1890s was $10 to $15 for a small pasture of ten acres.

Water was an essential requirement for all settlers. Of the nineteen Abernethy settlers who reported wells on their homestead applications, fourteen recorded values of $50 or less and the average valuation was $23.50. Others who were less fortunate were obliged to abandon their homesteads altogether for lack of water. Still, the variability in these cases is so great that they cannot reasonably be averaged with the costs incurred by those individuals who did strike water. Assuming that the settler did not need to hire a professional well digger, he could avail himself of a government-owned digging apparatus for a nominal fee. One homesteader at Morse, Saskatchewan wrote that he spent seventy-five cents per foot to dig a well in 1912.[37] Another source for 1912 placed the cost of well-digging at $1 a foot for the first fifty feet. Thereafter the cost increased steadily until it reached $2 per foot at a depth of a hundred feet.[38] The writer observed that in wooded areas water could generally be found at a depth of twenty to forty feet. On the open prairie, despite the presence of occasional springs, a well digger would usually be obliged to go to a much greater depth. Assuming that a settler struck water at twenty feet, as several Abernethy settlers appear to have done, the cost of digging a well, in 1900 dollars, was about $14. To crib his well the Morse homesteader estimated that he needed three hundred feet of lumber, costing $9.[39] If this estimate is converted into 1900 dollars, the overall cost of constructing a well of twenty to forty feet was $21 to $35.

Livestock constituted another homesteading cost that was highly variable. Professor Ankli has estimated the cost of horses to have been $75 to $100 at the turn of the century, and bases this figure on the 1901 census, which gives the average value of horses in Manitoba and the North-West Territories as $96 and $62 respectively. Since the census average includes horses of all sizes, the figures are misleading. The basic operations of breaking sod and pulling implements required heavy draught horses, which were more expensive than the average. In 1881 a settler in Manitoba estimated the cost of a team of horses and harness to have been $325.[40] Walter Elkington, who settled north of Fort Qu'Appelle in 1891, wrote that teams of horses ranged between £30 and £60, or $150 to $300.[41] In 1912 a Dutch homesteader at Morse, Saskatchewan purchased a team of large sorrel horses for $450.[42] The following year, Boam's *The Prairie Provinces of Canada* included an estimate of $360 for a team of "good horses."[43]

Alternatively a farmer might purchase oxen to pull his implements. While a team of oxen sometimes cost $200 or more,[44] most estimates in the 1880–1900 period range between $100 and $130.[45] Costs varied according to the age and quality of the animals and to the time of year in which they were purchased. Oxen sold in the spring, for example, would command a higher price than those sold in the fall.[46] The usual practice was to purchase a yoke of oxen for the purpose of prairie breaking and to exchange them for more expensive horses a year or two later (Fig. 10).

One homesteading cost that few settlers could avoid was the outlay for a wagon. Wagons were essential to the hauling of farm products to market centres and supplies from the town back to the homestead. They were used in the collection of wood for building and fuel, and were an essential means of transport to social and religious activities. Settlers could, and often did, begin farming with Red River carts worth $10,[47] but once they had begun to transport large quantities of farm produce, wood, and hay, the purchase of a more serviceable wagon was mandatory. Four sources for the 1880s give $80 as the standard price of a farm wagon,[48] a figure that does not seem to have changed for the entire period leading up to the First World War. These estimates pertain to the entire unit, including a four-wheel chassis and removable wooden box. For winter travel, settlers were also obliged to purchase a set of sleigh runners on which to mount the wagon box. One Manitoba resident wrote in 1883 that the price of a pair of runners was then $30.[49]

Expenditures for farm implements could exceed $1,000, or several thousand dollars, if a settler purchased a threshing machine. For purposes of initial farm making, however, most settlers' requirements were much more modest. In most accounts, a small investment of $40 was sufficient to purchase a prairie-breaking plough, a stubble plough, and a harrow.[50] These simple implements permitted the breaking and preparation of several acres for crop. Cultivation was of necessity very crude as the prohibitive cost of sophisticated implements required that the poorer farmers broadcast seed, stock, harvest, and thresh their crops by hand, using such primitive tools as cradles and flails.[51] Professor Ankli's estimate of $330 as the cost of implements for "substantial settlers" does not seem unreasonable, although it is difficult to accept an overall estimate without some indication as to the size of farm for which these implements were considered sufficient. Obviously, too, a farmer's expenditure for implements was related to the kind of crops he or she cultivated and to the extent to which they raised livestock. If farmers raised cattle, for example, it was incumbent that they invest

Fig. 10. "Breaking sod in the Bracken district, 1911." Saskatchewan Archives Board, Regina, Photo no. R-B1410.

in a mower and rake for purposes of haying. Table 2 provides price quotations for implements essential to farm 160 acres.

Most settlers did not incur expenses for seed in the first year of settlement, as few were able to prepare any of their newly-broken land in time to put in a crop. In the study areas of Abernethy and Neudorf, only forty-five or about 10 per cent of the 461 homestead applicants reported that they had placed one or more acres under cultivation in the first year. Sixty-two per cent reported no cropping, and the remaining 28 per cent did not answer the question. It should be noted that before 1900 comparatively few settlers began farming on improved land. Purchasers of CPR and colonization company lands were similarly obliged to break virgin prairie prior to cultivation. Most sources suggest that 1.5 to two bushels of seed grain were sufficient to sow one acre of wheat. At $1 per bushel, seed sufficient to sow ten acres could be purchased for $15 to $20, but even this small expense could be deferred if the settler obtained a seed grain advance from the Dominion government.

TABLE 2. Price quotations for selected farm implements on the Prairies, 1884–1915.*

INTERVAL	1884[52]	1889[53]	1897[54]	1909[55]	1914[56]	1915[57]
Breaking Plough (Walking)	22	22	20	22	20	24
Stubble Plough (Walking)	16	20	18	22	14.5	24
Iron Harrows	18	20	15	20	18.5	20
Drill Seeder	55	65	90	88	85	100
Mower	80	63	55	52	56	52
Binder	290	150	155	150	145	150
Horse rake	35	20	28	33	33	33
Sleigh Runners	30	28	25	37	35	38
Total in Current $	546	388	406	424	407	441
Total in 1900 $	494	371	477	369	310	324

* Price quotations were deflated using the J1 Wholesale Price Index, after Mitchell, 1868–1925 (Urquhart and Buckley, *Historical Statistics of Canada*, p. 291). Since each set of price quotations relates to a different location and documentary source, allowance should be made for variations in freight rates and retail mark-up. An effort was made to ensure that quotations pertain to the same kinds of implements, but it must be recognized that technological changes had modified some of these implements by the end of the period.

If the essential tasks of wood gathering and house building limited the amount of acreage cropped in the first year, these factors also served to curtail the extent of breaking for most settlers. Anglo-Canadian settlers at Abernethy broke an average of fifteen acres in the first year, compared with an average of eight acres broken among the German settlers at Neudorf. If a farmer elected to contract his breaking out he usually paid from $1.50 to $3 an acre for his work,[58] and an additional $2 if he also wished to have the land backset in the fall.[59] Ankli is correct in his statement that by 1900 steam ploughing offered a cheaper alternative than breaking with horses or oxen. Yet the continuing debate in the agricultural press in the early 1900s suggests that steam ploughing was slow to supersede horses and oxen as the preferred method.⁶⁰ The relatively small extent of land clearing in the Abernethy area suggests that the majority of settlers performed their own breaking with mould-board ploughs.

A factor which should not be overlooked in assessing homesteading costs was the role of co-operative activities among settlers. While the principal mode of land disposal was that of individual "free" homestead, as opposed to group settlement, settlers tended to select homesteads close to their friends or relatives, to permit the pooling of implements, livestock, and labour. Pooled labour was common not only among continental European immigrants, but in the Anglo-Canadian communities as well, as the memoir of an English settler near Fort Qu'Appelle indicates:

> The people in Canada are very good in helping one another; if a man wants a new house or stable put up, he gets the material ready and goes around among his neighbours and asks them to come and help him on a certain day; this is called a "bee" and it is not only done for buildings, but for ploughing, sowing, reaping, or if a man has had a misfortune and is behind in his work, the neighbours go in together and give him a day's work.⁶¹

Similarly, expensive implements, which were beyond the means of many farmers, were sometimes loaned by more prosperous settlers in exchange for a day's labour from the borrower.⁶² Alternatively, some setters found that they could still save money by contracting various farming operations out. During the first five years, for example, W.R. Motherwell hired a neighbour to do his binding.⁶³ Settlers also defrayed capital costs by purchasing implements and livestock in partnership with their neighbours. John Teece, who homesteaded near Abernethy in 1883, held a one-fourth interest in a team of horses, wagon, and plough during the first four years. Thereafter, he possessed his own yoke of oxen, wagon, and plough, but purchased a one-fourth interest in other implements. Shared ownership does not seem to have impeded Mr. Teece's progress as a farmer; in 1913 he reported that he was in possession of nine quarter-sections of land and $10,000 worth of livestock.⁶⁴ Another way in which partnerships operated was to cultivate crops "on shares."⁶⁵ One Manitoba settler related that farmers commonly worked out arrangements whereby one settler would perform

Estimates of Homesteading Costs

the cultivation duties for two farms while his partner continued to work on the railway to support both homesteads.[66]

There is little doubt that the majority of settlers needed to borrow or hire the use of draft animals initially. Fewer than 25 per cent of the German homesteaders recorded having owned cattle in the first year of farming and only 10 per cent stated that they owned horses in the same year. At Abernethy, approximately one-third of the homesteaders reported that they had owned cattle in the first year, compared with fewer than one-fourth reporting the initial ownership of horses. It is possible that some settlers misread the question on the form and therefore failed to answer it completely. Such a small percentage of respondents, however, leaves little doubt that a substantial portion of this homesteading population did not own oxen or horses at the time of entry. Reliance on their neighbours was therefore axiomatic.

It is important to recognize that farm making was usually a gradual process in which homestead development was phased in over a period of several years. Settlers normally made only a partial beginning in the first year, and gradually improved their holdings over the "proving-up" period of three years or more, and beyond it. In the first year a typical pattern was to select land early in the spring and put up a tent or very rudimentary shelter. No fieldwork could be performed until after the spring run-off, but once it was dry enough to plough, the settler would break five to fifteen acres. If his land was stony, rocks would have to be picked, and the rate of breaking would be slowed. Working industriously, the settler could harrow a few acres in time to plant small amounts of wheat, oats, and root crops sufficient to provide him with flour, livestock feed, and vegetables for the winter. Late June and July was customarily devoted to haying. Homesteaders who settled within reasonable proximity of sloughs would be able to cut enough hay to meet their own requirements. Any surplus could generally be sold in the towns or to other farmers. During the harvest months of August to October, the new settler would often find employment with threshing crews harvesting in the developed farm communities in Manitoba and districts along the CPR main line. For the duration of the winter, young settlers frequently hired on with other farmers, or found work with the CPR or other employers.[67] Those who stayed on their homesteads tended to such tasks as fencing and other farmstead improvements. Commonly, they worked in the bush most of the winter, cutting wood for their own needs and marketing surplus cords in town

Fig. 11. "Harvesters gathering wood for fuel in a ravine, Pinto Butte, Saskatchewan," ca. 1900–1920. Saskatchewan Archives Board, Saskatoon, Photo no. S-B 9283.

to support themselves.[68] Figure 11 depicts several Saskatchewan settlers gathering wood for fuel which they obtained in coulees or ravines in the early twentieth century. This income might be supplemented by sales of milk, eggs, and other livestock products. Generally, farming in the first year of settlement was of a subsistence nature and was carried out chiefly to offset costs while homesteaders prepared their farms for future market-oriented agriculture. Only after the passage of several years, and the accumulation of the necessary buildings, tools, prepared acreage, and other improvements, did many settlers begin to treat their farming as a full-time occupation.

In his report on the Canadian North-West in 1904, James Mavor observed that "the amount of capital necessary for the establishment of a colonist varies with the district, with the kind of cultivation he intends to adopt, and with the standard of comfort of the colonist himself."[69] For the present study three ranges of farm-making costs have been tabulated: the minimum, the average, and the substantial (Table 3).

To conclude, the costs of prairie homesteading between 1882 and 1914 varied greatly among settlers of different economic and cultural origins. Costs also varied according to a host of individual variables, including the size of the settler's family, the availability of indigenous building materials, water, and wild game, his working relationships with the neighbours, and opportunities for co-operative ownership of implements. A critical variable was the timing of settlement. In the 1880s incoming settlers usually possessed a good choice of free grant and CPR lands with easy access to timber, water, and unoccupied pasture lands. After 1900 the price of purchased lands increased rapidly and the selection of homestead lands of high quality diminished.

Despite these differences, it is apparent that a prairie settler could get started with a small investment of about $300 to $550, albeit in a rudimentary way. If the settler made a more independent beginning, he would spend $600 to $1200. Alternatively, the settler with a large accumulated capital could spend $2100 or more to begin farming on a substantial scale. These figures tend to support Ankli and Litt in their tabulation of farm-making costs for substantial settlers, but suggest a somewhat lower limit to cost estimates for average settlers. Where this study departs significantly from Ankli and Litt is in its conclusion that a settler could, if necessary, begin farming with a very small investment.[70]

The low minimum capital requirement implies a degree of economic democracy in the prairie settlement experience. Throughout the period 1883–1914 the standard wages for experienced farm labour remained fairly constant in current dollars, including board. Even in the context of inflation at the end of the period, a hired man usually earned enough to start farming on free-grant land with a year's accumulated wages. At the same time it must be acknowledged that homesteaders who arrived early and claimed the best lands were in a comparatively better position than later arrivals to turn a small investment into a small farm. Farmers who arrived with substantial capital also had a head start in bringing their farms into market-oriented production. Poorer settlers often spent an inordinate amount of time establishing themselves. The homesteader who began with little was more vulnerable and had fewer options than his more comfortable counterparts. Any single economic disaster, such as the loss of his buildings to a prairie fire or the failure to find water on his land, could bring an early end to his homesteading venture. These qualifications notwithstanding, homesteading was open to most settlers in the earliest period.

TABLE 3. Estimated homesteading expenditures for three categories of settlers, 1882–1914 (converted to 1900 dollars)*

	No. of Observations	Minimum Homesteader's Expenditures	Average Homesteader's Expenditures	Substantial Homesteader's Expenditures
Entry Fee	(N/A)	$8–12	$8–12	$8–12
Dwelling	(303)	5–75	75–250	250–1,500
Stable	(147)	5–30	30–100	100–725
Granaries	(81)	2–20	20–75	75–400
Well	(19)	21–35	21–35	42–70 (2 wells)
Fencing	(111)	5–20	20–100	100–500
Provisions	(3)	90–130	90–130	130–217
Stove	(4)	14–23	14–23	28–46 (2 stoves)
Furnishings	(3)	22–34	22–34	44–68
Implements	(6)	13–17 (2 ploughs, ½ int.)	94–131 (2 ploughs, 1 mower & rake)	310–494 (full complement)
Small Tools	(3)	14–31	14–31	14–31
LIVESTOCK				
- yoke of oxen	(6)	47–77 (½ int.)	93–153	–
- team of horses	(4)	–	–	338–696 (2 teams)
- dairy cattle	(3)	–	–	138–288 (6 cattle)
HARNESS				
- for oxen	(2)	10–19 (½ int.)	93–153	–
- for horses	(1)	–	–	76–76
BREAKING				
100 acres	(3)	–	–	305–517
SEED				
150 bushels	(N/A)	–	–	65–151
Wagon	(7)	35–41 (½ int.)	70–82	70–82
TOTAL		$291–564	$590–1,193	$2,093–5,873

* Due to the great variability in reported values of structural improvements, estimates for the cost of dwellings, stables, granaries, wells, and fencing have been left in current dollars.

The analysis of homesteading costs in Abernethy's settlement period goes far to explain that community's subsequent economic development. Abernethy, comprising an early group of Anglo-Canadian homesteaders arriving mainly from Ontario and Britain, was settled in a period of expanding economic opportunities. Settlers benefited from the comparatively low costs of setting up a homestead, and particularly from the low cost of land. Those who began farming with a small amount of capital often phased in their homestead development over a number of years. With their accumulated acreage in cultivation and other improvements, they were in a good position to reap the benefits of higher wheat prices after the wheat export boom in the late 1890s.

Quantitative tabulations of homestead improvements at Abernethy and Neudorf show that most settlers before 1900 began farming with a modest outlay. The accumulated improvements at the time of patent application indicate a slow rate of homestead development in both areas. Anglo-Canadian settlers at Abernethy had broken a slightly greater acreage than their German counterparts, but possessed fewer cattle. Since Abernethy settlers took a longer period of time to fulfill their homestead obligations, their annual rate of breaking was no greater than that of the Germans. This evidence would seem to refute the proposition that the subsequent comparative success of the Anglo-Canadian group was necessarily attributable to a higher initial investment. This conclusion implies that other factors were at play, and these are examined in the next chapter.

An essential aspect of early farm making was the co-operative pooling of labour, shelter and implements among settlers. Co-operative activity worked well in the early period when most settlers were obliged to share their resources. Settlers who arrived after this initial period entered a situation in which the co-operative principle had already begun to erode.

3

Economic Development of the Abernethy District, 1880–1920

While settlers prepared sufficient land, built the necessary shelter, and amassed the materials of farm production, they confronted the problem of converting their investments into viable economic operations. The transition from the early homesteading to profit-oriented agriculture was often slow. The settler's chances of success depended on a multitude of factors, many of which were beyond his or her control. A continually unpredictable factor was weather and, more generally, climatic conditions. While Abernethy settlers had settled outside the semi-arid region of the Palliser Triangle, their lands were not beyond the reach of drought in a dry year. Early frost was an ever-present danger, particularly in the early period. Crops were frequently invaded by infestations of gophers, locusts, and crop diseases. Prairie fires, too, could destroy in a few minutes what the settler had taken years to produce. Returns were further imperilled by an unstable market and frequent fluctuations in the price of farm products. On the other hand, there existed a number of ways of reducing the risk in prairie agriculture. As farmers observed the effects of various agricultural techniques on production, they were presented with a number of options to increase yields and a degree of reliability in revenues. Since the margin between profit and loss was often very small, farming demanded astute management to turn the anticipated returns into a reality.

The economic history of the Abernethy district, 1880–1920, divides into two distinct periods, although there was some overlap in terms of historical processes. The initial settlement period, 1880–1900, corresponding to the use of nineteenth-century technology and farming techniques, was labour- rather than capital-intensive. Homesteading represented a small investment for most newcomers who were able to begin cultivation with a yoke of oxen and a few simple implements. The abundance of unclaimed lands presented the settler with the option of choosing his homestead near wooded areas, and thus reduced his necessary expenditure for building materials. Land was cheap, but required a considerable labour investment to prepare the virgin grassland for cultivation. In this early period, the farmer also spent an inordinate amount of his time in marketing his grain. Reflecting a general initial dearth of adequate transportation facilities on the prairies, the Abernethy district did not obtain rail linkage with outside markets until after 1900. In the first twenty years, farmers hauled their wheat as far as thirty to thirty-five miles to the CPR line – a process consuming much of their labour time during the fall and winter months.

In the second phase of settlement, after 1900, rapid increases in the price of land and other farming costs necessitated a shift to a predominantly capital-intensive agriculture. Land prices jumped dramatically, as the range of accessible homestead lands diminished and incoming settlers were obliged to purchase farms in settled areas. The costs of motive power increased, as oxen were supplanted by horses, and horsepower in turn was replaced by steam and gas. Building costs went up as the supply of lands close to wooded areas diminished and settlers were obliged to purchase the materials of construction. Accompanying this process was an eventual increase in the size of farms and a reduction in the farming population. Settlers who were able to persist eventually reaped the benefits of large-scale production. In the process, however, many fell by the wayside.

EARLY FARMING HISTORY, 1880–1900

The Qu'Appelle Valley farming region established itself very early as a grain exporting area. The earliest references to grain shipments appear in a December 1884 issue of the *Regina Leader*.[1] Rail linkage had been established for the Indian Head district in 1882 and Regina in 1883, but it was not until 1884 that the CPR built a grain terminal at Port Arthur and grain shipments proceeded immediately upon its completion. Nevertheless, the scale of production remained small. At

Abernethy, and in surrounding rural communities, most new settlers devoted the first year or two to the construction of farm buildings and sod breaking. Before they had prepared sufficient acreage to undertake large-scale wheat production, the settlers sold limited quantities of root and cereal crops at local markets at Fort Qu'Appelle and the nascent villages and towns that sprang up along the CPR main line as railroad construction progressed. These centres, including Indian Head, Sintaluta, Wolseley, and Grenfell, also constituted a market for cords of wood and bales of hay, which the settlers could collect in the wooded and marshy areas to the north and east of Abernethy. The mode of exchange for these transactions was often barter and not cash, as little money was in circulation in the initial period.[2]

Historians of settlement in the American Midwest and eastern Canada have interpreted the phenomenon of barter as representative of a subsistence or self-sufficient agriculture. Yet, as V.C. Fowke has noted, even these early non-monetary transactions were connected to commercial agriculture and markets.

> In stressing "cash products" as we have, it is important not to fall into the error of dismissing barter transactions as of no commercial importance, or of assuming that the scarcity of money in frontier communities was a proof of self-sufficiency. Much of the produce disposed of by the pioneer was bartered, but it was nonetheless disposed of commercially and constituted effective demand for capital equipment and consumer's goods of non-agricultural origins.[3]

Cash continued to be a scarce commodity throughout the early period of Abernethy's history, but this fact should not obscure the essential market orientation of its economy.

Wheat quickly established its economic importance as the principal cash crop in the Abernethy area, and generally in the District of Assiniboia. Mixed farming was practised in several areas in which the soils were less conducive to crop than to pasture, such as the Pheasant Hills to the east of Abernethy and the Moose Mountain, Touchwood, and Beaver Hills districts. These areas also possessed extensive marshy lands that afforded an accessible supply of hay for cattle feed. Most settlers in the District of Assiniboia, however, had settled on the flat open prairie. They tended to raise only a few livestock for home consumption and a local market, while concentrating on wheat production for export.

WHEAT PRICES, 1880–1900

Given the heavy reliance on wheat as a cash crop, the importance of wheat prices to the economic fortunes of Abernethy farmers is evident. For this study, the relevant figure is the local or farm-gate price, representing the actual amount paid to the farmer on delivery of his wheat to the elevator. Theoretically the local price should have been equal to the world or Liverpool price, less the costs of moving grain from the local market to Liverpool. Similarly, the price of wheat at the principal port for east-bound grain, Fort William, should have equalled the farm-gate price, less delivery charges. In practice, however, the farmer often received a much smaller figure, particularly in the latter years of the nineteenth century, before government regulation of the grain handling business removed the worst abuses of the system.

As Figure 12 indicates, the local wheat prices fluctuated greatly in the 1884–1900 period.[4] Frosted wheat, if a buyer for it could be found, brought less. Before 1900 the unfortunate combination of late ripening strains of wheat and recurrent summer frosts commonly caused frost damage and concomitant low grading. Grain prices were also adversely affected by the existing system of grain dealing, which farmers saw as conspiring against their interests. Often only one buyer could be found at the market centre, and farmers frequently complained of a lack of competition, with the attendant threat of short weight and low prices. If the grain buyer had already filled his weekly requirements he might not take any new grain, or he might offer to purchase a load at a considerably reduced rate. Having hauled his wheat thirty miles or more at considerable time and expense, the farmer was often obliged to accept whatever price the dealer offered. The alternative was to return empty-handed with wagon full. Figure 13 shows a farmer unloading his grain at a local elevator after agreeing to sell it to the buyer. The farmers' suspicion of price fixing was echoed in some local newspapers. In 1885 the *Qu'Appelle Vidette* related that an Abernethy area

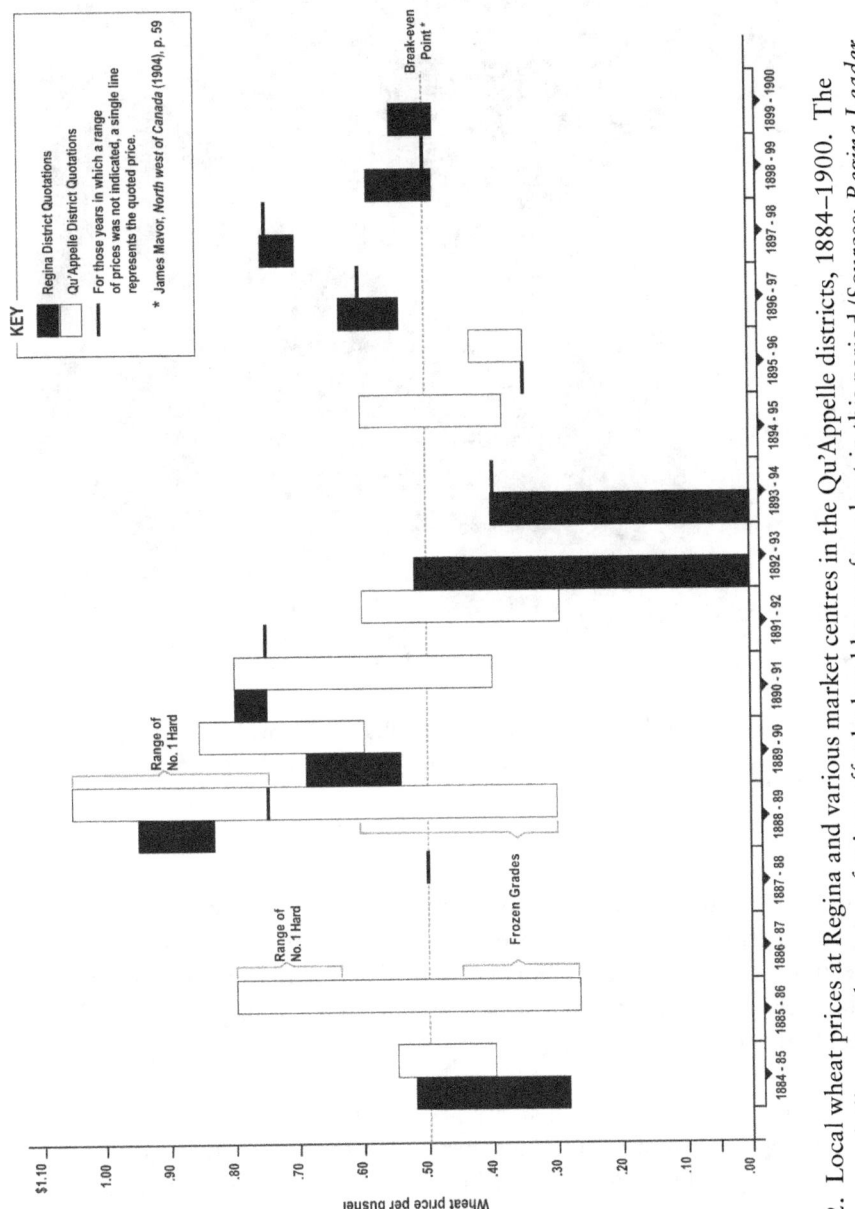

Fig. 12. Local wheat prices at Regina and various market centres in the Qu'Appelle districts, 1884–1900. The graph illustrates the range of prices offer by local buyers for wheat in this period (Sources: *Regina Leader*, 1884–1900, *Qu'Appelle Vidette*, 1884–1900, and *Qu'Appelle Progress*, 1884–1900.

Fig. 13. "Dumping grain in an elevator, Andy Blair of Lumsden standing," 1900. Saskatchewan Archives Board, Saskatoon, Photo no. S-B 447.

farmer had been offered only fifty cents per bushel for wheat equal in quality to grain that had brought fifty-eight cents a few days earlier. Since one of the two grain buyers at Indian Head was out of town, the farmer had no choice but to sell to the remaining purchaser. The *Vidette* commented, sarcastically:

> This action clearly illustrated the beauty of monopoly and we believe it would be much better for those buyers to join the "ring". Then they could buy wheat at their own prices from those who were compelled to sell and there would be

no grumbling and farmers would not expect any competition in the market....[5]

Hard evidence that might verify the charges of monopolistic practices remains elusive. It is perhaps worthy of note that freight rates comprised less than half the difference between the quoted local and Fort William prices. For example, assuming that a farmer received the maximum price (quoted in Qu'Appelle district newspapers in 1890–91) for No. 1 Northern he would have earned an average of fifty-five cents a bushel for wheat that sold in Fort William for eighty-eight cents (Fig. 14). Immediately prior to the initiation of Crow's Nest rates in 1897, the freight charge at Qu'Appelle was approximately thirteen cents per bushel.[6] This left twenty cents per bushel for elevator storage charges and the grain buyer's profit. Farmers clearly believed this margin to be exorbitant.

If farmers wished to avert the problems associated with the sale of their wheat to local elevator companies, they could direct the elevator officer to load their grain and ship it directly to Fort William. In this way, the price to the producer could never have differed substantially from the Fort William price plus freight charges, except under circumstances of chronic congestion of handling and transportation facilities. Such a situation did not occur until the unprecedented wheat harvest of 1901. The business of shipping carloads of track wheat presented other obstacles, however. A farmer had, first of all, to have produced 1,000 bushels of wheat of the same grade in order to fill the car, or 1,500 bushels for the larger cars. Before the 1890s few farmers had prepared sufficient acreage to produce such a large amount of wheat.[7] Even into the 1900s, the logistical problems involved in hauling 1,000 bushels within the fifteen-day waiting period proved insurmountable for many.

Before the building of the Kirkella CPR branch line in 1903, Abernethy area settlers were forced to haul grain distances of twenty to thirty miles to the elevator. This trip normally took at least a day, so no settler could make more than three round trips per week. Since most farm wagons held a maximum of forty bushels of wheat, a farmer could only hope to meet the deadline with the help of hired "teamsters." But short-term labour was hard to find and was costly in the initial period, when grain prices remained depressed. Other expenses associated with grain hauling included the hotel and meal costs

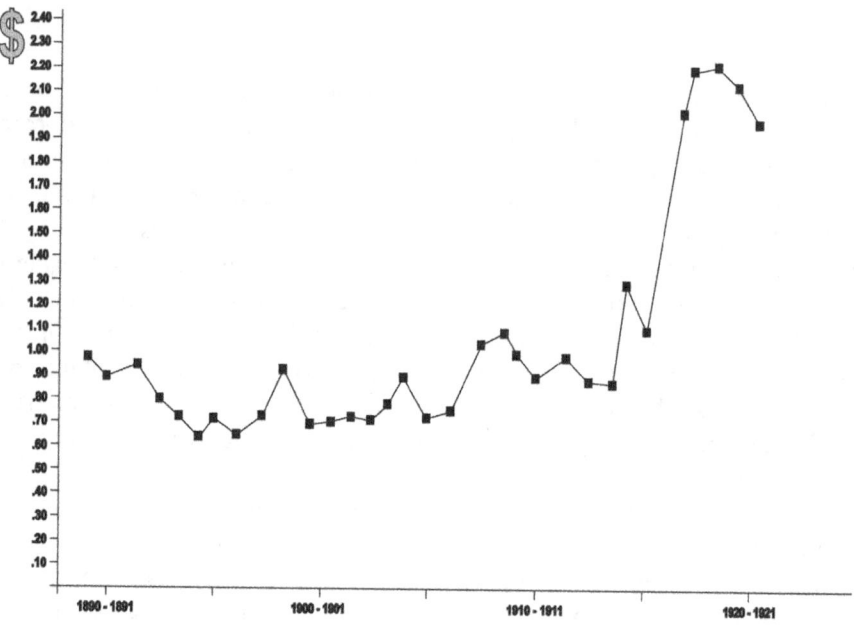

Fig. 14. Wholesale wheat prices at Winnipeg, 1889–1920. This graph illustrates estimates for the wholesale market price of No. 1 Northern wheat at Winnipeg, 1889–1904, and the Winnipeg cash closing for No. 1 Northern, basis in store, Fort William, 1905–1920. Adapted from M.C. Urquhart and K.A.H. Buckley, *Historical Statistics of Canada* (Cambridge: University Press/Toronto: Macmillan of Canada, 1965), pp. 359–60.

for the drayman and a livery charge for their horses. Together, these expenditures exceeded $2 per trip. Abernethy farmers estimated that the cost of hauling wheat was ten cents a bushel, a figure that does not seem at all exaggerated.[8]

In 1893, John Nicholls, a Fort Qu'Appelle area settler, estimated the revenues and expenses settlers on free-grant lands would encounter in raising a hundred acres of wheat:

REVENUE	
16 bushels per acre at 40 cents	$640
EXPENDITURES	
2 bushels seed at 40 cents	80
Threshing at least	100
Twine and hired help in harvest alone	60
Repairs, wear and tear of implements and teams at least	$150
Cash expense for marketing	20
TOTAL CASH OUTLAY	$470

In Nicholl's tabulation the farmer was left with a net revenue of $170, "out of which he has to provide flour, groceries, clothing and furniture, for himself and his family, and to make provision for sickness, accidents and old age."[9] For those settlers who lived twenty miles or more from market, farming was an even more marginal proposition. If the cost of marketing each bushel of wheat were ten cents per bushel, the farmer's net revenue, using Nicholl's other estimates, would be a paltry thirty cents per acre, or $30 for a hundred acres.

YIELDS, 1880–1900

The other major factor bearing on farm income was yield. Statistics on crop yields were not compiled by the government of the North-West Territories before 1898. In their absence, incomplete estimates may be gleaned from the scattered reports of local immigration and agricultural officials and in local newspapers. These sources must be approached carefully. Local newspaper editors and immigration officials, anxious to promote settlement, tended to emphasize the successes, while paying scant attention to the failures.

Crops in the North-West in the 1880s were extremely variable, and farmers suffered recurrent failure in many districts. In 1883 the Dominion Immigration Agent at Qu'Appelle reported an average yield of eighteen bushels of wheat per acre in that community.[10] The following year he reported:

The crops has not been so abundant as was expected from the Rapid Growth and appearance through the summer – the lateness of the spring season, besides being very dry – together with the frost in fall has been the cause of a slight damage to some fields of grain, which did not vegetate and ripen equally."[11]

In 1885 the *Qu'Appelle Vidette* reported that Indian Head area farmers suffered "great losses" from prairie fires. Drought was another source of crop damage.[12] The exceptionally hot and dry summer of 1886 resulted in an almost total loss for Abernethy farmers. Samuel Chipperfield, who reaped two bushels per acre, also recorded the yields of his neighbours: Wallace Garratt, seven bushels per acre; F.S. Evans, three bushels; W.H. Garratt, five bushels; C.S. Dickenson, 2.5 bushels; Dixon Brothers, ten bushels. Chipperfield attributed the higher yields registered b the Dixon Brothers to the greater precipitation received by their fields.[13]

Despite the general destruction of crops, a few fields recorded surprisingly high yields. Abernethy-area farmers empirically deduced that the higher-yielding fields were the lands that had been left fallow the previous year by farmers who had dropped their seeding in favour of "teaming" supplies to government troops during the North-West Rebellion. Some farmers, such as W.R. Motherwell, did not actually leave for transport service but instead rented their horse teams and wagons to the government. Unable to work their lands, they nevertheless reaped an unexpected benefit.

In 1887 the wheat crop at Qu'Appelle averaged twenty bushels per acre, compared with twenty-five bushels at Fort Qu'Appelle.[14] But at Abernethy John Teece reported that in that year his crop was "almost a failure and the same is true of most in this part of the district."[15] Grasshopper and gopher infestation approached epidemic proportions, and early frosts in 1888 also wrought a near-total destruction of wheat in the Abernethy district. W.R. Motherwell, who had sowed sixty-five acres, reaped a paltry three hundred bushels, or less than five bushels per acre.[16] Such crop failures inspired various petitions to the Territorial government for seed grain assistance. Many of the losses resulting from drought and gopher destruction approached 90 per cent in some areas.[17] Beyond the seed grain question many settlers found themselves unable to make the payments for their pre-emption entries, which had come due that year. They faced imminent cancellation and the loss of half their homestead lands. To cap a difficult decade, farmers reaped widely divergent yields in 1889, ranging in one report as low as twelve bushels per acre, which Samuel Chipperfield obtained on summer fallowed land,[18] although only five miles away, W.R. Motherwell reaped thirty bushels to the acre.

In June 1890 crop yields improved overall, but with mixed results for different localities. An indication of the great variation in yields may be drawn from a series of crop and weather bulletins commissioned by the federal Department of the Interior in 1890.[19] Average yields recorded for the three principal grains in different districts were:

Place	Yield		
	Wheat	Oats	Barley
Qu'Appelle	23	45	30
Fort Qu'Appelle	35	70	35
Moose Jaw	15	17	nil
Wolseley	18	30	-

The year 1891 was described as a "banner year" for wheat farmers in the Blackwood district south of Abernethy.[20] Farmers took advantage of the large harvest to purchase additional lands and implements. Yet the early nineties were marked by continuing problems with early frost and irregular yields. To compound these difficulties, mixed farmers lost from 40 to 75 per cent of their cattle in the severe winter of 1892–93.[21] The cumulative impact from this series of disasters was considerable, as was revealed in a petition sent by a group of Saltcoats settlers to the Governor-General in 1894.

> This the Petition of the residents of Saltcoats, Assiniboia, humbly sheweth.
>
> That families who came from the British Isles to this district under Colonization Company's influence, are in need of relief from the heavy and grievous burdens which oppress them.

That at great cost of Toil and Money these people have improved the lands they took up from the Government.

That in the present Financial position of this district the settlers are in such a state of uncertainty as to what these Colonization Companies will do; that they are afraid to expend further capital and labor in improving these Homesteads.

That the margin of profit of late years is so small, on account of low prices, loss of cattle, by the thousand in this district alone last spring through a long winter, scarcity of hay and good wells of water, through losses by drought, gophers, hailstorms, and summer frosts, that not one of these husbandmen, can meet his indebtedness to these companies.

That fully fifty per cent of the families who came in under these auspices have abandoned the lands they settled upon, that the obligations are now so large of those that remain to the Implement Firms, and Storekeepers, that they most certainly cannot repay principal or int. to said Companies.

That owing to lack of a home market, heavy freight rates, scarcity of money in circulation in this district and that our business has to be done by Barter there is no chance whatever of farmers making money, (mortgages are about as plentiful as the dollar with the farmer).

That these men, on the strength of immigration literature, published by Governments and Land Companies, gave a mortgage to these Co's. on their homesteads and improvements, in good faith of being able to redeem the same, and obtain the title deeds of their Holdings.

That consequently Your Excellency and the Gov't should see how very hard it is for these men and their families to be compelled to abandon their improved homesteads, after so big a struggle to make their own property.

That in the interest of the Dominion and especially those of the Territories, we are of the opinion that any further exodus of valuable settlers should be immediately arrested and their Homesteads saved from going as the others have gone to their original condition of wild prairie.[22]

There can be little doubt that many settlers suffered great hardships in the first difficult years. Numerous letters to local newspapers and petitions to the federal government document the various economic problems faced by the settlers. Frequently these letters were addressed to persons contemplating the North-West as a field for settlement, as disillusioned authors admonished readers not to believe the "glowing reports" in immigration literature. In a letter printed in the *Belfast Newsletter* in 1891, a Regina-area settler bitterly related his tribulations during seven years of struggle:

> There were only two years during the last seven that really should be called good years. These were 1890 and 1887. In the former the average yield of wheat per acre was about 15 bushels per acre, the general price is about 50 or 60 cents a bushel (a cent equal to a halfpenny). This, after deducting the expense of seed, labour, cost of cutting, and threshing – except the farmer has his own machinery, which is very expensive – is very small to the producer. Most years there is about a half or third crops, some years so bad that it's not worth cutting....[23]

The 1890s were particularly hard on newcomers, who had been unable to build up sufficient improved acreage to withstand the worst effects of crop failures and depressed prices. In 1895 the *Qu'Appelle Progress* related the plight of one young bachelor at Lorlie whose financial woes had forced him to subsist on a diet of bread and tea. Lorlie is located only about 10 miles northeast of Abernethy. According to the story this hapless homesteader had lost a yoke of oxen, suffered the strangulation of his mare, and witnessed the complete failure of his grain crop. Referring to other settlers, the author of the article wrote: "I am afraid there are a good many suffering this if they would only own up to it."[24]

At the same time, farmers who persisted through the period of low prices and crop failures were often able to progress. It is difficult to generalize about the individuals who stayed – quantitative comparisons of farmers who held their land for 20 years or longer do not reveal many characteristics that they shared in common. For example, no relationship between family size and longevity of tenure could be discerned. Yet, successful farmers appear to have taken similar approaches in critical aspects of their farming operations. Testimonials by successful settlers suggest a common tendency to diversify. During the 1880s, when grain crops were frequently destroyed, some settlers were able to market sufficient garden and livestock produce to be able to continue. In 1888, for example, W.R. Motherwell's frozen wheat crop yielded a negligible 300 bushels, but he was able to meet expenses by selling $200 worth of pork.[25] In 1893, while wheat prices plummeted, Motherwell expanded his herd of cattle to 50 head. This is not to suggest that a full-fledged diversification took place for most farmers but that some of the more successful producers were at least able to adapt to changing conditions.

Farmers who persisted also demonstrated a capacity to adjust to the exigencies of dry farming. After 1886 the practice of summer fallowing to preserve soil moisture spread rapidly throughout the Qu'Appelle Valley and neighbouring districts. An important impetus in the diffusion of summer fallowing and other new techniques was Angus MacKay, under whose superintendency the Dominion Experimental Farm was established at Indian Head in 1888. By 1897 the *Qu'Appelle Progress* was able to report that "summer-fallowing is now the order of the day."[26] Crop yields registered at Indian Head and at the Motherwell Farm demonstrated the enhanced returns of summer fallowing vis-à-vis continuous cropping.[27]

YEAR	RAINFALL	INDIAN HEAD DOMINION EXPERIMENTAL		MOTHERWELL FARM
		FALLOW Yield/Acre	STUBBLE Yield/Acre	FALLOW Yield/Acre
1891	14.03	35	32	30
1892	6.92	28	21	29
1893	10.11	35	22	34
1894	3.90	17	9	24
1895	12.28	41	22	26
1896	10.59	39	29	31
1897	14.62	33	26	35
1898	18.03	32	-	27
1899	9.44	33	-	33
1900	11.74	17	25	25

Individuals such as W.R. Motherwell continued to experiment with new techniques and approaches designed to reduce the risk and costs entailed by wheat farming. In 1894 Motherwell began to import brome grass from Austria, which he incorporated into his crop rotation program at his farm. In addition to its beneficial effects in enriching the soil and controlling weeds, Motherwell found that he could profit from selling brome for seed. Since brome seed required a much larger acreage than wheat to produce a load for market, Motherwell saved many trips and considerable expense. Instead of hauling to Indian Head the product of one and a half acres of wheat worth $20, Motherwell could haul the product of seven or eight acres of brome grass worth $200.[28]

In the late nineties, Abernethy farmers had also begun to benefit from a general reduction in the federal tariff on farm implements. At the time the area was settled in the early eighties, the Dominion government had raised the duty on American machinery to 35 per cent. While farm revenue remained at low ebb in the early nineties, the high cost of implements was surely an impediment to expansion. Between 1895 and 1900 implement duties were reduced to 20 per cent and Canadian implement prices dropped by a somewhat smaller margin. The retail price of mowers and binders was reduced 7.5 per cent; seeders, 15 per cent; fanning mills, 20 per cent; rakes and ploughs, 10 to 12 per cent; and wagons, 8 per cent. Notably, the price of the self-binding reaper had dropped 50 per cent since the early eighties.[29]

With the opening of new European markets in 1896, wheat prices recovered from a three-year slump. Concurrently yields increased dramatically. At Sintaluta, the principal market centre for Abernethy area farmers, wheat sales increased from a low of 80,000 bushels in the season of 1894–95 to 240,000 bushels the following year.[30] That prosperity had finally arrived for Abernethy settlers was evident in a tabulation of revenues in the Qu'Appelle Basin, published by the Winnipeg firm of Osler, Hammond and Nanton in 1897. In several cases revenues in excess of three to five times the original outlay for land had been recorded.[31] Even before the wheat boom, settlers in several farming districts, including Abernethy, had performed much better economically than their counterparts who had settled on poor land and with smaller amounts of capital. In his annual report for 1895, North West Mounted Police Superintendent A.B. Perry reported that while hardships in the early nineties had occasioned the wholesale desertion of townships in some areas,

> ... As a contrast I should mention that I have been astonished with the material progress made in many districts where the settlers are of the right class, the soil fertile and the surrounding conditions favorable. Take the settlements north of Pense, at Spring Bank, north of Qu'Appelle Station, the Wideawake settlements, around Kenlis, Abernethy and Indian Head, and you will find very marked signs of increasing wealth and prosperity.[32]

By 1900, then, the Abernethy agricultural economy was largely established, with heavy dependence on the wheat staple as its principal feature. Evidence of a developing prosperity could be discerned in the impressive fieldstone, frame, and brick residences that farmers were starting to build after the onset of the wheat boom in 1896. While most early homesteaders either had dropped out or would soon depart, the more successful settlers had acquired sufficient machinery and improved enough acreage to begin to cash in on the long-awaited gains.

DEVELOPING PROSPERITY, 1900–1910

The decade 1901 to 1910 saw the consolidation of the Abernethy economy that had been established in the 1880s and 1890s. With the arrival of local rail service in 1904, the townships to the north were repopulated, and thousands of acres were brought into cereal grain production. Abernethy farmers generally improved their economic position in this period. A major reason for their enhanced prosperity was a general increase in the price of wheat. Between 1901 and 1910 the Fort William price averaged 88.6 cents per bushel, or 12 cents higher than the average price between 1896 and 1900.[33] Most of the increase was registered in 1907 and 1908, when the price averaged $1.10 at Fort William. Meanwhile, freight rates at Regina dropped to 10.8 cents, a level at which they remained until the last two years of the First World War.[34] Having steadily expanded their improved acreage,

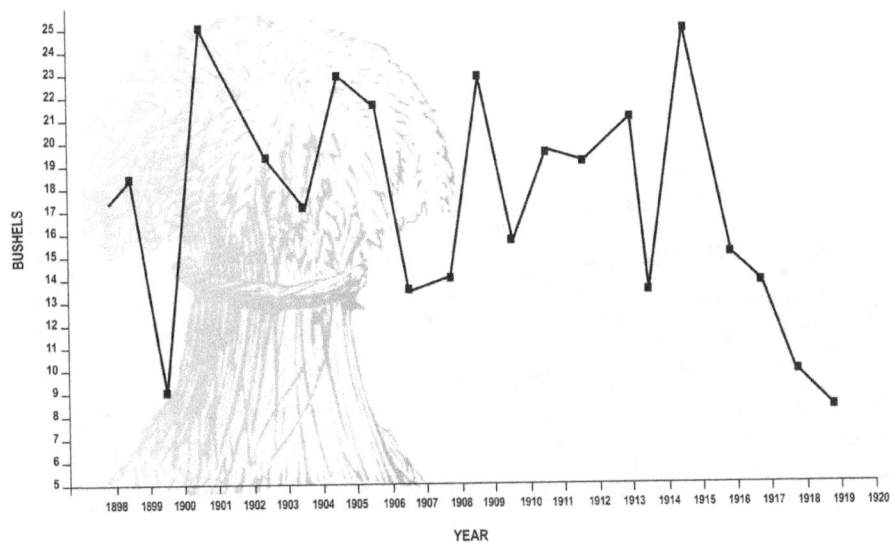

Fig. 15. Wheat yields in Saskatchewan, 1898-1920. Saskatchewan, Department of Agriculture, *Annual Report* (1913), and Canada, Dominion Bureau of Statistics, Agriculture Division, *Handbook of Agricultural Statistics* (Ottawa: The Bureau, 1951), Part 1, "Field Crops."

Abernethy settlers were now in a position to profit from the increased grain prices. By 1906, farmers cultivated an average of 151.85 acres in the townships surrounding the village of Abernethy.[35] With such a large extent of improved acreage, farmers could enhance their yields by leaving one-third of their lands fallow, while still placing a hundred acres under cultivation.

Crop yields continued to fluctuate wildly, ranging from the very dry year of 1900, when wheat averaged only nine bushels per acre throughout the District of Assiniboia, to the bounteous harvest of 1901, when yields reached twenty-five bushels overall (Fig. 15). Between 1898 and 1912, the average yield in the District of Assiniboia and the province of Saskatchewan was fifteen bushels per acre.[36] Yet yields in the rich agricultural district around Abernethy continued to outstrip the provincial average. Between 1898 and 1903, spring wheat

Fig. 16. "Postcard, Abernethy, Saskatchewan, 1905." Saskatchewan Archives Board, Regina, W.R. Motherwell Collection, Photo no. R87-219, No. 198.

in Crop District No. 4, comprising the Grenfell, Wolseley, Indian Head and Qu'Appelle Districts, averaged 20.4 bushels per acre, topped by a harvest of 28.5 bushels in the bumper year of 1901.[37]

A third reason for the developing prosperity of Abernethy farmers was the arrival of accessible rail linkage when the Kirkella Branch of the CPR was built north of the Qu'Appelle River in 1903–4. Instantly a series of market centres was established along this line, including the villages of Abernethy and Balcarres. Figure 16 is a photograph of the new village of Abernethy soon after it was established. Farmers now could save most of the time and expense of hauling grain thirty miles or more to market, and it was now possible to fill grain cars within the specified period of time. Farmers could also save the transportation costs entailed in travelling to distant supply centres in order to purchase material goods, as a full complement of retail outlets was established at Abernethy.

Clearly the principal economic advantage possessed by the settlers who arrived before 1900 was the capital gain that accrued from a rapid increase in land values after the turn of the century. In some

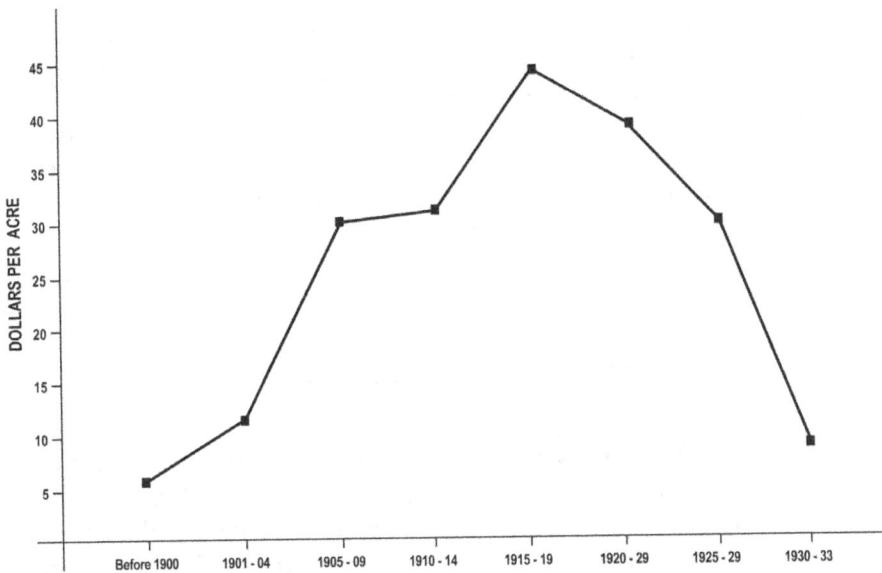

Fig. 17. Average price paid for land, Indian Head and Balcarrres districts, pre-1900 to 1933.

areas CPR lands that were purchased for $3 per acre in the 1890s had jumped in value to $30 or more in 1910 (Fig. 17).[38] In the Abernethy area, most settlers had already obtained at least 320 acres as homesteads and pre-emptions for a minimal outlay. In some cases an additional quarter-section, the so-called second homestead, was obtained by settlers who had arrived before 1886. In this way, W.R. Motherwell obtained 480 acres of Dominion land and an additional quarter-section of CPR land for a total outlay of only $1,200. By 1910 these lands were worth about $20,000. With such a cushion of real estate equity, farmers could mortgage their lands to obtain funds for various capital expenditures, including more land, additional machinery, or new farm buildings. In 1907, for example, W.R. Motherwell mortgaged one of his quarter-sections for $5,000 to finance the building of his barn.[39] Mortgages continued to be an important source of financing for farm implements throughout a farmer's period of tenure; in some cases as many as ten mortgages were registered against quarter-sections in the township around Abernethy.[40] In the context of variable grain prices and yields, mortgages also gave farmers access to cash for operating expenses in years of low revenue. In other words, they conferred a degree of flexibility on the farmer that newcomers without accumulated equity did not possess. Established settlers also received preferred treatment from lending institutions vis-à-vis interest rates, although rates rarely dropped below 7 per cent.

THE MIRE OF FARM DEBT

If post-1900 farming provided opportunities for larger profits, it also entailed the possibility of greater debts. A provincial Royal Commission on Agricultural Credit reported in 1913 that farm indebtedness in the province had reached $1,500 per farm unit. In the eighteen months preceding August 1913, more than 1,723 involuntary sales proceedings had been initiated for farms whose owners had defaulted on their mortgages.[41] A recurring problem in the settlement era was a general shortage of loan capital in western Canada. Many new farmers, already burdened with the large farm-making costs in the post-1900 period, could not obtain access to loans in periods of low revenue. The more prosperous farmers, including those in the Abernethy district, could compete more easily for the available capital. That the more successful Abernethy farmers experienced little difficulty in securing mortgage money is evident from a search of land titles in Township 20, Range 11, between 1883 and 1920, on which numerous mortgages were registered.

Another worry for many farmers was the threat of dispossession by implement companies. In 1913, the Saskatchewan Department of Agriculture conducted an investigation into the cost of farm implements to farmers during the preceding ten years. In its report, the Department observed that many farmers found the credit conditions surrounding the purchase of implements to be particularly stringent. While having the assurance of the implement companies' agents that they would receive fair treatment in the event of not being able to meet their payments, the farmers' experience "is somewhat different when the collector calls around."[42]

Much of the problem evidently stemmed from the nature of the agreements entered into by the farmers and the companies. Ordinarily, the farmer signed a lien note, by which he surrendered the right of ownership until the implement had been paid for. Notes were usually made out for several months to one year, although in some cases an agent would grant a buyer two years in which to pay. Farmers complained, however, of a clause in the agreement which permitted the implement company to declare the note due and payable if it considered it to be insecure. Failing payment on such demand, the company could seize the implement. One settler wrote:

> If you are unable to meet your notes, the company will send their collector out to get security of some kind. If you refuse, your life is made miserable by threatening letters. If you consent to give a mortgage you are charged ten per cent. Those who buy a threshing machine on time are especially hard pressed, as the banks are very tight with threshers and the companies are equally hard on them.[43]

In the event a farmer did not arrange a mortgage to cover the outstanding debt, the implement company or other interested party would register a caveat against his property.

Indebtedness was not limited to mortgages and implement loans. General stores frequently were obliged to extend credit, and the farmers' debts could persist over long periods of time. Even if the farmer's crop materialized, mortgage and implement companies often had first claim on the settler's revenues. In 1905 the three proprietors of Abernethy general stores, frustrated with long-standing debts, announced a new policy:

> To Our Customers
> A few words as to our views on the system of giving credits. We have found on referring to our accounts that we are carrying several from year to year which aggregate large sums with which if we had cash we could carry on business more satisfactory to ourselves and customers. We have realized that if we are to keep on doing business, we must have a better understanding in regard to the matter, and wish it understood that as this is the most bountiful crop that the

country has ever produced, and in view of the consequent prosperity and circulation of money attendant thereon, we must have old scores and open accounts, both large and small, settled in full.

There may be exceptional cases which deserve clemency, but we will have to secure such accounts as cannot be paid in full by mortgages on chattels or on other assets; for an account which cannot be paid in this year of plenty is a poor one for a merchant to carry.[44]

Farmers were perennially short of cash. To a large extent this was inevitable given the variability in annual returns and the irregular timing in which the farmer received his income. The Agricultural Credit Commission noted that "in most districts" farmers borrowed from banks to meet current expenses, particularly in the months preceding the fall harvest. Usually these short-term loans were made out for three to six months, but as the Commission noted, loans were perpetually renewed.

These features of our mortgage system are reprehensible. The mortgage is not calculated to develop business habits nor promptness. It is a document that places the farmer, from the beginning, in an impossible situation. It holds out to him the prospect of confronting a payment which he can never hope to meet. Under the guise of a short-time mortgage, there actually exists a system of long-term mortgages, but with this difference, that the farmer is compelled to renew every five years or lose his farm should he fail to meet his mortgage.[45]

Farmers also continued to suffer from problems in the grain handling system. The huge harvest in 1901 had produced more wheat than the CPR had been able to handle, and a wheat blockade developed at virtually all grain market centres along the CPR main line in Manitoba and southeastern Saskatchewan. With elevators and warehouses filled to capacity, farmers were forced to pile their bagged wheat on the ground along the rail line. A long-developing alienation found expression in the founding meeting of the Territorial Grain Growers' Association (TGGA) at Indian Head in 1901. Led by W.R. Motherwell, the co-organizer and first president of TGGA, farmers pressed for parliamentary amendments to the Manitoba Grain Act to redress perceived abuses of the system of apportioning grain cars. Yet many continued to complain of shortages of cars and discriminatory practices in grading by local elevators. As in the earlier period, new farmers experienced difficulties in raising sufficient grain to be able to ship in bulk to Fort William. Street sales of wheat usually netted the producer a much smaller amount than he could have received by loading his own car. Even in 1908, when the price of No. 1 Northern wheat exceeded $1, local elevator companies paid as little as twenty-five cents per bushel for lower grades.[46]

The plight of many of the post-1900 settlers might be best illustrated by a detailed example. In 1905, Georgina Binnie-Clark purchased 320 acres of partially improved land near Lipton, Saskatchewan, about thirty miles northwest of Abernethy.[47] She had been an English journalist and possessed no previous experience in farming. Lacking the capital to buy the land outright, she entered into an agreement to purchase for $5,000, at $15 per acre, with $800 down and annual installments of $1,000 at 6 per cent interest. From an outside perspective, the proposition seemed attractive enough; Binnie-Clark anticipated that she would be able to reap 2,500 bushels from her 80-acre wheat field which, at $1 per bushel, would bring $2,500 annually.

Binnie-Clark's anticipated gains proved illusory. While her harvest was relatively bounteous – about 22 bushels to the acre – the actual yield of 1797 bushels was considerably less than the anticipated 2,500 bushels. Of this, 160 bushels had to be held in reserve for seed. The remainder graded No. 1 Northern at the elevator and sold for $1,026. From this sum, expenses for harvest labour, binder-twine, threshing, hauling, and groceries had to be deducted. She had incurred heavy expenditures in her first "housekeeping bill" for such necessary chattels as a cook stove, wood, and various pieces of furniture totalling $519.30. Implement costs added another $700 to the yearly expenditures, leaving a net deficit of more than $200. To meet the land payment that was due on 1 January 1905, Binnie-Clark was obliged to send to England for an additional £200. At this time, she renegotiated her land purchase agreement and signed a three-year mortgage for $3,000 at 6 per cent.

In 1906 she engaged a farm labourer, who seeded the wheat crop on an unprepared seed bed. While yields in the Lipton district were generally fairly high that season, Binnie-Clark's 1,200-bushel crop was considerably smaller than the previous year and was further reduced by heavy dockage for wild oats. The 780 bushels that remained graded No. 1 Northern, and brought $1 per bushel, leaving her only $533 after threshing expenses were deducted. This return was not enough to cover the annual mortgage installment of $1,160, much less working expenses.

In a desperate attempt to increase revenues, Binnie-Clark paid $100 to have an additional 25 acres broken on her land, which she seeded the next spring. But owing to a wet spring seeding was delayed that year, with the result that crops matured late in the summer. With temperatures hovering near freezing in late August, Binnie-Clark had her wheat cut while still green. The loss in grade resulted in net proceeds of only 35 cents per bushel, after the deduction of freight and storage charges. Total receipts that year for sales of wheat, pigs, and butter amounted to less than $500, as opposed to $1,050 in working expenses. These included:

Feed and seed	$64.90
Breaking 25 acres	100.00
Wages	260.00
Taxes	15.00
Fencing	9.50
Fanning Mill	26.00
Binder-twine	27.00
Hail insurance	19.50
Horse	150.00
Grocery, flour and meat, repairs, veterinary attendance	245.00
Threshing and teams	133.00
Total	$1,050.60

In Binnie-Clark's view this outlay was nearly twice as large as the necessary average, as her grocery bills were too high, and certain capital costs such as breaking would not recur in succeeding years.

Whether the 1907 expenditures were an aberration or not, they had contributed to an ever-deepening mire of debt for the beleaguered Binnie-Clark. Obliged to obtain a deferment of her mortgage installment, she also had run up a debt at the local hardware store and an overdraft at the bank. Much depended, therefore, on her capacity to turn a profit in 1908. That year, however, Binnie-Clark was thwarted by the vagaries of the grain handling system. Since her crop was sowed on three different plots of land, including the newly broken twenty-five-acre field, she had reaped three different grades of wheat. Short of mixing the wild-oat infested wheat with the clean she did not have enough grain to fill a thousand-bushel rail car. Second, she had not ordered a car early enough to be sure of its delivery. Forced to sell "on the street" she received a poor price for the grain. Gross receipts that year amounted to $1,191, and net revenue, after labour and living expenses, was $700. Yet even with this profit, Binnie-Clark had fallen further behind in her mortgage commitments, which now amounted to more than $2,000, including the previous year's deferred payment. After further drawing on her reserves in England, she continued doggedly for two more years before concluding that farming without an accumulated capital of £5,000 was a futile proposition.

Yet other conclusions could be drawn from Georgina Binnie-Clark's experience, or more precisely, her lack of it. Unschooled in the principles of dry farming, she learned through bitter experience the inequities of the grain handling system. As a woman, especially one not used to manual labour, she was obliged to pay others to do most of the farming work, including grain hauling, which consumed a substantial share of revenues from the wheat crop. Her mistakes included an excessive reliance on wheat as a cash crop and financial over-extension.

To some extent Binnie-Clark's difficulties were beyond her control. Women, unless they were widows of homesteaders, were ineligible to apply for free-grant lands. After the revival of the pre-emption privilege in 1908, men over the age of eighteen could acquire 320 acres of homestead lands for $970 – one-fifth of the $5,000 Binnie-Clark spent for her half-section. Instead of acquiring an equity of several thousand dollars to assist in the financing of implements and other improvements, she was immediately saddled with the "killing weight" of a land payment. Doomed from the start, she joined the ranks of the thousands who vainly tried to establish themselves as farmers. Their failures gave the lie to the official predictions of easy profits.

THE FIRST WORLD WAR AND THE BOOM IN WHEAT PRICES

For many prairie farmers, the long-awaited promise of prosperity seemed to arrive in 1914. The First World War brought a renewed demand for exported farm products, and grain and livestock prices began a steep climb in 1914–15. By 1918 the price of No. 1 Northern wheat peaked at $2.24 at Winnipeg.[48] But while western Canadian farmers enjoyed unparalleled returns for three or four years, the war sowed the seeds of future financial trouble. Buoyed by their large returns, many farmers invested heavily in farm equipment and land, for which they paid record-breaking prices. By the early twenties, however, the price bubble of prosperity had burst and farmers found themselves burdened with heavy mortgages for land that earned less than half as much as it had in the late 'teens.

Meanwhile salaries of farm labourers, which had been virtually static in the pre-war period, increased steadily from $35 to $80 per month between 1914 and 1918.[49] The increased labour cost was partially reflected in higher overall production costs for wheat growing. In 1918 the Saskatchewan Department of Agriculture estimated that the cost of producing wheat had risen to $1.06 per bushel, or $14.85 per acre.[50] Their figures were similar to the results of a questionnaire distributed to thirty farmers throughout the province that year. The farmers reported that their costs were then $1.27 per bushel or $14.24 per acre.[51] This represented an overall increase of between $2 and $5 per acre in production costs since 1914.[52] Revenues had maintained a healthy margin over expenditures throughout the war, but the gap had narrowed considerably by 1920.

Despite the large increase in revenues and farm capital expenditures, no major trend towards mechanization occurred until after the end of the war. Farmers tended to replace old implements with

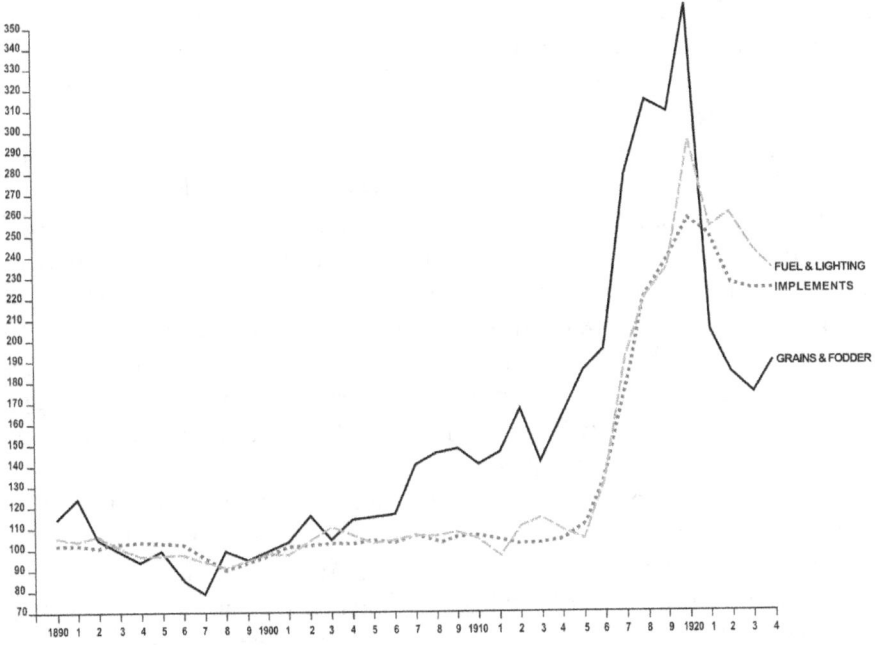

Fig. 18. Wholesale price indexes by commodity groups [Department of Labour], 1890–1924 (1890–99 =100). Source: M.C. Urquhart and K.A.H. Buckley, *Historical Statistics of Canada* (1965), p. 292.

new ones, but for the most part those purchases embodied traditional technology. Very few prairie farmers purchased tractors, as farmers clung to draft horses to provide their motive power.[53] By the 1920s the wholesale price index for fuels and implements had surged ahead of wholesale prices for grain products (Fig. 18).[54] Thus, farmers who had lagged behind in mechanization in the period of maximum farm revenues relative to costs were obliged to purchase expensive implements and gasoline in a period in which their income was declining relative to expenses.

The war had other, more serious, effects for the prairie economy. High wheat prices had encouraged over-cropping in many areas, resulting in soil exhaustion. After an outbreak of stem rust in 1916, yields declined steadily, reaching a pre-Depression low of eight bushels per acre in 1919. Yet, despite the reduced yield, unprecedented grain prices at the end of the war momentarily kept returns high. As a result the transition to diversification was further delayed, and the prairie economy remained tied to the wheat staple.

Inflated wartime prices had the additional effect of spurring the cultivation of marginal and pasture lands that previously had been unprofitable to crop. It also provided incentive to settlers to clear wooded lands for production. Overall wheat acreage in the prairie provinces jumped from 9.3 million acres in 1914 to 16.1 million in 1918, and 40,000 new farms were established between 1916 and 1921. A search of land maps in the Abernethy district shows a similar trend. During the four years of the war, the number of farms in Township 20, Range 11 increased from 46 to 63, reversing a decline in the number of farms between 1906 and the outbreak of the war.

The most comprehensive source of economic detail for the postwar period in the Abernethy district is a study on farm indebtedness prepared by the Saskatchewan College of Agriculture in 1933.[55] This study, comprising the Indian Head–Balcarres, Lemberg-Neudorf, and Grenfell-Wolseley areas, constituted a quantitative analysis based on interviews with 414 farmers in the three areas. While much of the information provided in this report related primarily to the early thirties, many of the trends in land acquisition and net worth of individual farmers shed light on the earlier economic development of the area.

One of the most striking revelations of the Farm Indebtedness study was the relatively large capital accumulation, particularly in the Indian Head–Balcarres district in which Abernethy is situated, in relation to the province as a whole. There the total average net capital per farm was $19,576 or more than double the provincial average of $9,261 (Fig. 19). At Grenfell-Wolseley, the average capital was $14,766, and at Lemberg-Neudorf, $12,487. These figures represented the total capital as represented in land, buildings, livestock, machinery, feed, seed, and supplies. While investments in each of these items exceeded the provincial average, the outlay for machinery and equipment was particularly large in the Indian Head–Balcarres area, where 65 per cent of farmers had invested in tractors. In the Lemberg-Neudorf and Grenfell-Wolseley areas, 50 per cent had purchased tractors up to that point. Other amenities such as modern equipment for lighting and water supply were reported as prevalent in the Indian Head–Balcarres area. Further evidence of prosperity lay in the fact that 22 to 24 per

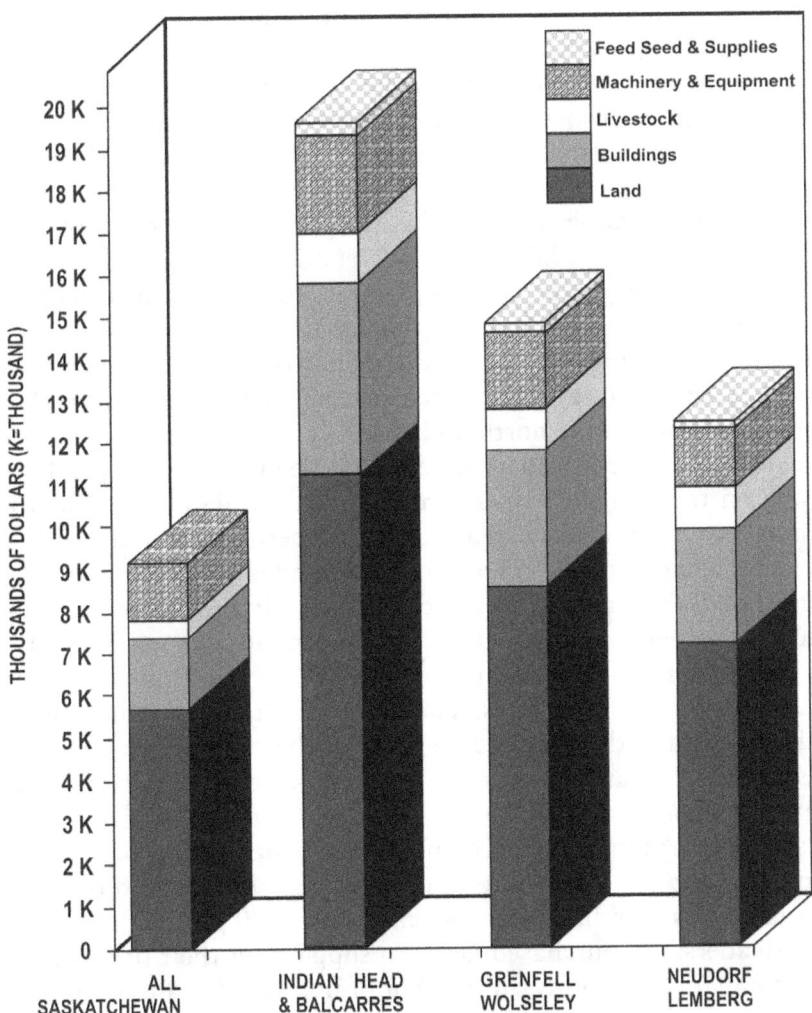

Fig. 19. Average capital per farm for the Province of Saskatchewan, 1931, and for farms surveyed in the Indian Head–Balcarres, Grenfell-Wolseley, and Lemberg-Neudorf districts, 1932–33 crop year. The data were derived from William Allen, E.C. Hope, and F.C. Hitchcock, "Studies of Financial Indebtedness and Financial Progress of Saskatchewan Farmers," Report No. 3, University of Saskatchewan, *Agricultural Extension Bulletin No. 68*, 1935.

cent of farm capital was invested in farm buildings, as opposed to a provincial average of 18 per cent.

In all three surveyed districts the majority of the farmland was devoted to grain growing, although the proportions varied rather considerably from one locality to the next. For example, Abernethy farmers averaged 470 acres, or 83 per cent of their farmlands, for crop or summer fallow, compared with only 268 acres of cropland, or 57 per cent of the available land, at Neudorf. The remaining acreage was consumed by waste, pasture, woods, roads, and farmsteads. To a large extent, the large non-crop areas at Lemberg-Neudorf can be attributed to the presence of uncultivable wooded and marshy areas. As late as 1920, surveyors reported that only 65 per cent of the lands in the townships to the north and south of Neudorf had been cleared for cultivation. In consequence, Lemberg-Neudorf farmers, while possessing on the average 83 per cent as much farmland, cultivated less than half as much wheat acreage as their Abernethy counterparts.

Farm tenancy in the Abernethy area tended to be more extensive than in the Lemberg-Neudorf and Wolseley-Grenfell areas. Forty per cent of Abernethy farm operators were tenants, compared with 19 per cent of farm operators at Grenfell-Wolseley and only 9 per cent of Lemberg-Neudorf farmers. The authors of the Farm Indebtedness study interpreted this fact as indicative of the profitability of the clay soils around Indian Head and Balcarres, which induced farmers to retain possession of their lands. The Farm Indebtedness study confirmed that farmers who had homesteaded continued to enjoy an economic advantage over those farmers who arrived later and purchased their land. In the Indian Head–Balcarres area, 90 per cent of surveyed homesteaders had purchased lands to supplement their original holdings, compared with only 38 per cent of non-homesteaders. Similarly, a far greater proportion of homesteading owners had made third and fourth additions of land subsequent to the first addition. In addition to owning larger tracts of farmland, homesteading owners also possessed a much smaller debt in relation to their overall assets than their non-homesteading counterparts. In 1933, farm owners in the Indian Head–Balcarres area who operated farms of three hundred acres or less possessed debts equal to 46.1 per cent of their assets, compared with debts of only 21.4 per cent of assets for owner-operators of farms exceeding 676 acres.

From the foregoing statistics, it can be inferred that the comparatively minor investment entailed in settlement prior to 1900 placed early settlers in a position of economic advantage that they never surrendered. Early settlers who persisted eventually tended to possess the largest farms, the most expensive farm buildings, the greatest assets, and the smallest debts in relationship to assets.

Even more striking than the economic differences between settlers on free-grant land and later settlers were the disparities between districts. In 1935 the Saskatchewan College of Agriculture published a study that calculated the approximate revenues of farms on varying types of soil in the province, on the basis of their previous performance.[56] Based on a comprehensive analysis of farm indebtedness in representative districts in nine distinct soil belts, the farm revenue study made separate tabulations of revenues and expenditures for 320- and 640-acre farms in each zone. Its conclusions demonstrated the marked inequalities in income that still existed among farmers on lands of varying quality. That income disparities were present was not surprising, but what was noteworthy was the virtual unprofitability of the quarter-section farm in six of the eleven identified soil zones. As the economic historian Robert Ankli has noted, if W.A. Mackintosh's cost figure of $9.81 per acre is correct, "then average prairie, fair prairie, inferior prairie, and poor prairie were either losing money or barely breaking even during the 1920s."[57]

Just as the Farm Indebtedness study had shown significant differences in accumulated capital between Abernethy and Neudorf farmers, the farm revenue report demonstrated markedly divergent incomes. Half-section farms on "very good park and prairie" soils at Abernethy earned cash incomes, on the average, of $510. After deduction of non-cash depreciation on buildings and machinery, farmers were left with a net of $307. But farms of 640 acres earned net incomes of $1,280, or more than twice as much per acre of cropland as the half-section farms. These figures demonstrated the marginal status of smaller farms which were, for the most part, still dependent on horse-power, even those in areas of higher quality soils. Operators of larger farms, who tended to use tractors, earned incomes sufficient to amortize lands to twice the value of the smaller operations. With such a large difference in income, the trend toward even larger farm units seemed an inevitability.

On the "fair to good Eastern Park" soils at Lemberg and Neudorf, similar differences were reported between 320- and 640-acre farms, except that the half-section farms in these areas were actually losing money. On these farms, incomes averaged only $134 in net cash revenue, but after deducting depreciation expenses, the farmers' returns evaporated into an annual deficit of $10 each. Farms of 640 acres on these soils, on the other hand, earned net incomes of $523. These figures suggest that the larger farms at Neudorf were modest, but viable operations in 1931. Yet most farmers in the area operated smaller farms. At the time the survey was conducted 50 of 70, or 71 per cent, of owner-operators at Neudorf possessed farms of 300 acres or less. The marginal status of their holdings was representative of broader inequalities in rural Saskatchewan society. To a significant degree these inequalities were the outgrowth of inequities in the original land disposition when these communities were first settled 40 to 50 years earlier.

Throughout its first fifty years the economy of the Abernethy district centred on wheat production and export. Despite recurrent anomalies in this predominantly single-crop economy, Abernethy never really experienced stages of agricultural development that had taken place in eastern North America. In those areas the early stage of subsistence-like agriculture had been succeeded by a period of wheat specialization followed by overcropping and concomitant soil exhaustion, and finally a diversification better attuned to the resource base of the particular area. At Abernethy and throughout most of the wheat belt of southeastern Saskatchewan, agriculture was never conducted on a purely subsistence basis, although many early settlers marketed wood, hay, and root products to provide a temporary livelihood in the farm-making period. Once they had cultivated sufficient acreage, however, they quickly concentrated on wheat production, and to lesser extent oats. While some farmers attempted to cultivate other crops, such as brome grass and flax, census statistics show that diversification never really took hold, even in the context of the economic depression of the early thirties.

Since farm income was predominantly dependent on the wheat staple, it was necessarily somewhat erratic. Many early settlers had been crushed by the unfortunate coincidence of low wheat prices, poor yields, and inferior grades in the 1880s and 1890s. Those who persisted, particularly early settlers who were able to reap a capital gain on their land, enjoyed a reasonable degree of economic success. By adopting new methods of dry farming they were able to reduce some of the risk and to regularize their income to some extent. But wheat farming, even in the relatively prosperous Abernethy district, continued to be a risky proposition. Farm debt became a common feature of economic life, particularly after the First World War price boom gave way to depression in the 1920s and 1930s. At the same time, the effects of debt were not uniformly felt. The earliest settlers, who had reaped a large capital gain by acquiring cheap land before the boom in real estate values, possessed the greatest assets and the fewest debts. Similarly, settlers who had occupied lands in areas of higher quality soils continued to augment their capital. Farmers on poor to average soils either continued to operate marginal operations or quit farming altogether to join the ranks of agrarian or urban labour.

4
Work and Daily Life at the Motherwell Farm

Since the publication of Marc Bloch's *French Rural History: An Essay on its Basic Characteristics*,[1] agricultural historians have increasingly turned to the study of daily life in search of a more rounded picture of rural society. Day-to-day activity can reveal a great deal about the objectives and material limitations of a society. The description of daily life specifically also serves as a basis for historical animation at the Motherwell homestead near Abernethy. Excepting some unique features, the processes of work and life at this farm paralleled those at other farms in the Abernethy district and throughout east central Saskatchewan.

Between 1882 and 1914 daily life on Abernethy district farms revolved around work (Fig. 20). This was particularly true in the early homesteading period when the tasks of developing a farm on unimproved land demanded long hours of hard toil by all members of the family. Even after 1900, when farmers like W.R, Motherwell had built up extensive operations, work continued to be the focus of activity. As late as the 1930s, the busy summer months were characterized by 18-hour days. Yet the Motherwell family, like other prairie Anglo-Canadian families, had experienced significant changes in the kinds of work routines they performed after the initial settlement period. To some extent these changes stemmed from the appearances of advanced farm machinery and domestic labour-saving devices in the district in the 1890s. In a broader sense they represented the imposition of a middle-class lifestyle on a foundation of hired labour. These farm families had graduated to a new set of social responsibilities that accompanied their new lifestyle.

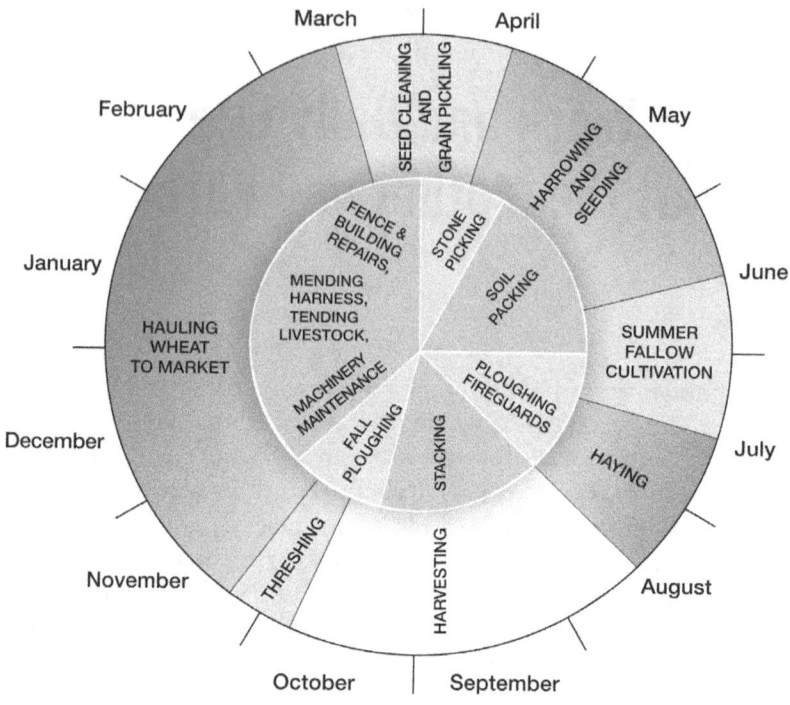

Fig. 20. The yearly cultivation cycle at the Motherwell farm.

In the period before the First World War, most of the manual labour on the Motherwell farm was performed by employees. During this time, W.R. Motherwell employed at least two "hired men" or agricultural labourers and one or two "hired girls" for domestic work. In summer one of the hired men was employed principally to do work within the farmstead. His duties included livestock maintenance and yard work – hedge trimming, grass cutting, hoeing around the shelter belt trees, and cultivating the garden and flower beds.[2] The other labourer was engaged for field work. At least one labourer was retained for winter work. Motherwell's son Talmage also worked in the fields with the hired men. Having completed a one-year course at

Fig. 21. The Motherwell barn. Saskatchewan Archives Board, Regina, Photo no. R87-219, No. 342.

the new Saskatchewan College of Agriculture in 1912,[3] he returned to the farm for one year to serve a brief apprenticeship. At the time of his marriage in 1913 Talmage received a dowry of two quarter-sections and was written out of his father's will.[4]

For farm staff a typical working day began early. Rising at 4:30 a.m. the hired men performed their ablutions,[5] dressed, and headed for the barn. There they cleaned the stables by guiding a horse-drawn stone boat through the corridor between the cattle and horse stalls. The draft horses were made to pause before each stall while the men shovelled the manure onto the stone boat. Immediately after, they drove the manure out to the fields where they spread it with hand tools, either a fork or a shovel.[6] The Motherwell barn is shown in Figure 21. After cleaning the stable the men fed the livestock, pumped water from the barnyard well into stock watering troughs, and let the animals out into the yard to drink. Usually they gave "chop" or milled oats, to the horses and a combination of raw oats and hay to the cattle. Accounts of two observers differ as to the form in which hay was given to the livestock and may suggest changing practices over time. Jack Bittner, who was Motherwell's neighbour after 1912, has stated that in this period, W.R. fed them brome hay from the sheaf.[7] Dan Gallant, who worked as a hired man at Lanark Place in 1920–22 and 1930–37, recalled having fed the stock loose hay.[8] In any event, the men produced the chop by operating a grain grinder in the loft. This machine was powered by a small stationary gasoline engine. The hired man shovelled oats into the implement's funnel-shaped hopper, which channelled the grain between the grinding plates or "burrs." The milled grain was then delivered along a chute and was shovelled into a bin.

The men groomed the horses with curry combs and brushes, a task described by a contemporary hired hand as "bothersome," because of the rambunctiousness of the horses.[9] During the cultivation season, they also clipped the horses' hair to help keep them cool for heavy draft work.[10] At the Motherwell farm they used a set of automatic clippers. One man powered this tool by turning a hand crank, while the other did the clipping. After clipping, they harnessed the horses in preparation for fieldwork after breakfast. Subsequently they gathered eggs from the poultry house in the east wing of the stable, milked the cows, and carried the eggs and milk containers to the kitchen.[11] Before breakfast at 7 a.m. one man emptied the bedroom chamber pots into a bucket from the upstairs chemical toilet and thence into the outdoor privy.[12]

Meanwhile the "hired girl," who also rose at 4:30 am, peeled potatoes, prepared breakfast, and set the kitchen table for the first of several morning sittings.[13] Breakfast was a large meal at the Motherwell farm, and usually consisted of potatoes and some form of protein – meat, eggs, or fish.[14] Serving the hired men, the hired girl cleared off the dishes and began preparations for the next round, as the family members rose in sequence. One hired man returned to the barn, where he hitched up the horse teams and headed for the field. The other stayed behind to turn the crank of the cream separator.[15] Depending on the amount of milk, this operation might last up to an hour. Dan Gallant, who was on the farm in the early 1920s, has related that he fed the hogs and chickens and cleaned out the hog manure after breakfast. The pigs were given chop dampened with milk, and chickens were usually fed wheat kernels.[16] Cleaning out the two hog pens was a particularly unappealing job. Since Motherwell had not widened the original door to permit a manure cart to be backed in, the farm labourer was obliged to shovel the hog manure twice. First he filled up his shovel and pitched it over the pen fences in the direction of the door. He then reshovelled the manure and heaved it through the opening in the door onto a wagon.[17] After cleaning out the piggery, the hired man distributed the load of manure on the fields. He then began the work of maintaining the farmstead grounds, including trimming shrubs, thinning the shelterbelts, cutting grass, and tending the garden and flower beds.[18]

Following the departure of the men, the hired girl continued to serve breakfast to other members of the household as they rose. Between servings she cleared away and washed dishes, and gave the upstairs rooms a light cleaning.[19] By mid-morning, she had begun preparations for the noon meal. While all meals on the farm were large, the noon meal, or dinner as it was called, was the largest. It lasted one hour or longer since the farm horses needed as much time to rest between shifts of fieldwork.[20] After dinner the hired girl repeated the cleanup of dishes and immediately began preparations for supper. Other afternoon tasks may also have included butter churning and the packaging of one-pound cubes in paper for market.[21] While the hired girls were spared most outside work, they were sometimes obliged to do the milking at the homestead.

At the Motherwell farm, supper was a meal wrapped in tradition and ceremony. All members of the household, including hired hands, were in attendance and all were expected to participate in its rituals. One of these was the "cent-a-meal" fund[22] for Christian missionary work. W.R. Motherwell kept a box into which each person was expected to contribute a penny. The other ritual was the daily religious service which Motherwell conducted after supper. After clearing away the supper dishes the hired girl was required to fetch the family Bible and to pass out hymn books.[23] Motherwell read a scriptural message, and then asked one of the assembled diners to select a hymn, which the group sang. Some of the hired men were less than enthusiastic about the enforced religiosity. One hired girl took advantage of the opportunity to select the longest hymns possible, while delighting in the suppressed grimaces of the victims.[24]

'Hired girls' in many respects had the most arduous daily schedules on the farm. Their working days often lasted from 5 a.m. to 10 p.m.[25] While male labourers could at least rest during the meal hours, female domestic staff were busy serving others at meal time. Two hired girls at the Motherwell farm have related that they never sat down at the supper table, but were in constant attendance.[26] Statistical studies of field and household labour on American farms in the early 1920s show that, overall, farm women worked an average of 11.7 hours per day, seven days a week.[27]

In addition to the daily regimen, larger housekeeping activities were organized on a weekly cycle. Different tasks were performed on specific days. At the Motherwell's and other Abernethy-area farms, laundry was usually done on Monday. Tuesday was reserved for ironing.[28] At the Barnsley farm, north of Abernethy, Friday was devoted to cleaning the upstairs, and on Saturday the downstairs rooms were cleaned in preparation for guests on Sunday.[29] Twice a week the Motherwell's hired girl baked bread. She was required to rise twice in the middle of the night to punch down the bread dough.[30]

Seasonally, the two busiest times of the year for farm women were spring and fall. Every spring the house received a thorough cleaning. In addition to cleaning the main and upper floors, the domestic staff also whitewashed the basement.[31] In the fall, farm women worked for weeks in advance of the harvest, preparing large quantities of food for hungry threshing gangs. In her autobiography, Nellie McClung remembered:

> It was for threshing that sauerkraut was put up in barrels (chopped by a new spade) and green tomato pickles were made; red cabbage and white were chopped up with onions, vinegar cloves and sugar, corn, scraped from the cobs and all kept in stone crocks. Every sort of cake that would keep was baked and hidden. The woman who had nothing ready for the threshers was almost as low in the social scale as "the woman who had not a yard of flannel in the house when the baby comes."[32]

During the actual harvest, farm women worked particularly long hours to provide the threshers with three square meals a day, while preparing basket lunches each morning and afternoon.[33] Their work began long before dawn and lasted until late at night. Figure 22 depicts a meal being served to harvesters in the field during Saskatchewan's settlement era.

Fig. 22. Harvesting meal served to threshers in the field, near Tiefenrgund. Saskatchewan Archives Board, Saskatoon, Photo no. S-B 4596.0.

The yearly cultivation cycle (Fig. 20) began in late March or early April, when the hired men cleaned grain in preparation for seeding. The first of two cleaning operations was the winnowing of the grain by means of a fanning mill (Fig. 23). This implement consisted of a square framework, a series of vibrating sieves and a revolving drum. The essential principle was for the grain to slide along the sieves through which extraneous particles of different sizes would drop, thereby separating out impurities. A wind current created by the revolving drum also helped blow dust and weed seeds away from the seed grain while it moved along the sieves. The drum was operated by a hand crank, which, through a system of chains and gearing, also agitated the sieves. While one man turned the crank, another kept the hopper at the top of the machine filled with grain.[34] As the cleaned grain emerged from the machine he shovelled it into a bin.

Fig. 23. Cleaning seed oats with a fanning mill on T.J. King farm, Carbon area, Alberta, 1920. Glenbow Archives, NA-2298-1.

The second cleaning operation was the "pickling" of grain, which entailed immersing it in chemicals, usually the day before seeding. Grain pickling was practised principally to make the seed resistant to smut, a fungal infestation, but this operation also served to remove impurities that had not been separated out during the fanning process. In the early period grain was usually treated with "bluestone," or copper sulphate, but by 1908 most farmers in the Qu'Appelle region were said to have switched to "formalin,"[35] a trade name for formaldehyde. The recommended solution was two pounds of formalin to 60 gallons of water.

Depending on the amount of grain to be treated, farmers used either simple or more elaborate implements in pickling. However, the basic principle, to coat each kernel with the solution, was the same. Georgina Binnie-Clark, who farmed about 30 miles from Abernethy, has related her method for treating the grain. She purchased a pickling barrel and immersed bushel bags of grain in a formalin bath for at least ten minutes.[36] While immersed, the grain was further cleaned as weed seeds and smut balls floated to the surface and were skimmed off. Binnie-Clark then lifted out each bag and left the grain to dry in the sack. She observed that more prosperous farmers often purchased two pickling barrels. Rather than use bags, they dumped loose wheat in the top barrel; when the grain had been sufficiently soaked, they pulled out a cork to permit the formalin solution to drain into the bottom barrel. The pickled seed was then thrown into a corner of the barn, covered with moistened sacks to prevent the formalin fumes from escaping, and left to dry. The process was repeated with the next quantity of seed grain.[37]

The Motherwell homestead, a substantial farm of 960 acres in 1912, required a somewhat more efficient apparatus for pickling larger quantities of seed grain. The remains of two seed grain picklers of the pre–First World War period have been found on the homestead. The first, a Dominion Specialty Works model, consisted of a table-like apparatus on which was mounted a funnel-shaped hopper and a galvanized barrel for the chemical solution. A hose ran from the barrel to a turbine placed directly beneath the hopper, which was punctured by numerous holes. The hired men shovelled the grain into the hopper and pulled away a stopper, causing the grain to fall on the turbine. The falling grain turned the turbine and was simultaneously sprayed with the solution. As the treated grain piled up on the floor it was shovelled away from the machine. The other pickler, an Acme model, was similar to the Dominion pickler, except that the grain did not drop through a turbine but was instead channelled into a horizontal chute. An endless wooden screw turned the grain over several times, so that it was well-coated by the time it was delivered to the end of the chute.

Meanwhile the hired men had begun to work the fields to prepare the seed beds. As one of the principal proponents of systematic "dry farming," W.R. Motherwell gave precise instructions to his men relating to the timing and techniques practised in field cultivation. As soon as the soil was in sufficiently friable condition after the spring thaw, the men harrowed the land. Motherwell insisted on an early harrowing so as to leave a surface mulch that would effectively check evaporation of moisture. Moisture retention also helped the soil remain warm, thus facilitating early seeding, a critical factor in western Canada's brief growing season, and especially in the short season of the second prairie level where the Abernethy district is situated. Harrowing also broke down clods of earth into a fine granular condition, suitable for a seedbed.[38]

Motherwell's preferred harrow was the simple iron-toothed drag harrow, which the horse teams, guided by a hired man, pulled across the fields. This was a dusty, dirty job that was not made any easier by Motherwell's insistence that his men walk behind the harrows.[39] Nor surprisingly, the men would have preferred to use a harrow cart, which would enable them to sit while harrowing. Motherwell discouraged this practice because he said that if his men were seated, they would not be inclined to clear away trash when it collected on the harrow teeth. One man who was a labourer on the farm in the 1920s has related that he used a harrow cart despite the orders.[40]

Spring tillage was not limited to the preparation of a seed bed. No doubt Motherwell also followed the advice of his colleague, Angus McKay, superintendent of the Indian Head Experimental Farm, in pursuing a ploughing regimen on his summer-fallowed land. He would have instructed his hired men to plough the fallow lands as early as possible after completing the seeding, so that the land would be "in the most receptive condition to fully absorb and save from waste all the early and later rains."[41] This work was performed with a gang plough (Fig. 24). The land was then immediately harrowed to create an insulating surface mulch that helped check the capillary action through which evaporation of valuable moisture resulted.

The men sowed the grain with a drill-seeder, on which they stood or which they followed on foot while it was pulled through the fields by two or more teams of horses. After sowing they harrowed the ground again to provide a better cover for the seed. Beginning in late April, seeding and related activities sometimes lasted until mid-June, depending on the moisture content of the soil and the type of crop sown.[42] In a wet spring the clay gumbo soils around Abernethy were impossible to work until they had dried sufficiently. Different maturation times determined the sequences of planting. Forage crops, such as brome, were seeded early to permit two harvests of hay during the summer. Wheat was also planted early, while early-ripening grains, such as barley and oats, were seeded last.

After seeding and harrowing the seed bed, many farmers packed the soil. After 1905, when Saskatchewan farmers began to purchase gas tractors, all three operations were often combined in one, with seeder, harrow, and packer arranged behind the tractor. Before 1900 they had "rolled" the land with a heavy cylinder of wood or iron, drawn by horses over the ground.[43] Rolling provided a means of smoothing

Fig. 24. "Breaking sod with a Hart-Parr tractor and six-bottom gang plough, Asquith District, Saskatchewan, ca. 1910." This outfit was very similar to W.R. Motherwell's Hart-Parr tractor and Oliver gang plough. Saskatchewan Archives Board, Regina, Photo no. R-A 12,082.

and solidifying the soil, and served to arrest the loss of moisture. It also helped to break up the clods, a result that was particularly desired in areas of heavy clay soils, such as Abernethy. Another effect was to level the land to create an even surface for harvesting equipment, such as mowers and binders. If, however, clay soils were rolled when still saturated with moisture, the soil would "cake" or form a crust and prevent the wheat plants from forcing their way to the surface. Hence, rolling was generally not performed until the plants were three or four inches high. At this stage, they were still flexible enough to spring back after being flattened. The small plants also prevented small particles of fine soil from drifting in the wind, according to a contemporary manual of prairie farming.[44]

Rolling was a controversial technique. In order to establish its utility, the Dominion Experimental Farm at Brandon conducted rolling tests in the 1890s. The experimenters showed that, while slightly increasing yields, rolling had a pulverizing effect on soils that contributed to the danger of blowing and drifting. As a result, its use had waned by the turn of the century.[45] From 1907 on, prairie farmers turned increasingly to soil packing as an alternative to rolling. Packers applied pressure to the lower part of the seed bed, and thus firmed the soil around the seed, while leaving the desired loose mulch on top. While tests initiated at Brandon eventually showed no clear increase in yields resulted from packing, it was a widely used technique on the prairies before the First World War.[46]

It is not known whether or not W.R. Motherwell used a packer on his farm, but his ministerial correspondence reveals that he approved of this method. Motherwell was of the view that all fall and summer ploughing was improved by packing.[47] However, he regarded the diamond-spiked harrow as a suitable substitute for the packer, and he may have instructed his men to use this implement instead.[48] He also recommended harrowing the grain plants as soon as they appeared above ground, and a second time when the plants were four to five inches high. Since harrowing at this stage was certain to pull some of the sown plants out, he instructed that the grain be sown more thickly than usual. To minimize damage to the young plants, Motherwell had his men use light wooden harrows with round teeth. The light harrows were still effective in removing most of the weeds and in creating a soil mulch.[49]

Since Motherwell cultivated extensive acreages of brome grass in this period, he frequently encountered the problem of eradicating it when tilling the land to convert its use to grain production or fallow. The process began in the previous fall, when the men ploughed the brome grass just below the root system or about four inches deep. In the spring the men disked and harrowed this fall ploughing to get it in suitable condition for the cultivator; then, in May, they went over the same land with a duck-foot cultivator. This operation shook the brome sods and roots, so that, exposed to the sun and wind, the grass could not re-root itself. The cultivating was repeated once a week throughout the month.[50]

Alternatively, when spring ploughing was required to eradicate twitch and other weed grasses, Motherwell recommended shallow ploughing very early in the season. This tillage was followed immediately by harrowing and seeding as soon as a half-day seeding was prepared, so that unnecessary moisture loss could be avoided. The early seeding was intended to give the grain plants a head start so as to choke out the weeds before they had re-rooted.[51] An alternative method recommended by Motherwell was to spring-plough, then leave the field to enable the first crop of weed seeds to germinate. About the beginning of June, his hired men would plough the young weed plants under, using a shallow adjustment on the gang ploughs. They immediately followed by working the same ground with a cultivator or disc harrows, and seeded the tilled land with barley.[52] Motherwell was of the opinion that these operations were best done simultaneously.[53] Since he possessed sufficient motive power in his Hart-Parr tractor, he theoretically had the capacity to hitch all of these draft implements in tandem. The cultivator differed from a plough in that it did not turn the soil over and did not till as deeply. At the same time the cultivator tended to pulverize the soil less than a harrow, and it was more efficient in cutting off weeds and breaking up the surface to permit rain water to soak in.[54] In practice, Motherwell's farm labourers usually carried out these field activities as separate operations, as his large Oliver gang plough proved unworkable for tillage in the heavy clay gumbo soils on his farm.

The purpose of summer fallowing was fourfold:

(i) to kill weeds by stirring up the soil so that weed seeds in it would germinate, and then to expose them to the air so that they would be destroyed;

(ii) by admitting air and moisture, to hasten the decay of the unrotted stubble;

(iii) to bring the raw subsoil to the surface to be pulverized and aerated; and

(iv) to pulverize the soil so that it would create a mulch surface layer to retain the soil's moisture.

Immediately after the initial summer fallow cultivation, the fields were harrowed lengthwise of the furrow. The land was then left for several weeks while the weed seeds germinated, after which it was harrowed crosswise.[55] Early July provided a lull in the cultivation cycle between seeding and harvest. At the Chipperfield farm six miles south of Motherwell's, men customarily used this period to build and repair fencing, cut wood, and cultivate the garden.[56] In the early 1900s prairie fires were still a threat, so the hands ploughed fireguards around the farm buildings, fields, and hay stacks.[57]

Meanwhile, by mid-July, farmers had begun the process of haying. Hay grass grew naturally around sloughs or was cultivated in fields as a forage. One man cut the grass with a mower, while another "coiled" it into rows with a horse rake.[58] After leaving the hay to dry and "cure" for a few days the hands returned with horse rakes to collect the hay into piles.[59] They then drove a hay rack to each pile on which they pitched the hay with a fork. From there they drove the loader to the barn and up onto the drive floor where they pitched the hay into the hay mows.[60]

Since it was carried out in the month of peak temperatures, haying was particularly unpleasant and tiring work. Not only was the air hot, the hay also became hot to touch. One settler who worked as a hired hand related his experiences in haying:

> After you have worked a whole day in that glowing hot hay, under the broiling sun, all alone, let me tell you, you lose quite a few drops of sweat. "Pails full," Uncle Lewis would say. In the evening you're dead tired, innumerable blisters on your hands, and your back feels broken from the continuous bending.[61]

In early August the hired men began the fall harvest. Starting with the early-ripening barley and oats, they used a binder or self-binding reaper to cut the grain. As the binder moved forward it bound the grain into sheaves and, when it had collected several bundles, the operator depressed a lever, causing them eject to one side (Fig. 25). Two men, following on foot, raced to "stook" the grain by arranging eight sheaves in pyramid formation.[62] The sheaves were stooked in rows north to south to permit the wheat to ripen evenly on both sides.[63] They were then left for several days to dry. To prevent rotting after a rainfall the men occasionally restooked the sheaves by placing the inner sheaves on the outside of the stook.[64] Since the stook often had to stand for many weeks before threshing, the stooker had to drive the butt end of each sheaf firmly into the ground.[65] (Fig. 26).

Fig. 25. "Harvesting machinery." Saskatchewan Archives Board, Regina, Photo no. R-B 8853-2.

Stooking was said to be one of the most undesirable jobs of the harvest season. Stookers complained of sores and blisters on their hands, which, despite the use of gloves, were rubbed raw by the straw. Their backs ached from interminable bending. Often they worked in hot weather. Regardless of the temperature the men could not roll up their sleeves or else their forearms would likewise be rubbed raw. Another problem was the early morning dew which "gets into your shoes to mix with the dust ... to make walking a trial by ordeal."[66] Moreover, since the grain had to be cut at the optimal moment, harvesting operations entailed particularly long working days. In his personal memoir of working on a western Canadian farm, A.G. Street has related that a normal stooking day lasted between 6 a.m. and noon, and then from 1 p.m. to 8 p.m., or 13 hours' actual labour.[67] An optional activity between stooking and threshing was grain stacking. Stacking was probably practised principally in the era of the large threshing

Work and Daily Life at the Motherwell Farm   115

Fig. 26. "British harvesters stooking, Okotoks, Alberta." Saskatchewan Archives Board, Regina, Photo no. R-B 7696.

outfits, when farmers often waited for several weeks or months before a threshing machine and crew became available. Figure 27 depicts a threshing crew posing at nearby Lorlie, just 10 miles northeast of Abernethy. In the meantime, stacking provided added protection from moisture for the wheat sheaves. A contemporary prairie agricultural manual suggested that the quality of grain stacked before threshing was superior to that of wheat threshed directly from the stook.[68] In his diary entries for 1902, W.R. Motherwell's neighbour Samuel Chipperfield recorded that between 17 September and 11 October his men spent about three weeks stacking oats and barley.[69]

Stacks were built in the field to remove the need to draw the sheaf racks across the rough land.[70] Farmers commonly built four stacks, in pairs six to eight feet apart. Initially they built a large circular stook base with a diameter of eight feet. The hired men then arranged two layers of sheaves, one on top of the other, with the heads angled

Fig. 27. "Harvesting machinery, showing threshing crew and equipment." Saskatchewan Archives Board, Regina, Photo no. R-B 8853-3.

inwards. One man stood on the stack building it while two others pitched sheaves to him.[71] He placed the butt ends of the sheaves so that they extended past the bands of the previous row. Gradually, this shingle-like formation diminished in diameter until the stack formed a rounded peak. At the peak a long stake was driven through a sheaf to fasten the stack securely. Stacks were built high – about 20 to 24 feet – to prevent rain water from seeping in.[72] Two pairs of high grain stacks from different farms are illustrated in Figures 28 and 29.

While the stacked grain was "curing," the hired men were principally busy with fall cultivation of the stubble fields. Other fall activities included the slaughter and butchering of livestock, such as pigs and cattle (Fig. 30). James M. Minifie has provided a very detailed account of the operations involved in slaughtering a pig on his father's homestead.

Fig. 28. "Threshing grain from stacks with portable steam engine, Moosomin District, ca. 1895." Saskatchewan Archives Board, Saskatoon, Photo no. R-B 574.

Fig 29. "Horse-powered threshing crew near Boggy Creek, 1896." Saskatchewan Archives Board, Saskatoon, Photo no. S-A 69.0.

Preparations were intensive. My father built a bench about eighteen inches high, a solid job with two-by-sixes, with two-by-four legs. We set up a barrel with a tripod and pulley and tackle above it. We had a container of water boiling on the stove. Preparations included a bucket for the offal, a couple of iron scrapers which had started life as garden hoes, and a lard-pail for the blood. It was my job to catch the blood and stir it well to prevent clotting, fetch scalding water and scrape the bristles off the hide. This meant dunking the carcase in the barrel as soon as the pig was dead; if there was delay, the bristles did not come out, and a dinner of roast pork was like eating hairbrush.[73]

The Minifies then, with much difficulty, slaughtered the pig, dunked it in the barrel to scald the skin, scraped off the bristles, suspended it by the hind legs, and disembowelled it.

The last stage of the grain-growing cycle was threshing. Ordinarily the grain was threshed in late October or early November, within a few weeks of the completion of harvesting and stacking. Before the advent of smaller portable separators around 1910, however, most Abernethy-area farmers did not own their own threshing machines and were dependent on the availability of hired harvest gangs. If the farmer participated in the co-operative purchase of a separator, he and his partners drew lots to determine the order in which his grain would be threshed.[74] In either case Abernethy farmers might have to wait until December or even January to thresh their grain. Threshing was a brief but intensive operation at the Motherwell farm. In addition to his regular hired men, W.R. Motherwell employed a large number of extra hands, including migrant threshermen who had come west from Ontario on harvest excursions. Figure 31 is a view of threshing operations being carried out in the farmyard of the Motherwell farmstead in the settlement era. Motherwell's daughter Alma remembered that on occasion at least 20 men participated in threshing operations at the Homestead.[75] Most other period accounts similarly place the number of required threshers at about 18 to 20 men.[76] Figure 32 shows a stooking crew from the settlement era near Davidson. By about 1910, W.R. Motherwell and other Abernethy area farmers had purchased their own smaller grain separators, which required fewer men to operate them.[77] The number of grain bundles to be pitched did not diminish,

Fig. 30. Butchering a pig on the A.E. Cox ranch near Pincher Creek, Alberta. The suspension of carcasses from a tripod apparatus for butchering was generally practised on farms across the prairies. Glenbow Archives, NA-2001-15.

Fig. 31. Threshing scene at the Motherwell farmstead. Saskatchewan Archives Board, Regina, R87-219, Photo no. 348.

Fig. 32. "Stooking crew on Wells Land and Cattle Company farm, Davidson, Saskatchewan." Saskatchewan Archives Board, Saskatoon, Photo no. S-B 2014.0.

however, and the time required for threshing lengthened, accordingly, from several days to perhaps two weeks.

The threshing machine or grain separator was driven by means of a belt from a power source (Fig. 33). In the 1890s, before straw-burning steam engines were prevalent, horse teams were used to power the machines.[78] A horse-powered threshing outfit in 1896 in the area of Boggy Creek, Saskatchewan is shown in Figure 29. By 1910 the steam engines, in turn, were beginning to be superseded by gasoline tractors (Fig. 23). While the separator was in operation, two drivers or "teamsters" drove wagon racks loaded with grain sheaves beside it. Two men climbed on to each wagon and, using a fork, pitched sheaves into the separator. Figure 34 shows sheaf wagons as driven next to a threshing machine in preparation for pitching the sheaves into the separator, during the harvest season at Ingleside, Saskatchewan. As the machine separated the wheat kernels from the straw and chaff, the threshed grain was discharged through an auger into a granary or the box of a grain wagon. If the grain was threshed into wagons, three or four other drivers hauled wagon loads, when filled, to granaries placed elsewhere on the farm. At the separator, meanwhile, as each rack was emptied of its sheaves, the teamsters drove their racks off to stack a new load. Two other wagons then drove beside the separator, and their contents were similarly emptied into the machine.[79] Meanwhile, in the field, pitchers used forks to throw sheaves onto the bundle wagons as they returned for a new load. These forks had three long steel prongs and round wooden handles about five feet long. As the sheaves were heaved to the middle of the wagon, the driver, using a fork, arranged them to permit quick loading and to distribute the weight evenly.[80]

Threshing gangs usually included seven or eight teamsters and four pitchers to keep the machine in constant use. Gangs also included an engineer for the machine, a man to keep watch over the separator and oil it periodically, a tank man to provide water for the steam engine, a stoker, and three or four men to transport the separated grain to granaries.[81] The water man was busy throughout the day hauling water from nearby sloughs and dugouts to the steam engine.[82] He and the engine man had to be vigilant for any outbreak of fire. The combination of dry straw, great heat generated by the engine, and the frequent need to load it presented a very real danger of igniting the prairie. If a fire started and got out of control, it could quickly destroy the entire year's harvest.

Fig. 33. "Threshing on J.R. Brown's farm, south of Qu'Appelle, early 1890s," Prior to the advent of the gas tractor, threshing operations were often delayed into the winter months. Saskatchewan Archives Board, Regina, Photo no. R-A2476.

If threshing operations were delayed until the winter months certain improvisations were necessary. The snow-covered ground obliged the teamsters to mount their sheaf-racks on sleigh runners.[83] Nightfall arrived early, necessitating the placement of a lamp or electric headlight on the engine. The headlight illuminated the "self-feeder" end of the separator to enable the pitchers to see where to throw their sheaves. A contemporary thresherman later recalled that "when you pulled out from the glare of it you were almost blind for a moment or two."[84] Aside from the obvious danger of slipping into the machine, he described other adverse conditions associated with the work:

Fig. 34. Siegert Dalen's threshing outfit in the Ingleside district south of Marchwell, Saskatchewan, ca. 1910–1930, Saskatchewan Archives Board, Saskatoon, Photo no. S-A 377.

> ... It was harvesting extraordinary. The stooks were just mounds of snow to look at. As the sheaves were pitched up, frozen ice and snow whipped across the loader's face. It was about zero weather and the cold made the engine difficult to start. This was worse than the work, as the cold got right into you as you were waiting about.[85]

The thresherman related that one of his fellow workers built a straw fire beside the separator to warm himself. Clearly, winter threshing entailed hardship for many seasonal workers. In another account a harvest hand described winter threshing as particularly miserable when snowfall alternated with melting, making the work "cold as the Dickens, especially at night." He stated that since there was insufficient room in the threshers' shack, he was obliged to sleep outside in the December snow.[86] He also wrote of "the ease with which you strain your muscles" while pitching sheaves in cold weather.

Like the proverbial ploughman, the western Canadian thresherman has been the subject of romanticized folklore. Numerous live demonstrations of threshing operations at country fairs such as the Manitoba Thresherman's Reunion are steeped in nostalgia. Possibly modern farmers, with their solitary practice of driving a combine through fields of swathed grain, long for the days when neighbours and travelling threshermen worked together at harvest time. What is easily forgotten is that work with threshing machines was an arduous, sometimes dangerous, often alienating labour. One observer in the period distinguished between the social aspects of threshing in old Ontario and what he regarded as the more mercenary character of prairie threshing after 1900.[87]

Threshing accidents were not uncommon. Occasional reports of accidents appeared in the local press and in pioneer reminiscences. In 1899 one Billy Ringrose, a young bachelor working on a threshing gang at Pheasant Forks, stepped on a belt feeding the separator. Simultaneously the horses which were providing the power for the separator began walking on the treadmill. The belt was icy; Billy slipped and was drawn into the machine. His leg crushed and severed at the groin, the young thresher bled to death.[88] In 1906 a young English hand was standing beside a threshing machine on the farm of Motherwell's neighbour Arthur Bearden, when the machine collapsed and crushed him. By the time the others had emptied the machine to release him, he had suffocated.[89] Even when the machines were not in operation, they posed potential hazards. At Kenlis, south of the Motherwell homestead, Archie Wright slipped while moving a separator. The wheel cut into his leg, breaking the bone.[90]

Harvest employees contended with other problems related to their transient status and lack of material comforts. James M. Minifie has described one difficulty that is not mentioned in the usual accounts.

> There is another hazard for the thresherman. The water at each farm varies slightly in mineral content, not much, but just enough to throw the body off balance, and produce a weakening looseness, which adds to the miseries of a sleepless night, even where adequate toilet facilities are available, an unlikely occurrence when casual labour is employed. Squatting at night in a field of stubble qualifies for Dante's Purgatorio.[91]

Grain hauling or "teaming" was the principal work activity during the winter months of November to March. This was particularly the case in the period before 1904, when most of the farmers' productive time was consumed by tasks relating to marketing wheat. As noted earlier, the lack of local rail service before 1904 obliged farmers to haul their grain 20 to 35 miles to market centres on the CPR line.[92] In addition to the distance factor, farmers north of the Qu'Appelle River Valley had to contend with the problem of negotiating its 300-foot or 400-foot banks. Since grain wagons could easily tip over on the descent, farmers were limited to loads of 50 bushels.[93] The danger of spillage also required that they bag their grain – another time-consuming process. Settlers were further impeded by the need of at least two teams of horses to pull their grain wagons up the valley slope. They therefore travelled in pairs and doubled up their teams for the ascent. After both teams pulled one wagon to the top, the teamsters unhitched them and returned to the valley floor to pull the other wagon up.[94] Figure 35 depicts horse drawn grain wagons as they arrive at an elevator at Yorkton, Saskatchewan.

Estimates vary as to the time required to make the trip to the rail centre. Obviously the time was affected by distance and by driving conditions. In the winter trails were often made "heavy"[95] by drifting snow and the teamster's progress was slowed. Late in the winter, melting affected the capacity of the trails to carry the load. Often, as the driver tried to negotiate the high drifts, one side of the wagon collapsed under the weight of the load. He had to throw the grain bags off the wagon to get the sleigh back on the trail. A teamster related that, when hauling grain in late March, he was once forced to load and unload three or four times in a five-mile stretch.[96] Most oral history sources state that the round trip took a minimum of 20 hours.[97] In addition, the teamsters needed to clean the grain before loading it. Fanning mills in use in the period before the First World War carried a maximum capacity of 25 to 35 bushels of wheat per hour. Hence, for each wagon load of wheat, teamsters spent about 1 to 1.5 hours in cleaning the grain, in addition to time spent in hitching and unhitching the horse teams before and after the trip.

Another task ancillary to grain hauling was the lifting of the grain wagon box on to the bob sleighs. First the front end had to be lifted on to one set of runners; then the back end was raised, swung around, and lowered onto the rear runners. A second tier was then fitted onto

Fig. 35. "Horse-drawn wagons hauling bags of grain at the Lake of the Woods elevator in Yorkton," 1905–7. Saskatchewan Archives Board, Saskatoon, Photo no. S-B 4265.0.

the wagon box. As James M. Minifie has noted this was a "tricky job" for one person to perform; swinging the box around always carried the danger of a "pulled back" or "slipped disc."[98]

In his social history of a Saskatchewan rural municipality, Bruce Peel provided a vivid account of the arduous process of grain hauling.

> In preparation for the journey the farmer loaded the wheat on the wagon the night before. Standing in the granary, ankle-deep in the wheat, he scooped up shovelful after shovelful and, with a rhythmic swing, tossed it into the wagon. Poets have waxed eloquent over the "golden grain"; but no poet ever shovelled the stuff. Dust filled the air, irritated the shoveller's nose and throat, and itched his sweaty cheeks. Perspiration trickled a smeary path down his dusty cheeks. He puffed with exertion. His back ached from stooping and straightening as he shovelled the innumerable scoops

needed to fill a wagon. Sooner or later, when he straightened his back to rest it, he caught it on one of the taut strands of twisted wire used to brace the granary. With dogged determination he stooped again to shovel. The wagon had to be loaded. Between two and five o'clock in the morning the farmer arose. Lantern in hand, he went to the stable to feed and harness his horses, while his wife prepared breakfast, and packed a lunch. The half-frozen lunch would be eaten when the farmer stopped along the trail to feed his horses.

The four horses were hitched to the wagon. To facilitate the loading of the wagon little hollows had been dug into which the hind wheels had been lowered. Now the horses tugged and strained to pull the wagon out of the hollows. The silence of the morning was broken by the shouts of the farmer urging his horses on, by the crack of the whip, by the metallic clanking of whipple trees and traces, and then by the crunching of wagon wheels on frozen ground.[99]

On reaching town the teamster drove the wagon to an elevator to sell his load. When the wagon was in place on a large set of floor scales, the elevator agent weighed the loaded wagon and graded a sample. In the period before 1904, when the grain was bagged in heavy cotton sacks, the teamster then dumped each bag's contents through a hatch into a hopper. The building of local elevators at the time of the founding of the village of Abernethy obviated the need to bag grain, since farmers then did not have to cross the Qu'Appelle Valley or Pheasant Creek Coulee while hauling their loads to market. Thereafter grain wagons were filled in bulk and weighed at the elevator before the wagon was lifted mechanically at the front end to permit the bulk loaded grain to spill through a hatch at the rear. Once all the grain had been dumped, the elevator agent again weighed the empty wagon, to arrive at the net weight of grain. After calculating the value of the grain he gave the teamster a cheque in exchange.[100] The teamster then drove to local retail outlets, to load up with groceries and supplies, before beginning the return trip.

In his diary entries for 1902–1903, Samuel Chipperfield recorded that between 10 November and 24 March his hired man and another seasonal labourer made 40 trips to market. These trips consumed two days each – one day outbound and another for the return journey. On the average the men made slightly more than one return trip per week, usually to Sintaluta on the CPR main line. In early January, after the bulk of the hauling had been completed, Chipperfield laid off the seasonal hand. His regular hired man hauled the remaining stored grain for the balance of the season.[101] Other winter activities included cutting and hauling wood, building and repairing granaries, and fencing.[102] Winter generally afforded a somewhat slower pace of activity, enabling farmers also to mend harness and repair machinery. As in all other seasons, feeding and watering livestock, and cleaning out stables, were ongoing responsibilities. Milking of cows, cream separation, and butter preparation as well as the collection of hen's eggs, continued throughout the winter.

By 1914, for its part, the Motherwell family was preoccupied with tasks relating to its role as a leading middle-class family in the Abernethy district. As provincial minister of agriculture, Motherwell himself was living in Regina and returned home only on weekends and holidays. Catherine, his wife, acted as a surrogate farm manager, and communicated Motherwell's instructions to the hired men. She also held sway in the domestic sphere, supervising the work of the hired girls and her sister, Janet Gillespie, who now lived on the farm. Mrs. Motherwell did domestic accounts[103] and, with the hired girl, worked out daily menus. As a prominent member of local women's groups and in her role as a politician's wife, she hosted large gatherings at the farm throughout the year.

For the Motherwell family, daily roles in the pre–First World War period increasingly assumed the form of prototypes that have been described in Thorstein Veblen's *Theory of the Leisure Class*[104] and other sources. Between 1841 and 1914, as Ann Oakley has shown, middle-class wives in England assumed roles of leisure.[105] Their dependence on their husbands was elevated to a sought-after ideal, since this very dependence became a mark of status for their husbands. "The successful businessman delighted to show off his wife and daughters expensively clad, living a life of ease and elegance."[106] In a study of Edwardian women, Kate Caffrey described the typical English middle-class life of leisure of that era:

> Even the middle classes could keep up a style of living that surprises the present-day reader.... They went to luncheon parties, tennis parties, garden parties, and like the aristocracy, subscribed to the elaborate ritual of afternoon tea. It was, perhaps, the halcyon time of suburbia. Trim villas with evocative names like The Laurels, standing in neat gardens, contained a firm family life, gradually relaxing to permit croquet on Sundays with an indoor variation on rainy days played on a green baize cloth spread on the dining table, and church connections encouraged amateur dramatics, music, tennis and badminton.[107]

This picture of Edwardian England might in many respects equally be applied to the Motherwells, a middle-class Anglo-Canadian prairie family aspiring to social prominence in their community. In imitation of his English counterparts, Motherwell established an elegant late Victorian home and pastoral estate with an evocative name; he called it "Lanark Place" in fond memory of the Ontario county of his birth. Contemporary photographs show the Motherwells entertaining large groups on their lawn, sometimes under the huge canopy of an open-air tent. The Motherwells also hosted lawn-tennis parties, and possessed the only tennis court in the district[108] (Fig. 36). The Motherwells even went so far as to emulate a particularly fashionable practice among the English middle classes. They sent their daughter to study in Germany for a year. Her stay was cut short by the outbreak of the First World War in 1914.

Further indications of "conspicuous consumption" among the various members of W.R. Motherwell's family were revealed in their clothing. As Thorstein Veblen has noted, the function of middle-class dress was to convey the impression that one did not work for a living. While Motherwell's wife and daughter were fashionably clad in current vogue that seemed to proclaim their leisured status, so Motherwell himself donned the bowler hat, vested suit, and prominently displayed gold watch chain that marked him as a gentleman farmer and politician (Fig. 37). Perhaps the ultimate expression of Motherwell's adoption of Edwardian dress codes was his first class Civil Uniform, replete with cocked hat, embroidered closed-front tailcoat and sword, which he wore to the coronation of King George V in England in 1911.[109]

Fig. 36. Lawn tennis game, Motherwell house in the background. Saskatchewan Archives Board, Regina, W.R. Motherwell Collection, R87-219, Album, p. 37.

This is not to suggest that Motherwell or even his wife, Catherine, led lives of leisure; both were particularly active and contributed much to community affairs. Motherwell also continued to labour on his farm into his old age. It does, however, reveal the importance to the Motherwells of assimilating aspects of the lifestyle of the leisured classes in the parent societies of eastern Canada and England. By virtue of not being employed and of being largely free of housework, Catherine Motherwell could evoke the image of a leisured matriarch, even if this image was not really accurate.

Fig. 37. W.R. and Catherine Motherwell on their front door step at Lanark Place. Saskatchewan Archives Board, Regina, W.R. Motherwell Collection, R87-219, Photo no. 378.

5

Abernethy's Social and Economic Structure

Writers of western Canadian history have been slow to analyze rural prairie social structure in vertical or hierarchical terms. Seymour M. Lipset, in his study of the C.C.F. in Saskatchewan, stressed what he saw as the "classless" nature of that province's rural population.[1] In Lipset's view the scattered distribution of the population, dependence on the wheat staple, and other factors made Saskatchewan farmers less differentiated economically than the population in most other rural areas. While acknowledging gradations in income and property among farmers, Lipset believed that these differences "exist mainly *between* areas rather than within individual rural communities."[2] Within each district, differences were minimal. Similarly, in *Democracy in Alberta: Social Credit and the Party System*, C.B. Macpherson showed an awareness of variations in income and capital among farmers, but stressed that these differences were not so great as to create a clear demarcation of economic interest within this group.[3] Neither Lipset nor Macpherson gave much attention to socio-economic groups other than the broad class of farm owners/proprietors. To an extent their analyses reflected an emphasis on sociological realities in the middle decades of the twentieth century, including the accelerating disappearance of hired labour on prairie farms.

Before the First World War, prairie society, particularly in the older areas of Anglo-Canadian settlement such as Abernethy, exhibited features of a more stratified social structure. As elaborated by anthropologist Morton Fried, stratified – as opposed to egalitarian – societies are characterized by an unequal access to available resources.[4] With specific reference to nineteenth-century Canada, the most comprehensive historical analysis of social structure in this era was the study of social organization in eastern North American cities, including Hamilton, Ontario, by Michael Katz, Michael Doucet, and Mark Stern.[5] Katz and colleagues put forward a two-class model, which assumes that in a capitalist society, most people share common relationships to one or both of its two central aspects: the private ownership of capital and the sale of one's labour as a commodity. Most people who do not own capital sell their labour and those who own capital are purchasers of labour. The problem in applying such a model to Abernethy's social structure is to determine to what degree it applies to independent commodity producers in rural areas of this period.

It would first be useful to define the term class. Most theories of social class have identified common properties of a group that must exist if its members are to be considered to belong to a class. These include a vertical ordering of people, an element of permanence in the class's interests, a shared sense of class identity, and a relative degree of separation of individuals from different classes. In a behavioural sense this last criterion implies minimal personal contacts between classes and little crossover in terms of one's class membership.[6]

An examination of socio-economic relations at Abernethy in the late nineteenth and early twentieth centuries shows some degree of conformity to the standard definitions of class. Farm owners persisted as a socio-economic group during the first 40 years after settlement. They showed a collective awareness of their class position in the context of agrarian protest movements after 1900. Abernethy's society was vertically ordered, stratified not only between socio-economic groupings, that is, farm proprietors and labourers, but within the farm proprietor group itself. This model is tempered somewhat in terms of the separation of and interaction with other social groupings. Farmers and labourers often did work together in the fields, and there was the possibility of social mobility for at least some labourers to move into the proprietor group.

What must be recognized is that Abernethy, like all other prairie farm communities, was in a transitional stage in the settlement era. A substantial degree of mobility still existed among homesteaders, at least initially. By 1920, a trend toward consolidation of homesteads into larger farms was already well underway, as many early settlers sold out or went bankrupt. Despite this instability in their situation, Abernethy farmers overall showed considerable continuity in their position by maintaining their individual ownership of land. This group of farm proprietors may reasonably be termed a class, albeit a transitional one, or at least a class segment. On the other hand, farm labourers did not conform to the usual definitions of class. They were too geographically dispersed and their working situation too transient to permit them to form organizations or to act in common in their own interest. Moreover, many farm labourers did not view their situations as permanent, but as a prelude to becoming farm owners themselves.

It would be shortsighted to treat Abernethy's social structure as a self-contained unit, without reference to outside influences. The lives of local farmers were heavily affected by a business class, but this class consisted of corporate transportation, financial, and manufacturing interests based in Winnipeg and central Canada. Abernethy's economy, like that of other prairie farm communities, was inextricably integrated into the Canadian commercial framework on which farmers relied to purchase and ship their grain, manufacture farm implements, and finance their land or equipment purchases. Farmers tended to identify themselves as independent entrepreneurs operating within a free enterprise economy. The work of economic historians Harold Innis and Vernon Fowke, who explored the implications of the price system for Canada's staple industries, suggests that such notions were largely illusory.[7] Unrestricted competition between producers in a price system they were powerless to control could only serve to erode their socio-economic structures in the long run.

MAJOR SOCIO-ECONOMIC GROUPINGS

Abernethy's population before 1914 was divided between a broad class of independent farm proprietors and a more transient group of wage labourers. A similar dichotomy existed between the proppertied group and the propertyless. In modern capitalist societies ownership of property, both capital and land, has been a major determinant of social and economic position. Property presents options to the holder that are denied to the propertyless. As noted in Chapter 3, Abernethy farmers were able to use their ownership of land equities for various economic purposes. Property ownership also granted the holder access to social and political power in the community.

Within the dominant class of independent farming proprietors, there existed considerable economic differences. The comparison of farm sizes indicates progressive differences between farmers' holdings throughout the first 35 years after initial settlement. In terms of capital accumulation, a study of the Abernethy district carried out in the early 1930s[8] showed that the upper 15 per cent of the groups of farm owners possessed 30 per cent of farm wealth; the lower 30 per cent possessed only 18 per cent of the wealth. Farmers in the highest category with farms of 676 acres or more recorded an average net worth of $37,580 per person. The lowest third of farmers, with 300 acres or less, had an average net worth of only $7,579. Hence, the ratio of capital between the highest and lowest levels was five to one.[9] What was striking was the extent to which the gap had widened since those in the Indian Head–Balcarres sample had started on their farms. Owners in the lowest category had begun farming with a net worth of $7,782. Twenty years later their average net worth was $7,579, a loss of $203. The wealthiest group of farmers started with an average of $17,930. Twenty years later, their net worth had increased by almost $20,000 to $37,580.[10]

Abernethy society was further differentiated by distinctions between landowners, part-owners, and tenants. In 1933, only 60 of 122, or slightly fewer than half the farms in the Indian Head–Balcarres study area, were owner-operated. Part-owners operated 14 farms, or 11 per cent of the total, and 48, or 40 per cent of the farms, were operated by tenants. Owners and part-owners were in a demonstrably superior position to tenants vis-à-vis their net worth. In 1933, farm owners possessed total assets of $24,723 on the average and about $16,000 after deduction of liabilities. Part-owners possessed assets of $23,607 per person and $14,000 each after deducting liabilities. Tenant farmers had average assets of only $2,800 and an average net worth of $1,134.[11]

Significantly, the emergence of a large proportion of tenant farmers pointed to the formation of a new *rentier* component within the farm proprietor class. Evidence for the *rentier* thesis exists in the fact that W.R. Motherwell, and others of his socio-economic stratum, indulged in land speculation outside their home district. By 1906 Motherwell had purchased sections of farmland in the newer areas of settlement near Markinch, Loon Lake, and Davidson, Saskatchewan.[12] In 1909 he again speculated in improved land in the Outlook district.[13]

It would appear that the general trend towards farm tenancy had begun much earlier than the 1930s. The 1926 tax assessment roll for the Rural Municipality of Abernethy indicates a high proportion of absentee landlords, comparable to the 1933 figures.[14] Tax rolls for the pre–First World War era have not survived, but demographic trends in this period are suggestive. In the years 1901–6, during which the village of Abernethy was founded, the townships adjacent to and south of the village were significantly depopulated.[15] Many of the new residents of the village were farmers from the surrounding district. It is probable that many of these farmers began to rent out their lands when they moved into town. As the authors of the Farm Indebtedness study noted, the high tenancy rates in the Indian Head–Balcarres areas partly reflected the high value of the farmlands in these districts, and partly stemmed from "the desire of the original owners to retain possession of their farms which have been generally decidedly renumerative."[16] A large proportion of Abernethy's population, therefore, lived to some degree off the economic rents that accrued from their ownership of property.

At the opposite end of the economic spectrum was the non-propertied group. In the period before 1914 a fairly large proportion of Abernethy's population consisted of female domestics as well as hired hands. They both occupied social positions even lower than that of the tenant farmers. Where farm tenants at least had access to the use of property, hired men and women could sell only their labour. Some insight into the insecurity of the non-propertied classes in the nineteenth and early twentieth centuries was provided by a contemporary observer:

> A propertyless person is one without any economic reserve power. He is in no position to ward off the sufferings which must frequently come to most persons depending wholly upon their ability to labor and upon the demand of the community for their services....[17]

In periods of labour shortages, which were frequent, farm hands could command fairly high wages. Their position was, however, dependent upon market conditions beyond their control. When the price of wheat dropped, as in 1913, farm wages accordingly went down.[18] Since they worked for wheat farmers whose incomes were erratic, the hired men often experienced considerable difficulty in collecting their wages.[19] Hired men could also be dismissed summarily with little recourse. Sometimes farmers would start a quarrel with their employees or make life difficult enough for them that they would leave without collecting their wages.[20] This is not to suggest that most farmers treated their employees unfairly, but periodic reports of these practices underscored the farm labourers' lack of status or power in society.[21]

STATUS AND LEADERSHIP

Property relations, while central to the Abernethy social structure, were not the sole determinants of position in the community. Since Max Weber, sociologists and anthropologists have often emphasized the role of status in conferring social position.[22] In rural societies, where economic differentiation is less pronounced than in cities, community prestige may assume a special importance. This is particularly the case, given the predominance of primary group relationships, that is, "intimate face-to-face association," in rural areas[23] and the resulting "short status ladder."

A context for interpreting social status in pre–First World War Abernethy may be derived from John Bennett's *Northern Plainsmen*.[24] This book comprised an anthropological analysis of farmers near Maple Creek, Saskatchewan in the 1960s. While relating to a much later period than the Abernethy study, the Maple Creek analysis showed some traces of earlier social constructs. Bennett identified six distinct status-giving criteria in Maple Creek farm society: "social credit," ethnic prestige, occupational prestige, settlement prestige, hardship prestige, and economic and political power.

Bennett used the category "social credit" to denote one's status as "a good human being, manager, homemaker." He viewed this criterion as the most important of the six. By "ethnic prestige," he meant the status that accrued to one's belonging to a preferred ethnocultural group, in this case, the Anglo-Canadians. "Occupational prestige" was an important factor in the Maple Creek social structure, since two large occupational categories, ranching and farming, co-existed, and ranching generally carried more prestige than farming. Farmers who immigrated early into the district qualified for "settlement prestige," or status based on the date of one's arrival. In the 1960s, when Bennett compiled his data, he concluded that this factor was of minimal influence. "Hardship prestige" was to a large extent analogous to "settlement prestige," but reflected the hardships experienced by farmers during the drought and depression of the 1930s. "Economic and political power" accrued to those farmers who were wealthy or possessed connections with the provincial government.[25]

At the turn of the century, economic power was one of the central criteria in determining one's status in the prairie Anglo-Canadian communities. In a society of self-made men, one's accumulated wealth was a testament to one's initiative, business acumen, and managerial skills. In the novel *Fruits of the Earth*, Frederick Philip Grove charted an Ontarian settler's progress to material prosperity and status. While fictional, Grove's account was based on his own observations and experiences as a farm hand in Manitoba before the turn of the century. He wrote of his protagonist, Abe Spalding:

> Abe's prestige had grown enormously. He owned the biggest holding not only in the ward but in the municipality. He paid the highest taxes....[26]

Like Abe Spalding, Abernethy's most prosperous farmers enhanced their prestige as they augmented their capital. Possessing six quarter-sections by 1906,[27] W.R. Motherwell was one of the largest landowners in the Abernethy district. Among the positions he held in the 1890s were the chairmanship of the local school board, elder in the Presbyterian Church, and Justice of the Peace.[28] Prosperity also freed farmers to participate in community service. By 1901, when Motherwell instigated the Territorial Grain Growers' Association, he employed at least two full-time hired men.[29]

Conversely, subsequent settlement groups were often constrained by their marginal economic position from achieving a significant degree of status. As a result of their late start, some of the settlements tended to be in a relatively primitive state of development at the turn of the century. At Neudorf and other eastern European settlements in the District of Assiniboia, farmers were preoccupied with the basic problems of subsistence – feeding and clothing their families, while gradually expanding their holdings. As noted earlier, land clearing in the Neudorf district was not carried out on a significant scale until the settlers were able to afford access to sophisticated clearing technology in the 1920s. While property qualifications for candidate in North-West Territorial elections were removed in 1888, a $100 deposit was still required. Few settlers in the recently settled districts, even if backed by their neighbours, could afford to risk such capital on a candidacy. Since so much productive effort was consumed in the process of eking out an existence, manifestly little surplus was left to be devoted to politics.

Abernethy residents were also differentiated by occupational status in the settlement era. Michael Katz has identified the great gulf in status between the entrepreneurial and labouring sectors in nineteenth-century Ontario.[30] Abernethy and other rural prairie communities had only two significant occupational categories. The dominant group of independent farm proprietors was guaranteed a degree of status inaccessible to the wage labourers. In fact the farmer's occupational status was derived to some degree at the labourer's expense. One Anglo-Canadian prairie farmer portrayed the independent farmer as a responsible entrepreneur just as he regarded the hired man as irresponsible.

> I resent the idea that there is a parallel between the man who shirks his task and the farmer who finds that he cannot raise an abundant crop. The farmer did not put in his crop for some other man; it was his own crop and it was his own failure if he did not reap the harvest he expected. He did not accept any man's wages for which he was expected to raise a full crop and then only try to raise half a crop.[31]

In a similar vein, a contemporary farm labourer noted that his employer, a "small-time" farmer, derived a personal sense of superiority by virtue of having a hired man. He wrote: "he knows damn well that he has a hired hand, and he's proud that he can afford one now."[32] Despite the economic deprivation of many farmers in the period, the fact of employing staff served to reinforce their self-image as independent entrepreneurs.

Ethnocultural and religious association was also an important prerequisite to status in Abernethy's settlement era. Throughout the first 40 years, Abernethy was an almost monolithically Protestant

Fig. 38. Exterior of Knox Presbyterian Church in Abernethy, ca. 1909. Saskatchewan Archives Board, Regina, W.R. Motherwell Collection, R87-219, Photo no. 165.

Anglo-Canadian community. In 1901, 84 per cent of Abernethy's population belonged to one of the Protestant churches – Presbyterian, Anglican, Methodist, or Baptist, with Presbyterians predominating. Figure 38 shows Knox Presbyterian Church in Abernethy shortly after its completion in 1909. As well, 92 per cent of the community's residents in 1901 were of English, Irish, or Scottish descent, including the dominant Ontario-born settlement group.[33] Most non-Anglo-Canadians in the district were either Ukrainian Orthodox or Russian-German Lutheran seasonal labourers from the newer settlements such as Ituna and Neudorf to the east and north.[34] Few members of these ethnocultural groups purchased farmlands near Abernethy; few, indeed, could have afforded it.[35]

Non-Anglo-Canadians had little access to status and leadership roles in rural Saskatchewan society. The prairie Anglo-Canadians were wary of eastern European cultures and feared that these "aliens" were not capable of being assimilated. In a historical study of ethnocultural groups in Alberta, Howard Palmer observed that between 1880 and 1920 most leaders of opinion in Alberta voiced doubts that European peasant settlers would conform to Anglo-Canadian political institutions, owing to their presumed lack of experience with representative government.[36] In the Qu'Appelle district of the North-West Territories similar sentiments had currency in the settlement era. In 1899 the *Qu'Appelle Vidette*, published at Indian Head, approvingly quoted one Father Morin's comments about the "alien" immigrants. Among other things, Morin stated that Galician setters (meaning, Ukrainian immigrants from the province of Galicia in the former Austro-Hungarian Empire) were "from the point of view of civilization, ten times lower than the Indians...."[37] Whatever this comment reveals about the status of Aboriginal people in the period, it also indicates how eastern European newcomers were received by some within the dominant central Canadian group.

Status also accrued to men and women who were respected for their human qualities. In this respect Bennett's criterion of "social credit" seems particularly applicable. As a family the Motherwells were held in high esteem not only for their formal involvements but for the personal assistance they provided to others of the community. A reading of W.R. Motherwell's ministerial correspondence indicates that he gave local farmers advice on all aspects of prairie agriculture, including field crop techniques, the planting of shelter belts, and even appraisals of land values.[38] Even those actions of Motherwell's which appear to us to be paternalistic were often received warmly as expressions of his personal interest in community affairs. As the prime mover behind the building of a "temperance hotel" in Abernethy, Motherwell received accolades even from political foes. In 1908 Conservative Senator W.D. Perley praised Motherwell "on questions of morality and the practical manner in which he demonstrated his faith in ultimate success of the temperance cause."[39]

Catherine, Motherwell's second wife, also achieved considerable status through her community involvement (Fig. 39). Before her marriage, Mrs. Motherwell had achieved recognition as a teacher and Presbyterian missionary among the File Hills Cree north of Abernethy.[40] After her marriage she continued to give Cree women on the reserves advice on homemaking questions.[41] Mrs. Motherwell also spoke frequently on a variety of women's issues at meetings throughout the province. At the Regina founding convention of Saskatchewan

Fig. 39. Catherine Motherwell. Saskatchewan Archives Board, Regina, W.R. Motherwell Collection, R87-219, Photo no. 161.

Homemakers in 1911, alongside such luminaries as Lillian Beynon and Nellie McClung, Mrs. Motherwell gave a lecture on "domestic book-keeping."[42] In 1915, she spoke on the woman's franchise before the Equal Rights League at Lemberg, Saskatchewan.[43]

In the Abernethy area, Mrs. Motherwell was a leading member of the Women's Missionary Society of Knox Presbyterian Church. A view of the Whitewood, Saskatchewan chapter of this society is shown in Figure 40. After 1916, she assumed a prominent role in the local Women Grain Growers Association. Further evidence of the stature of the Motherwell family in the community emerged over the issue of church union. In 1920 the Abernethy Presbyterian congregation was deeply divided on the question. The church appointed Mrs. Motherwell and Englehart Stueck as a committee of two to study the issue and make recommendation.[44] They recommended that the church participate in the union, and the congregation supported this recommendation.

American settlement historian Allan Bogue has postulated that since newly settled areas on the frontier lacked an existing political structure, aspiring leaders jockeyed for power in this early period.

> ... In a period when groups are being formed, competition exists among potential leaders to a considerable extent, creating a situation which is much more unstable, than is ultimately the case when the group has shaken down and the members have come to know the virtues and deficiencies of their fellow more thoroughly. An important function of the leaders is to regulate the membership of the group.[45]

Bogue provides an interesting model for interpreting the establishment of social structures in Abernethy's formative period. Prospective leaders may have spent up to two decades establishing their respective claims to position and status in the community. Leadership was initially comparatively fluid. For example, of the first four justices of the peace to be appointed in the Abernethy District (at Pheasant Forks to the northeast and Kenlis to the south), at least three had disappeared as landowners by 1906.[46] Obviously a settler needed to establish himself economically and occupationally before laying claim to long-term influence in the community.

Those early settlers who did persist were best placed to assume leading positions in Abernethy society. For the decade before the First

Fig. 40. Presbyterian Women's Missionary Society delegates at Whitewood, Saskatchewan, ca. 1907–8. The middle-class composition of this and other women's missionary groups and their emphasis on conspicuous consumption is suggested by the members' dress, and particularly, their hats. Saskatchewan Archives Board, Regina, Photo no. R-A 7394-1.

World War, an analysis of the officers of local organizations, including the local improvement district, various school boards, church groups, and the Abernethy Agricultural Society shows the dominance of farmers who had arrived in the 1880s. The same names recur in different contexts, for example, the Motherwells, the Chipperfields, and the Morrisons. Wives of prominent farmers also tended to belong to the principal women's organizations and to wield considerable influence in these groups, that is, women's missionary societies, the Women's Christian Temperance Union and the Abernethy chapter of the Women Grain Growers Association.[47] If one examines the movers of key motions in various groups, the influence of a few key families, particularly the Motherwells, is readily apparent.

6

Social Relationships in the Settlement Era

Apart from socio-economic structure, social relationships provide important insights into the character of a community. A measure of the community's openness and democracy is the extent to which persons from different socio-economic strata interact with one another. Generally speaking, it is assumed that informality in social relationships suggests a degree of egalitarianism. A tendency towards formality and exclusiveness in social interaction, on the other hand, indicates a stratified or elitist orientation. These are some of the criteria that were applied to the study of social relationships in Abernethy before the First World War.

SOCIAL INTERACTION IN THE EARLY SETTLEMENT PERIOD

Abernethy's population in the early settlement period consisted primarily of unmarried male settlers, a demographic imbalance typical of the North-West Territories, as males greatly outnumbered females throughout the region in this era. The 1891 Census of Canada reported that within the District of Assiniboia West, which included Regina, males between the ages of 20 and 74 comprised about 65 per cent of the population, or nearly twice as many residents as females in the same age brackets. A large proportion of this overwhelmingly male population was unmarried. For example, in the Regina and Moose Jaw districts, 2,402 people, comprising only 31 per cent of the population were married in 1891, the remainder consisting of unmarried or widowed residents or children. As settlement progressed, the disparity of numbers increased, as in 1911 males outnumbered females in Regina by 13,616 to 6,020.[1]

In the first few years on their homesteads, many of these men evidently experienced great loneliness. By moving west they had cut themselves off from the network of social relationships that had sustained them in their places of origin.[2] Not only did many lack romantic relationships, they were also separated from neighbours by miles of treeless prairie. Winter months were described as particularly monotonous, as settlers found themselves surrounded by barren wastes of snow and frequently confined to their tiny shacks by blizzards or severely cold weather. As well, winter work was often solitary. Settlers had not yet developed their homesteads into farms, and they spent much of the winter cutting wood alone in the bush. Walter Elkington's account of his first weeks on his homestead illustrates the new settlers' solitude:

> The first night in the place was very cold, for winter had not yet passed, and in the morning I found a piece of plaster had fallen out of the wall just above my head, letting in more fresh air than was necessary. For the next few days the weather was very severe, and I frequently had to leave the house and go into the snug little stable, where the horses were, to keep myself warm. All this time I seldom saw a soul, and having very little to read, found it very lonely....[3]

In their personal accounts of the settlement experience bachelors sometimes reported that they lived in less than ideal conditions. One Manitoba settler wrote, "The bachelor lives on pork and bannocks, as a rule, never sweeps his house out, or very seldom; generally hoes the floor once a month."[4] Another settler from Hayfield, Assiniboia reported that of all the tasks in the daily regimen, washing up was the worst for the bachelors. Hence, "... things are always used as often as possible, and then piled up until everything is dirty, when a big washing takes place."[5] The decline in diet and domestic order was perhaps inevitable given the long hours of farm work on the homestead and the comparative inexperience of many young settlers in housekeeping. A correspondent to a contemporary Anglican journal in the District of

Fig. 41. "Bachelor in his palace (Tim Starks?), Prairie Rose, Saskatchewan," 1906–1910. Saskatchewan Archives Board, Saskatoon, Photo no. SB 1062.0.

Assiniboia asked: "Is it to be wondered at that very soon the refined and even cleanly habits of home are forgotten and with them I fear often religion is forgotten too...."[6] Figure 41 shows a bachelor settler of the settlement era who apparently tidied up the interior of his modest dwelling for the photographer.

In these circumstances it was not surprising that settlers attached great importance to the re-establishment of human contact on the frontier. In rural areas, settlers frequently lived with other men while developing their homesteads into farms. Sometimes, the cohabiting males were listed themselves as "partners" on the census returns; in other cases, male settlers cohabited with hired labourers. For example, in the rural district of North Qu'Appelle, the 1901 Census indicated that two male newcomers formed a farm household. It comprised Charles S. Thrina, aged 28 and listed as the head of the household, and Walter F. Sheppard, aged 27, denoted as his "partner." Both gave "farmer" as their occupation. Both men originated in England, although Thrina had emigrated in his twenties, while Sheppard had been in Canada from the age of six. In the same district, Archibald Matheson, a 32-year-old farmer, was then living with Samuel Paul, listed as a domestic, with the occupation of "farm hand." In South Qu'Appelle, Thomas Clarke, a 38-year-old farmer, headed an all-male household that included two farm labourers – Ernest Jones, 23, and Bert Grant, 18, both listed as lodgers. In 1906, in the Qu'Appelle district, Herbert Deker, a 27-year-old farmer, was living with George Penfold, his 17-year-old farm labourer.[7] Not all cohabiting males were linked to their companions by occupation. As reported in the 1901 census, J.R. McMeakin, a 47-year-old single farmer, was listed as the head of a household consisting of him and John Tuffnel, a 30-year-old single "pedlar."[8]

At Kenlis, just a few miles south of Abernethy, David Gibbons, a 35-year-old farm proprietor, was listed as cohabiting with Richard Penny, a 20-year-old farm labourer in 1901.[9] Also at Abernethy, in 1911 John Vellicott, a 26-year-old divorced head of household, was listed as living with David Jones, a 25-year-old labourer.[10] As well, Rory McKenzie, a 33-year-old farm proprietor and household head, was then living with Roy Dickson, a 23-year-old farm labourer.[11] In the same year, in the nearby Indian Head area, Ernest Devitt, a 26-year-old farmer and household head, was then living with Alexander McClintock, a 24-year-old farm labourer.[12] In addition to such shared living arrangements, many other all-male households included bachelor farmers who lived alone as the sole enumerated members of their households.

Since neighbours were few and scattered, the bachelor settlers tended to welcome any new addition to the community and could not afford to be too selective in their social relationships. In *Fruits of the Earth*, a fictionalized account of an Ontarian prairie settler, Frederick Philip Grove presents Abe Spalding in an initially isolated situation. Despite attitudes indicative of racial prejudice, Abe is shown to be pleased to receive a Ukrainian neighbour. In the novel's dialogue, Abe states, ""I'd like to have men of my own colour about. But rather than stay alone, let niggers and Chinamen come.""[13]

While the bachelor settlers often lived together for reasons of mutual enterprise, companionship, and economy, they encountered problems of adjustment that people generally experience when living with others. Often, a certain laxity in meeting domestic responsibilities set in. Ferdinand David, an English settler at Ellisboro (15 miles to the southeast of Abernethy), wrote in his diary entry for 1 January 1894 that he and his roommate resolved to turn over a new leaf:

> ... Tim and I made New Year's resolutions vis. – We keep our diggings clean and in order. I do the housework. Tim does the chores. We have kept these resolutions splendidly today and are therefore feeling very self-satisfied. How long will it stay thus? Don't ask.[14]

In other cases the settler's roommate in the small shanty was his hired man. In this situation of informal accommodations and limited human contact, ties between farmer and employee were often close and fairly unstructured. A Manitoba settler in the early 1880s has related that he and his hired man worked together in all farm operations, including cutting wood in the bush and field work. They also ate together and slept in the same room.[15] Figure 42 depicts two bachelor homesteaders hauling food purchased in town to their homestead in the settlement era.

Sometimes neighbouring settlers arrived unexpectedly at dinnertime. Abernethy pioneers recalled that invariably the uninvited guests were asked to stay for supper. Walter Elkington wrote that one Sunday bachelors from a nearby ranch showed up at his door carrying a few eggs. The impromptu arrivals proceeded to make some special "slap-jacks."[16] In the period before 1904, when Abernethy settlers hauled their wheat to the CPR line 20 to 30 miles to the south, they were often obliged to stop at farmsteads along the route for food or a night's accommodation. According to oral history sources they were never refused.

To alleviate the pain of isolation settlers tended to socialize as often as possible. Sometimes they congregated at another bachelor's shack for a game of cards or simple conversation and tea. Social intercourse was relatively informal in this early period, even for married families. In the winter, surprise parties were a frequent occurrence. Within a radius of 15 miles people travelled by sleigh to their friends' homes, announced a party, and stayed the night. Despite frostbitten fingers and toes that resulted from these trips, an Abernethy pioneer woman

Fig. 42. "Dave Campbell and Bill Spedding hauling food from Crooked River to their homestead near Speddington with oxen and sleds," 1913. Saskatchewan Archives Board, Saskatoon, Photo no. S-B 7177.0.

later related that "it was worth it, there was so little fun or entertainment."[17] Games at these parties included "charades," "upset the fruit basket," and "musical chairs."[18]

Indeed, settlers seized on almost any pretext to participate in social activity. A weekly ritual in one settlement was to congregate on Friday evenings at the post office.[19] The collection of one's mail provided an opportunity for human interaction. Even funerals provided an outing and were thus well attended. An English settler was somewhat perplexed, at a country funeral near Fort Qu'Appelle, when he witnessed the spectacle of mourners dressed in their gaudiest clothes, forming a procession of grain wagons to the cemetery.[20]

The first priority for most bachelors, however, was to find a spouse to provide companionship and essential domestic support. As a Manitoba settler wrote:

> ... my advice to the settler is marry. Every girl is pounced on directly she puts her face inside the settlement. Young fellows get so sick of the monotony of baching. I hope to get Frank out, after a year or two, to help me, or marry some young lady well versed in scrubbing, washing, baling, dairying, get up at 3:30 in summer, 5:30 in winter, strong nerves, strong constitution, obedient, and with money. Where can I find this paragon?[21]

Numerous sources suggest that fierce competition ensued among the overwhelmingly male population for the few women who established residency in these settlements in the homesteading phase. Usually the most eligible young women were schoolteachers, many of whom were soon wooed by bachelor suitors.[22] Bachelors' balls became a common fixture of Abernethy life throughout the period. These events permitted bachelors to meet prospective sweethearts in a socially respectable way. In some prairie communities an event that combined community service with social intercourse was the "Box Social." For these dances the young women of the settlement prepared boxes containing sandwiches, cakes, and pies, which they wrapped with ribbons. The bachelors bid on the boxed prizes, and the proceeds went to a fund for building a new school or church in the community, while their bids provided them with an introduction to the young woman and a dance.[23] The absence of women did not necessarily deter the men from convening or attending social events largely populated by males. In 1895, "Bachelor Billy," a correspondent from the Spencerville area to the *Qu'Appelle Progress*, wrote that he and 'a friend' had attended a social event at "Blakeney Castle." He wrote: "Much to our disappointment, the fair sex were in the minority." That did not stop the festivities, however, and "dancing was indulged in at intervals" until dinner. Afterwards, the men danced again until the party broke up at

3:30 a.m.[24] Bachelor Billy's account suggests that, at least within this group of men, there was no apparent stigma involved in men dancing together. It was a situation that would rapidly begin to change as the easy familiarity of friendship between young men in the nineteenth century became more constrained by emerging social conventions and fears of same-sex sexuality after 1900.[25]

Other bachelor amusements found less favour among the middle-class farming group that was gradually establishing its dominance in the Qu'Appelle region. These farmers or their spouses associated "bachelor dens"[26] with a host of evils, including alcohol consumption, card playing, and sexual licence. Reading between the lines of contemporary memoirs, one gains an impression of many early settlers spending as much time in the town saloons as they did working on their homesteads. Walter Elkington wrote that, along with tea, which was drunk at every meal, whisky was the most popular drink.[27] On another occasion he wrote that the crowd in Fort Qu'Appelle was "very orderly," as a result of raids carried out, under the prohibition laws, against whisky smugglers.[28] In her autobiographical memoir, one Saskatchewan woman recalled the experiences of an English "remittance man" east of Round Lake in the Qu'Appelle Valley. Having been established on a "dude" ranch by his father, this young fellow "kept a lot of hired men, gave stag parties, where beer flowed from 5 gallon kegs...."[29] While taken at a later date, Figure 43 shows two bachelor setters partying outside a homestead shack in the settlement era. Indeed, of 94 cases tried before Mounted Police justices in the Qu'Appelle District in 1882, the first year of settlement, 55, or 60 per cent, were for gambling or alcohol-related offences.[30]

The sexual behaviour of bachelors was similarly placed under increasing scrutiny in the early settlement period. James Gray has written of the red light districts in frontier communities in the Canadian West, including thriving houses of prostitution in the major centres, of the North-West Territories, including Regina and Moose Jaw.[31] A reading of local newspapers confirms that the Qu'Appelle district was similarly not immune to the attractions of prostitution. In 1885 a correspondent from Qu'Appelle wrote the *Vidette*, demanding to know "whose duty it is to rid the town of its present harlot pests?"[32] Complaining that the Mounted Police officers had instructed that the "unfortunate" women were not to be molested as long as they behaved themselves, the indignant writer asked,

Fig. 43. "George Warnock (left) and Alonzo R. Prestley partying outside Prestley's homestead shack at Crystal Hill," ca. 1925. Saskatchewan Archives Board, Saskatoon, Photo no. S-B 8301.0.

> ... are our women folk to continue to be daily shocked by the flaunting of these prostitutes, be it on horseback or on foot? Where are those respectable church going citizens, who are supposed to voice public opinion, that such a living disgrace to our hitherto fair town as a house of ill fame is permitted under their very noses?[33]

Evidently pioneer men, as well as women, felt threatened by the possible consequences of sexual laxity, especially adultery. It was perceived to threaten the family, and since women were considered indispensable to successful settlement, the whole social and economic basis of society. In 1886 another correspondent to the *Vidette* issued a stern warning to an alleged offender:

> We have here a young bachelor in lower town, and a married "lady" in upper town, who are working everything too much on the free-love plan. Why the husband of the lady don't put a stop to this sort of thing is more than I can tell; but I can assure the young bachelor that he will get a coat of tar and feathers if he don't cease to bestow his attention so openly. If the husband is foolish enough to put up with it, the public are not.

The letter was signed "DECENCY."[34]

Yet while the bachelor frontier had the appearance of an unstratified, relatively egalitarian society, this situation was only transitory. Bachelors were prepared to put up with material and social privations for a few years while they developed their rough homesteads into farms. Few regarded these circumstances as permanent; rather, they were simply a necessary penance that they experienced while establishing themselves. Similarly, the seemingly unstructured nature of social relationships in the early settlement period did not really signal a permanent democratization in social interaction. Spontaneous interaction was a necessary pragmatic response to pioneer isolation. It also reflected the relative youth of the population. Most settlers wished, however, to find spouses to establish familial ties to replace their looser network of frontier relationships.

THE MATURE SOCIETY, 1900–1920

As Abernethy evolved from the "pioneering" period into a settled community, social relationships reverted to more traditional patterns. By 1900 most of the bachelor homesteaders had married and begun to raise families. These settlers now endeavoured to build their own self-sufficient farm communities. In doing so they emphasized the household as the centre of social life. When financial resources permitted they built large Victorian houses that fostered a more dignified, middle-class familial interaction.

Another possible reason for the establishment of a more traditional lifestyle was that the pioneering process was selective. While remittance men, adventurers, and other confirmed bachelors dropped out in the initial period, the persistent, more ambitious – and conservative – settlers remained. Willem de Gelder noted these differences between two such types of settlers he encountered in Morse, Saskatchewan in 1911:

> You meet them here in the hotel, men of all nationality and ages.... Swells and dandies talking about their homesteads with little cigars in their mouths. I don't think you ever see one of those swells survive a second winter on their homesteads. And next to them, you see others on whose faces you can read, the resolution, ones who are frightened by nothing in this world and are determined. You can read the success on their faces.[35]

Like Grove's protagonist, Abe Spalding, the latter type of settler was preoccupied with developing his farm and had little time for social life. Moreover, at Abernethy and other settlements in the North-West, the mean age of the population increased in the first decades of settlement. Settlers who were in their early twenties in the 1880s had reached middle age by 1900. Obviously, farmers in this later stage in the life cycle had developed social roles and responsibilities different from those of the predominantly youthful population of the early period.

In keeping with the new Victorian middle-class order, social activities needed to be purposeful. In 1893 Kenlis residents founded a local Royal Templars of Temperance.[36] It provided the best of all possible activities – a heavy dose of Christianity, a sense of mission, a practical purpose, and, not least, a social outing. Women, too, were not long in forming their own chapter of the Women's Christian Temperance Union (W.C.T.U.). Young people also were encouraged to join their own Christian organizations. A chapter of the Epworth League of Christian Endeavor was formed by Kenlis residents in 1896.[37] Public lectures also combined the need for human interaction with social utility. In 1897, the *Qu'Appelle Progress* reported that John Allen of nearby Pheasant Forks was to deliver a lecture at the Abernethy school house on the subject "Harmony and Diversity."[38] Another lecture had as its title "Leadership Values." At nearby Lemberg, members of the local literary society heard a public debate on the topic: "Resolved that professionalism in sport is conducive to its highest interests."[39]

The middle-class farmers' desire for stability and permanence in social relationships found expression in the formal organizations they established after 1900. Abernethy men founded chapters of the major Anglo-Canadian lodges: Masons,[40] Oddfellows,[41] Canadian Order of Foresters[42] and the Loyal Orange Lodge.[43] A group photograph of Oddfellows from the Moosomin Lodge is shown in Figure 44. In the early period, social intercourse on the frontier had been spontaneous. Later, regular lodge meetings assisted in institutionalizing and regulating social behaviour. Lodge rituals, steeped in ceremony and hierarchy, symbolically expressed the established settlers' desire to entrench their vertically-ordered notions of social organization and served to inscribe a pecking order of status and privileges. The fraternal societies also afforded an element of exclusiveness in social relations, as men could mix with other men of similar ethnocultural background or socio-economic position. The formation of such exclusionary organizations appeared to reflect an attempt to emulate hierarchical social conventions of the settlers' societies of origin. Georgina Binnie-Clark commented on this tendency when she wrote of the Fort Qu'Appelle Tennis Club:

> ... [it] has all the distinction of tradition defended by the force of exclusion. At one time, to be known as a member of the club gave much the same cachet in the district as presentation at Court during the Victorian era. Also, in common with many of the great persons of the Victorian era, face value went for little. None would guess from a glance at the club enclosure with its distinctly primitive pavilion the important part it played in the creation and preservation of a social atmosphere in the little village....[44]

Abernethy women also participated in a developing more formal approach to social interaction after the period of economic consolidation than existed previously. In the early years of settlement, homesteaders' wives had been preoccupied with heavy housekeeping duties, including cleaning, cooking, and raising children, often combined with agricultural labour. Their social activities often flowed from their domestic work and were fairly unstructured as, for example, in quilting bees.[45] By the 1900s, the pioneering period had passed and women graduated to a new set of responsibilities. Their involvement in various women's

Fig. 44. Independent Order of Oddfellows, Moosomin, Saskatchewan, pre-1897. The ceremonial dress was characteristic of Oddfellows lodges in Saskatchewan's settlement era. Saskatchewan Archives Board, Regina, Photo no. R-A 3301.

groups was now regulated and goal-oriented by agendas and established meeting times. Refreshments were also formalized and pre-arranged, as the hired girl at the Motherwell farm served tea in a sterling silver service to the guests.[46]

Private entertaining among Anglo-Canadian middle-class families also took on a more formal character after the passing of the "pioneer era." Where surprise parties had been the norm before, Edith Stilborne of Pheasant Forks remembered that after 1896 "invited parties" became more general.[47] At Lanark Place, as soon as the Motherwells had built a large Victorian "new house," they began to host ceremonious dinner parties. In January 1898 the Qu'Appelle *Vidette* reported that "Elder" W.R. Motherwell had celebrated New Year's Day by having 25 couples to his home for dinner. The dinner was "given in a style and manner which reminded some of us of the more formal gatherings of the Old Country..."[48]

Social Relationships in the Settlement Era   159

The most striking evidence of the new middle-class approach to social interaction was in the material culture of the farm community. The early settlers had erected simple utilitarian log or sod houses that often contained only one or two rooms. After a period of economic consolidation, the Ontario-born farmers built handsome stone, brick, and frame farmhouses that seemed to proclaim their newly won social status (Fig. 45). Not only did these houses emulate the exterior design of mid-Victorian, Ontario middle-class dwellings; they imitated the interior spatial organization of their central Canadian antecedents. The new layouts provided distinctly specialized spaces that accompanied a more stratified conception of proper social intercourse.[49]

To provide comparative material, 13 large stone houses built around 1900 were examined in the districts of Abernethy, Sintaluta, and Arcola in southeastern Saskatchewan. The floor layouts for these houses reveal highly specialized, and compartmentalized, interior spaces. Emphasizing the change in lifestyle that accompanied the move from the one-room or two-room log dwellings, the new houses contained separate rooms for food preparation, family activities, formal entertaining and dining rooms, and most included front halls that provided buffer zones between exterior and interior spaces. Domestic servants could use such spaces to channel guests into the appropriate room, while preventing unwanted visitors from intruding on the interior living spaces of the family. Of the 13 houses examined, eight dwellings featured rear staircases leading up to the farm labourers' accommodations. In some houses the employees slept in an annex separated from the family's quarters that emphasized the inherent social divisions between employer and employee.

In W.R. Motherwell's house, the Victorian formality of Ontarian settlers was taken to an extreme. In addition to the provision of a parlour and dining room for the formal reception of guests, the Motherwell house contained a "lobby" or sitting room for more informal visits and an office for farm accounts. The house was divided into two wings: a front formal section, and a rear service annex (Fig. 46). On the main floor a rear hallway and rounded arch provided a buffer zone between living, business, and entertainment functions and the service section. The noise and odours of the kitchen were thus prevented from intruding upon the formal areas. At the same time, Motherwell built a narrow rear staircase to link the employees' quarters with the kitchen. In the daytime a door between the employees' rooms and

Fig. 45. W.R. Motherwell's stone house before the First World War. Motherwell's daughter Alma poses with a friend at the front gate. Saskatchewan Archives Board, Regina, W.R. Motherwell Collection, Photo no. R-87-219, Photo no. 3.

the family's chambers could be unlocked so that the removal of slop pails and chamber pots could occur unobtrusively. At night this door was locked. Conversely, the front stair permitted "dramatic descent to meet family and guests."[50]

The farm employees' quarters were small and spartan. Initially the hired men's room at the Motherwell house, in which two men were crammed together, was a meagre 60 square feet. It possessed no closet and was furnished with a single bed, a washstand, and a chair. The hired men hung their clothes on a series of hooks outside the room adjacent to the rear staircase. Next to the men's room was the hired girls' room, which, at 80 square feet, was somewhat larger. Its furnishings were similarly utilitarian, consisting of a bed, dresser, and chair.[51]

On the other hand the family's "chambers," while not large, provided an average of 113 square feet, including closet space. The largest was

Social Relationships in the Settlement Era

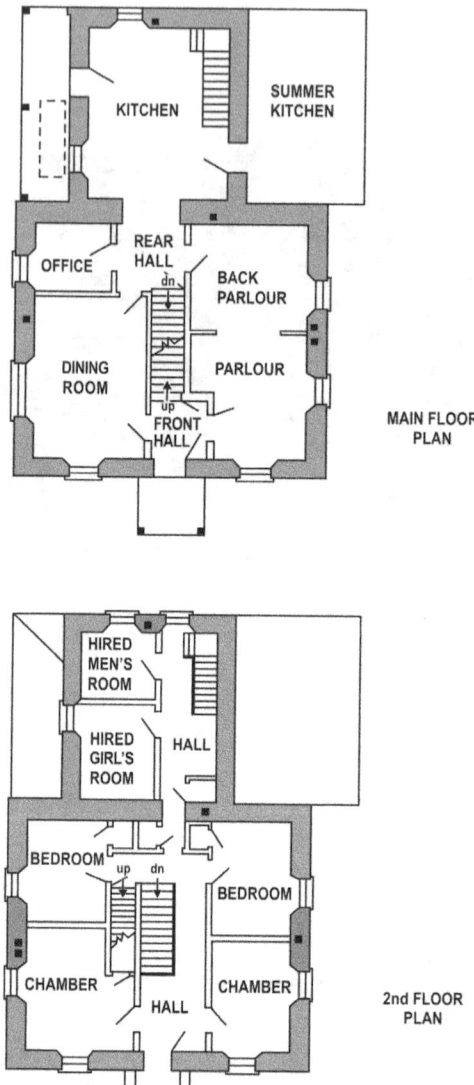

Fig. 46. The Motherwell house floor plan, showing the original hierarchical organization of space and compartmentalization of room functions, indicative of the elite social aspirations of its occupants. Plan adapted from the *Nor'-West Farmer*, 5 May 1900.

the bedroom of W.R. Motherwell and his wife, totalling 150 square feet. What distinguished these rooms principally from those of the employees was the nature and quantity of material possessions and objects they contained. If the architectural designs of the Victorian Anglo-Canadians' permanent farm homes provided a notable contrast between the families' and employees' quarters, an even more striking dichotomy existed between these accommodations and the lodgings of seasonal or transient staff. In the log house days, harvest labourers commonly slept and ate in the settler's kitchen, but after the building of the new houses, they were rarely invited inside. Instead they had their meals outside, and slept in tiny shacks, called "cabooses," which were crammed with beds. One settler who worked on a harvest gang has described these accommodations as

> ... about three-quarters the size of a bath house, lined with 8 bunks each capable of holding 2 men. Everybody softened his bunk with some hay or straw, but it was hard and stayed that way. These were rather strange surroundings for me; in the middle of so many rough harvest hands in the caboose you could find every kind of person....[52]

As the building of large Victorian homes signalled the establishment of a new middle-class hegemony, pioneer hospitality waned or disappeared. It was perhaps somewhat ironic that as soon as Abernethy residents had begun to inhabit much larger homes in which others could be more easily accommodated, they showed a reluctance to share their living spaces. In 1907, a correspondent to the Abernethy *Abernethan* complained that women teachers were unhappy with their rooms in the local King Edward Hotel but had been unable to secure alternative accommodation. Due to the unavailability of suitable lodging, the school had to be closed before the end of the school year. In the writer's words,

> ... while it speaks volumes for Abernethy that in these hard times, the people of our village are one and all too wealthy and aristocratic to take in a boarder of the social standing of a teacher, yet it is a fact to be deplored.[53]

Contemporary western Canadian fiction also linked the building of extravagant prairie houses to a loss in human interaction. Both Frederick Philip Grove and Nellie McClung depicted Ontarian settlers' permanent houses as representative of economic prosperity, but nonetheless symbolic of spiritual emptiness. In *Sowing Seeds in Danny*, McClung described the house of a fictional family called, appropriately enough, the Motherwells. It is not intended to make too much of this coincidence of names, but McClung's Motherwells live in a large stone house, "square" and "gray, lonely and bare."[54] In Grove's *Fruits of the Earth*, Abe Spalding also builds a palatial house, but discovers that the house only accentuates the estrangement of the family members living within it.[55]

Accompanying the formal Victorian layouts of the Ontarian settlers' "new houses" was a formalization of relations between farmers and their employees. In her study of domestic servants in Canada between 1880 and 1920. Geneviève Leslie has noted that household employees occupied positions of low status.[56] She quotes from the contemporary book on *Domestic Engineering* that domestic servants formed "the class that Society has relegated to the lowest place of human beings claiming respectability. Housework and houseworkers are classified at the very bottom of industrial occupations...."[57] Like their urban counterparts, rural domestics had experienced a reduction in status in the context of Canadian industrialization. Leslie argues that since rural domestics often shared work with their mistresses, particularly in the busy season, they were often in a comparatively better position socially than urban servants. Yet "hired girls" on farms were generally paid low wages, worked long hours, and experienced great isolation.[58]

At Lanark Place the Motherwells often recruited hired girls from the nearby German-speaking settlement of Neudorf or from the File Hills Cree reserves north of Abernethy. These women tended to be quite young, in the 18–20 age bracket.[59] Nina Stonechild, who had been both taught by Mrs. Motherwell at File Hills and then worked as a "hired girl" at Lanark Place, is pictured in a photograph from her wedding in Figure 47. She and the others were paid comparatively low wages; one former hired girl remembered having been paid $5 a month in the 1930s.[60] At least one of the Motherwell's former hired girls remembered her life at the farm as a very lonely time. She was not included in family activities.[61] Another recalled that for social

Fig. 47. "Stonechild-Pratt Wedding," File Hills, n.d. Both Nina Stonechild and Catherine Motherwell appear in this photograph. Saskatchewan Archives Board, Regina, Photograph no. R87-219, File 103-142, Album, p. 19.

interaction she usually went over to the hired men's cottage to visit with the Motherwells' hired man and his wife.[62] Most of the hired girls stayed at the farm only a few months, although one woman, "Lizzie" Lutz, lasted two years on the job. Generally speaking, they could be classed as transient labourers – young women who had few opportunities other than to work in domestic service prior to marriage. Their transient status is confirmed by the fact that these women seldom appear in the more than 300 photographs in the Motherwells' family albums. Moreover the term "hired girl" was itself indicative of the subordinate status of women in this period; male farm labourers were "hired men," but female domestics were "girls."

W.R. Motherwell's views regarding the social place of the hired men on his farm are well documented in a bulletin issued under his direction as Saskatchewan Minister of Agriculture in 1915. Entitled

Social Relationships in the Settlement Era

Practical Pointers for Farm Hands, the bulletin provided no fewer than 200 recommendations for persons contemplating seeking work as hired men.[63] The Foreword to this document explained the Department's purpose in publishing it:

> At this time a great many men, for one reason or another, are taking up farm work who perhaps, have had no previous experience of it and never expected to be engaged in it. The more such men know about their work the better they will like it, and the sooner they acquire the knowledge. The more pleasant it will be both for them and their employers.

The Foreword also states that most of the bulletin's contents were prepared by E.A. Blakesley and "are the fruit of twenty years' experience and observation." To the extent that they were followed, the pointers would "make for a better understanding between employers and employees on our farms, and help to solve what is apt to be one of the most vexatious problems on the farm – the labour question." The tone in which the 200 "pointers" are delivered is paternalistic; they are written as orders rather than suggestions. Little is left to the imagination or intelligence of the reader, as even apparently trivial matters warrant a stern admonition from the writer. The paternalistic tone is evident from the following excerpts:

> 133. Be a leak-watcher. Don't waste the hay. Don't waste the grain. Don't lose things. Don't cause accidents. Prevent breakages by keeping things well inspected. Keep things greased and oiled and mended.

> 168. Get up in the morning when you are called. It is only a habit to be called twice. Good riddance to a man who has to be called more than once. It is childish.[64]

With respect to proper social conventions the booklet recommended that hired hands be appropriately deferential to all members of their employers' families. In addition to the required obedience to the farm proprietor, hired men were instructed to be polite and not to "call the grown daughters by the first name at first sight.... It is the mark of a gentleman to call the eldest daughter Miss Susan and Miss Maud until they request you to drop such formalities and be more like one of the family." Above all, the hand was never to call the farmer's wife by her first name, "no matter how young and pretty she is ... If she likes to have you call her by her name, so much the more you should refrain. She needs a little training herself."

Quite apart from the assumptions implicit in these statements, the booklet offered guidance to farm hands regarding proper deportment and conduct on the farm. Hired men were instructed not to lie, to be "decent" and "neat."

> The man who bathes regularly, cleans his teeth, grooms his nails, keeps his shirt buttoned, hangs up his clothings, stacks up his books and papers, cleans his boots in muddy weather, etc., etc., is one of the men the women folk will want to keep when the extra men are dismissed.[65]

While the *Practical Pointers for Farm Hands* was evidently intended to be a definitive statement on relations between employers and hired hands, it must be recognized that their actual relationships depended on the personalities and relationships of individual farmers and hired men. In keeping with the tone and advice of the handbook, Motherwell's approach was patriarchal and somewhat authoritarian. He insisted – on pain of dismissal – that his farm hands eat with the family on most occasions.[66] As noted earlier, all persons including the farm labourers, were expected to participate in the religious service following each evening meal and to contribute to the cent-a-meal fund. Moreover, if found smoking near the barn, employees were fired immediately. Three former hired men at Lanark Place remember W.R. Motherwell fondly. Given that oral history informants were generally reluctant to criticize other people, it is nevertheless apparent that Motherwell treated his staff fairly and was concerned about their welfare. At the same time, his paternalistic approach signified a marked change in Abernethy's social relationships from the informality of the early pioneering era to the hierarchy of the settled agricultural community.

CO-OPERATIVE ENTERPRISE

In the first few years of settlement Abernethy farmers relied heavily on the co-operation and assistance of their neighbours. Homesteaders often lived together and shared the ownership of implements. They helped each other build shanties, barns, and fences. They also provided extra labour for harvest and other farming operations. It is important to realize, however, that these instances of mutual assistance were not really "co-operative" in the sense that they represented a communal effort from which rewards would be mutually shared. Pioneer co-operation was a pragmatic response to a situation of scarce resources.

A series of questionnaires drafted by the Saskatchewan Archives Board and filled out by the settlers in numerous districts of the province in 1955 provides an indication of the kinds of co-operative activity existing in the settlement period. Questionnaires suggest that the initial period of co-operation was of comparatively short duration and was largely limited to the process of farm making. George Hartwell, who settled at Pheasant Forks in 1882, recalled that farmers helped each other in building, haying, and harvesting "bees" until 1886.[67] Another account places the end of the "pioneer era" at about 1896.[68] This is not to suggest that the tradition of co-operation evaporated completely after the early period of economic consolidation. W.R. Motherwell's large barn was built in a one-day "raising bee" in 1907.[69] By this time, however, such events were relatively infrequent.

In addition to pooled labour, farmers also participated in co-operative ownership of implements. The example of John Teece, who started farming with a one-fourth interest in a yoke of oxen, plough, and wagon and graduated to shared ownership of other implements, has been noted. James M. Minifie has related that at Sintaluta his father and seven other homesteaders shared ownership of two of the most expensive implements – a seeder and a binder.[70] The group drew lots to determine the order of access to the implements. Minifie states that this arrangement was unsatisfactory since "the man at the foot of the list could not sow at the critical time when optimum conditions of land and weather prevailed, nor could he harvest the crop at peak." After the first year the co-operative plan was dropped, and "each man equipped himself as well as his funds, the bank and the implement companies would allow."[71]

Due to the great expense of threshing machines, co-operative ownership of these larger implements was relatively common. In 1891, Abernethy-area farmers formed a joint stock company to purchase, manage and operate a steam engine, separator, and appurtenances.[72] Capital stock of $2000 was issued in 40 equal shares of $50. Problems emerged in 1901 when a shareholder, John Teece, complained of too much grain being thrown out of the separator into the stack. He refused to settle his account with the company and sued for damages.[73] The case underlined the rather fragile basis for co-operation in a society in which an individualistic and competitive outlook prevailed.

Land relations between farmers also demonstrate the myth of a communal pioneer spirit. Co-operation stopped at the farmer's fence. Under the Herd and Estray Animals Ordinances,[74] farmers were entitled to arrange for the impounding of roaming livestock found grazing on their property. The animals would be kept impounded until their owner had settled for damages with the proprietor of the violated pasture. In a lighthearted vein the *Abernethan* admonished a "certain exalted personage" for having turned out his herd of swine "to fatten at his neighbours' expense."[75] In an obvious reference to W.R. Motherwell, the editorial continued:

> To allow one's pigs to soot up one's lawns and arboretums may in a year like this be a very praiseworthy act of economy, but to turn them out as a practical test of the soundness of one's neighbours' graneries [sic] and stooks is what even the scientific mind of a Minister of Agriculture in his thoughtful moments would depreciate [sic].[76]

In her autobiographical account of Anglo-Canadian life on the prairies, Nellie McClung has succinctly placed "pioneer co-operation" in its proper perspective.

> The farmers in these days [in 1895] were rugged individualists. They changed work whenever necessary but each kept his affairs to himself quite jealously.[77]

In support of this statement, McClung related the outcome of a community "beef-ring," which was the first formal co-operative undertaking in the community. Every week, under the terms of the ring, an animal was butchered and the meat distributed to the participating members. Butchering was performed by different farmers in sequence, and each distribution was weighed so that eventually all farmers received equal amounts of the meat. According to McClung, the "beef-ring" worked well initially, "but with the coming of better crops and prices the farmers began to realize that they were foolish to bother with their own meat when they could buy it at the butcher shops...."[78]

Thus the beef-ring concept, like the other co-operative plans among farmers, dissipated with the arrival of economic prosperity and greater independence. Co-operation, in the sense of a mutual activity directed at producing rewards to be mutually shared, was not really practised at Abernethy. In their family, work, and leisure activities, Abernethy farmers pursued relationships within an individualistic, middle-class context.

7

Abernethy's Social Creed

> ... It is a society of men united together for the attainment of a twofold object – moral and social – the moral standing first as out of moral statutes social consequences flow....[1]

In stratified societies, dominant groups develop cultural frameworks, incorporating ideologies, value systems, and codes of conduct to channel, motivate, and justify socially-sanctioned behaviour.[2] In Canada of the late nineteenth and twentieth centuries, such mentalities were integral to structuring relations of property, religion, gender, sexuality, among other aspects of social experience. Abernethy and other Anglo-Canadian prairie settlements were not exceptions. Beyond entrenching political, legal, and social structures, members of these recently-established communities sought to govern themselves by imposing a common set of values. Having recently established their ascendancy, Abernethy's middle-class settlers were anxious to not only to codify acceptable social relations but also to set benchmarks against which the behaviour of community members could be evaluated. This is not to suggest that all members of this community held identical values or views on all issues but in terms of prescribed patterns of thought and behaviour valorized by its leading citizens, a high level of conformity to established norms was expected of all members. The set of values promoted at Abernethy and other Anglo-Canadian prairie settlement communities of this period shared many affinities with anthropologist David Bidney's category of "moralistic culture," according to which culture is "lived and organized according to some dominant idea of the good for a given society."[3]

As the opening quotation of this chapter suggests, residents of the central Qu'Appelle region were aware of the instrumental role of "moral statutes" in regulating society. While referring specifically to the Independent Order of Foresters chapter at Indian Head, this passage could equally be applied to local churches and other organizations fulfilling similar roles in shaping and disseminating values in this era. A comprehensive treatment of the mentalities of Abernethy farmers is beyond the scope of this study. What is offered is an account of aspects of their shared world view that can readily be correlated to their social and economic structures, group interactions, and social behaviour. Abernethy's social structures were expressed in institutions imported and further developed in this community, governing a variety of social, religious, and family traditions and interactions. Overt statements by Abernethy farmers or their families that might illuminate their operative value structure are comparatively sparse. Occasionally, they can be found in letters to the editor, articles in period newspapers, minutes of local organizations, personal correspondence or reminiscences. It has sometimes been necessary to reach beyond the boundaries of the actual Abernethy district to refer to evidence generated in other Anglo-Canadian farming communities across the Canadian prairies. These communities were settled in the same era by agricultural entrepreneurs from central Canada or Britain sharing similar assumptions about society and how it should operate. As well, the surviving evidence of its material culture provides significant evidence of operative cultural values in this era, supplementing and enhancing the extant textual and photographic documentary record.

At the centre of the value structure held by Abernethy farmers was a deep-rooted belief in individual enterprise. Historian Allan Smith has shown that Canadian historians, in emphasizing the collective nature of the Victorian Canadian identity, have overlooked a strong undercurrent of individualism, "the myth of the self-made man."[4] Nellie McClung discerned this trend in her observation of prairie Anglo-Canadian society around 1895: "The farmers in these days [in 1895] were rugged individualists."[5] It is arguable that the individualistic ethic was even stronger in a frontier context than in established communities in the East. As agricultural historian Allan Bogue has suggested, the settlement process may have been selective, attracting a certain breed of settler possessing attributes uncharacteristic of the general population.[6] The point is that the individualistic ethos of many settlers was probably exaggerated by their perceptions of the role of individual strength, initiative, and perseverance in laying claim to and remaining on their homesteads. Obviously luck was also an important component of success. In the mindset of the settlers, however, luck played no role alongside the pre-eminent individualistic virtues. The settler was the maker of his own destiny.

A second key element to the Abernethy settlers' value structure was its materialistic orientation. That this should be so is not surprising; market societies, in which wealth or prosperity is an important measure of one's status, tend to promote materialistic values. However, the preoccupation with material expansion and accumulation appears particularly evident in some memoirs and other evidence left by Victorian Anglo-Canadian settlers. In the passage quoted at the beginning of this book, John Teece described his having "made good" solely in terms of material success, that is the acquisition of land, livestock, and cash. Other testimonials of the period defined success largely in economic terms.[7] Physical and iconographic evidence of the material culture of early Abernethy settlers is also strongly suggestive of a preoccupation with material accumulation. Following the period of homestead consolidation, the practice by many early settlers of erecting large showpiece masonry houses, either of stone or brick, seemed to proclaim to their contemporaries that they had "made good."[8] That these large dwellings were far from representative of the domiciles of the great majority of Saskatchewan's settlement era families is evident in census consolidations of the period. At the end of the settlement era in 1931, fewer than 1 per cent of Canadian prairie farm dwellings were built with brick or stone. As well, even at that late date, more than 60 per cent of prairie farm dwellings comprised four rooms or less,[9] indicating the modest living circumstances of the majority of prairie farm residents in that era. These figures suggest that the material culture patterns of Abernethy's successful farm proprietor class expressed an acquisitive orientation not shared by subsequent settlement groups. As referenced in the discussion of "conspicuous consumption" in Chapter 4, the elaborate dress conventions and extensive collections of furnishings and other material goods assembled in the Anglo-Canadian settlers' permanent homes revealed an emphasis on material acquisition and its display unrepresentative of the overall farming population.[10]

Operating in tandem with the Abernethy settlers' focus on material success was a shared belief in the value of the work ethic. In various testimonials published in the territorial press, settlers such as John Teece never ceased to stress the value of hard work. Allan Smith has shown how in the late nineteenth century prevailing conceptions of success in Anglo-Canadian society shifted from an emphasis on wealth to work itself. As the central provinces became more urbanized and stratified, and hence less upwardly mobile, ideologists of the self-made man were obliged to find a new rationale to keep the working masses happy with their lot. The principles of the Protestant ethic met this requirement; while divesting success of its materialist content, it offered moral virtue as an alternative incentive to hard work.[11] The predominantly agrarian western Canadian society was less stratified than the East. Nevertheless, in settled communities upward mobility diminished steadily as land values and the required capital outlay for farming increased. In such a situation of reduced mobility Abernethy settlers, like their eastern counterparts, were obliged to fall back on the work ethic as a means of motivating their employees to greater efforts. Hence in a contemporary booklet for farm labourers, the author could write:

> Be honest. Do as much work, and do it well, in the absence of the boss as when he is with you, and in due time you will reap your reward. Many a hired man has been turned into a son-in-law by this one trait alone.... Get a move on you. Act as if you want to get your piece of work done and out of the way. It will help you to enjoy your work and will be a pleasing sight to the boss, the wife, the children and the neighbours.[12]

Thus, without the promise of material rewards, the hired man was exhorted to work hard for work's own sake.

However committed the Anglo-Canadians were to the pursuit of material gain, they sought it less for comforts and luxuries than for the respectability it carried. W.L. Morton has noted that Victorian Canadians had three ideal images:

> ... the gentleman, respectable industrious, well bred and well read; the honest businessman and honest worker; the prosperous farmer. The common element was that of respectability, the possession and evidence of those personal and social traits that were generally commended.[13]

Respectability also remained a preoccupation with the newly established agrarian middle class at Abernethy. As if to proclaim their arrival in the new society, settlers in the more prosperous farming districts such as Abernethy erected large, elegant Victorian houses. Settlers also appropriated the symbols and rituals of the parent society. At the Motherwell farm, in emulation of the Anglo-Canadian model, tea was served in a sterling silver service after every evening meal. As we have seen, the Motherwells reinforced their respectable position in their public dress, which marked them as a family of prosperity and standing in the community. On Sundays they rode to church in a surrey, or phaeton, and occupied their own pew near the front of the sanctuary of Knox Presbyterian Church in Abernethy (Fig. 48). Yet, for all the emphasis on the material and the worldly, Abernethy, no less than other Victorian Anglo-Canadian communities, was dominated by a religious creed and accompanying code of morality. Paul Rutherford has described the role of the Christian church in imposing social authority and regulating moral behaviour in Victorian Canada. Anglo-Canadians believed that the church was a kind of "national cement" that gave cohesion and purpose to society. In the words of one Ontario journalist, "Our improved circumstances – the higher civilization of the present age – the comforts of the present and the prospects of the future are the results of religious influence, enlarging and developing our better nature, and lifting humanity higher day by day."[14]

Abernethy was not simply a Protestant community; its particular religious tone was predominantly Presbyterian and, to a lesser degree, Methodist.[15] The sociologist of religion Ernst Troeltsch has shown that both the Calvinist and Methodist creeds required the active participation of individuals in Christianizing the community. In the Calvinist-Presbyterian paradigm, the doctrine of predestination was paramount. On earth, men were divided between the "elect," whom God had chosen to be saved, and the "damned." In granting election to certain people, God assures forgiveness for their sins. This assurance frees the individual to serve as an agent of Divine Will. Indeed, through their election, individuals may become Christ's warriors to carry out His Purpose in the secular world. A second central element of Calvinism was its emphasis on the inward and individual character of piety. Since the Calvinist believed his election was sure, he did not need to be concerned with self-preservation or with winning salvation; he or she was free to concentrate on shaping the world according to

Fig. 48. Interior of Knox Presbyterian Church, 1913–14. Saskatchewan Archives Board, Regina, Photo no. R87-219, Photo no. 246.

Christian purpose.[16] The other aspect of Calvinism relevant to this discussion is the concept of the "Holy Community," dedicated to the glorification of God. Calvin did not envision the church merely as an objective means of salvation, but as a dynamic agent in Christianizing the community. The church's role was to develop institutions through which the Divine Spirit could mould both the secular and the religious spheres, family and social relationships, and even economic life.[17] Richard Allen has identified the Calvinist inheritance of Presbyterianism as one of the principal supports for the social gospel movement in western Canada. Where Canadian Calvinism had earlier seen its social role in more defensive terms, that is, in preventing society from disintegrating, late nineteenth-century Presbyterianism was largely transformed into an evangelical force.[18] On the prairies, a clear statement of evangelical Presbyterianism may be discerned in an 1898 sermon delivered by the Rev. A.T. Murray at an ordination ceremony in Minnedosa, Manitoba. Murray stated:

> The minister's stand upon all questions of public morals must always be decided, and always against the evil. Abraham was a soldier ready to do battle on the right side. So we are on a battlefield. We have much evil to contend against. With God we may be more than conquerors. Abraham, by erecting his altar, claimed the land for God. So we must claim our land for Christ, place missionaries in every district, and stop not until we have won our community, our province, our Dominion for our captain.[19]

The references to the need to "claim the land for Christ" and to be "more than conquerors" were revealing. Having appropriated the prairie lands from their former native inhabitants, Anglo-Canadians saw the need to invoke religion as both a justification and means of maintaining territorial control. As Abraham had erected an altar in "the promised land," so missionaries as harbingers of the Anglo-Canadian group needed to build churches throughout the West. Imperialism needed to be buttressed spiritually (Fig. 49).

Viewed in this context, the preoccupation of individuals such as Catherine and W.R. Motherwell with moulding Abernethy society to Christian purpose becomes comprehensible. As Catherine Motherwell's early career as a Presbyterian missionary among the File Hills Indians attests, the process of Christianizing the world entailed an active assimilation of non-Christian cultures.[20] Similarly W.R. Motherwell's involvement in the temperance movement represented an attempt to eliminate the "un-Christian" influence of liquor. His personal financing of the building of a temperance hotel in Abernethy, that is, a hotel serving only non-alcoholic beverages,[21] demonstrated how far he was prepared to go in pursuing his perceived responsibility in Christianizing the community. Another example of the extension of Christianity into community affairs was the Sabbatarian movement, which gained currency in Ontario in the late nineteenth century.[22] On the prairies, the North-West Territorial Assembly passed an ordinance prohibiting the performance of any labour, commercial activity, games, or amusements on the "Lord's Day."[23] It is interesting to note

Fig. 49. "Rev. T. De Witt Talmage, D.D." Frontispiece from Mrs. T. De Witt Talmage, ed, *T. De Witt Talmage As I Knew Him* (London: John Murray, 1912), p. i. The New York celebrity evangelist was a favourite in the Northwest Territories in the late nineteenth century, as his sermons were serialized in various territorial newspapers in this era. For a capsule biography, see *The Encyclopedia Britannica*, 1911 Edition. <http://www.1911encyclopedia.org/Thomas_De_Witt_Talmage>

that in the Abernethy area, the first prosecutions under this ordinance were carried out against several Métis teamsters, who were charged with transporting goods on a Sunday.[24] This is not to suggest that the Lord's Day Act was selectively applied, although further research in this area might support such an inference. Nor is it claimed that all Anglo-Canadian residents necessarily agreed with the draconian application of the Sabbath law. One Anglo-Canadian settler wrote to the *Qu'Appelle Progress* in protest over the Lord's Day prosecutions.[25] Sabbatarianism was a tool of social control that was available to the Anglo-Canadian leaders when required. It takes little imagination to see that this activist orientation of Calvinism could reinforce its advocates' social position. Leading citizens appropriated the power and status that accompanied their roles as pillars of the Christian church. As elders and leaders in church-related organizations such as the women's missionary societies and the W.C.T.U., they invoked a higher purpose in support of their objectives. The Protestant creed, then, assisted Abernethy's Anglo-Canadian elite in establishing and maintaining its leading role and influence in the community.

The religious orientation of Abernethy society was accompanied by a rigid code of morality. Since moral life entailed such a struggle between temptation and the exercise of will, Victorian Canadians created many taboos of unacceptable behaviour, including card playing, swearing, and alcohol consumption. Most of these activities were associated with the early settlement period when young bachelors sought relief in beer parlours and brothels from the monotony and loneliness of homesteading. The imposition of prohibitions against perceived immorality represented an attempt by the new middle-class order to entrench its authority. At Lanark Place, Catherine Motherwell was said to have disapproved of card playing and refused to participate in it.[26] In questionnaires filled out in the 1950s, a number of early settlers also drew attention to the taboo of card playing, particularly among Methodists.[27]

Victorian prudery found its most ardent support among settlers' wives, since they perceived that they had the most to lose from the sexual and other indiscretions that had been the preserve of men. Nellie McClung is quite clear on this point in *The Stream Runs Fast*.[28] Other evidence is comparatively rare but may be read between the lines in settlers' accounts. One homesteader, at Ellisboro, southeast of Abernethy, related in his diary that his bride of three months had banished him from the house. From his place of exile he wrote: "Ructions at night about swearing. I am evicted...."[29] Men also played a role in preserving community standards from moral decay. One Abernethy resident has related an incident that illustrates W.R. Motherwell's contribution. One weekend Motherwell returned to his farm via the Grand Trunk Pacific rail service to Lorlie, northeast of Abernethy. Arriving in the early hours of the morning, Motherwell was driven by horse team and cutter to Abernethy, where the driver stopped briefly, before proceeding to the farm. Motherwell heard sounds from a late-night revelry at the local pool hall on Main Street. He proceeded to register a complaint with the town authorities; all merry-making henceforth ceased.[30]

Like many other prairie communities of the era, Abernethy was beginning to come under the influence of the "social purity" movement of the late nineteenth century, whose adherents included leaders of both the Methodist and Presbyterian churches.[31] The social purity advocates' promotion of the nuclear family was consistent with the Dominion government's objectives of fostering stable settlement in the West. Federal policy in the late nineteenth century was itself suffused with notions of patriarchy, as women were generally excluded from eligibility for free-grant lands, which were solely reserved for male applicants over the age of 21 or the heads of families, defined by default as males except in the case of widowhood. The establishment of family farms adhering to the patriarchal heteronormative model, then, was a policy objective of the Dominion government, whether or not it was explicitly articulated. The assumed model unit would consist of a male household head and female spouse, who together would not only cultivate grain crops but also propagate children to populate the initially sparsely settled region. On the prairies, the preoccupation with promoting marriage probably also assumed greater urgency owing to the federal government's failure to realize anticipated levels of western settlement under the National Policy. In 1895, the thinly-populated settled areas west of Manitoba were still largely confined to the areas bordering the Canadian Pacific Railroad and a few smaller railway lines through central Saskatchewan, a settled strip between Calgary and Edmonton, and a few other isolated pockets of settlement.[32]

In early Saskatchewan, the legal framework was socially reinforced by both incentives and sanctions, as ideologists of the family used both the carrot and the stick to try to influence the bachelors to marry. For example, in 1893 the *Qu'Appelle Progress* carried an article entitled "Marriage an Aid to Longevity," which began with the admonition, "Bachelors, beware..." In the same year another article concerned agitation in both Manitoba and the North-West Territories to levy a special tax on unmarried males.[33] At Indian Head and Qu'Appelle, the *Vidette* extolled the benefits of marriage, as it opined: "The Good Lord meant marriage for youth and youth for marriage." In 1895, the *Leader*, in publishing an article entitled "The Decline of Marriage," lamented a reported rise in the average age of marriage from 25 to 32 years, a source of concern for advocates of population and economic growth in the West.[34]

According to the prevailing paradigm of socially structured gender relations, the Euro-Canadian newcomers assumed that heterosexuality within marriage was normative, defined in opposition to any alternative forms of sexuality. Their operative moral ideologies generally connected to a movement to defend marriage and heteronormative sexuality across Canada in the period 1890–1914. Spearheaded by the middle class, which felt threatened by the impact of industrialization and rural depopulation on the family, this movement influenced the Canadian government to sponsor legislation to reinforce traditional sexual mores, by providing stiffer penalties for non-familial, non-procreative sexual activities.[35]

Indeed, the legal prohibitions extended to all expressions of sexuality outside the heteronormative nuclear family. For the upstanding citizens of rural prairie communities, the initially large group of predominantly young bachelor settlers comprised a target for an emerging discourse advocating supplanting their homosocial environments with heterosexual marriage and procreation.[36] Moral critics associated the bachelors with various types of proscribed behaviour, including drinking, card-playing, gambling, and sexual licence, all considered to threaten religion and the traditional family, the twin pillars of the new moral order. Reflecting a widespread belief that society's morals were in decline, all Christian denominations framed issues of sexuality in terms of a mighty struggle between good and evil. The evangelical churches, especially the Methodists, railed against "diseases, deaths, crimes, vices, miseries, and losses," all thought to derive from the liquor traffic, for which its "total suppression" was "the only true remedy for the losses which it inflicts on the nation." Revealingly, a contemporary report of the Methodist Church Committee on Evangelism stressed "maintaining those ideals of Christian citizenship which must obtain if our civilization is to be permanent," suggesting a fear of the loss of civilization in the absence of a crackdown on moral decay. Of particular concern was the sexuality of teenaged boys, thought to be particularly susceptible to moral corruption. In 1893, the Methodist Church in western Canada formed a special "Boys' Brigade," intended "to preserve the boys during the critical age from sixteen to twenty-one."[37] It was organized by the Epworth League of Christian Endeavour, a branch of the Methodists which established a local chapter at Kenlis, just south of Abernethy, in 1896.[38]

By the 1890s, social purity advocates found ready vehicles for their moralistic campaigns in the territorial popular press, which periodically carried articles inveighing against non-procreative forms of sexuality. In an extended piece entitled "Social Purity," published in March 1895, a Reverend Dr. Douglas inveighed against the 'dark record of solitary vice,' the "crime ... that launches emasculated ruin into asylums of hopeless insanity." In 1891, the celebrity evangelist Reverend H.T. Crossley addressed an audience of men in Regina in an effort to turn them towards the path of righteousness. Focussing on "swearing," "idleness," "licentiousness," and the "secret vice," he painted a bleak picture of their consequences, saying: "the wages of sin is death." The *Regina Standard* commented: "So minutely did he describe and instance the shocking results of sinful indulgence, that there was not the remotest appearance of livity [sic] on the part of the young men." The point was clearly made that procreative heterosexuality within marriage was the only legitimate and moral alternative to a life of sin and degradation.[39]

In the 1890s, several territorial newspapers, including the *Regina Standard*, the *Saskatchewan Herald* and the *Qu'Appelle Progress*, also regularly carried serialized sermons by Dr. Thomas De Witt Talmage, the best-selling New York Presbyterian evangelist. Talmage was particularly fond of the cautionary tale, which he frequently directed towards younger readers. In a sermon re-published by the *Saskatchewan Herald* in 1896, Talmage warned young men about the dangers of falling in with the wrong crowd of males in the city, making them susceptible to the "monstrosity of wickedness." Talmage did not specify what he meant by that term but in other writings was more specific. In his tract *The Night Sides of City Life*, re-published in Canada in 1878, he heaped scorn on the wealthier classes, which he particularly associated with male homosexuality: "It is the iniquity that comes down from the higher circles of society that supports the haunts of crime, and it is gradually turning our cities into Sodoms and Gomorrahs waiting for the fire and brimstone tempest of the Lord God who whelmed the cities of the plain."[40]

In Saskatchewan, the campaign against non-procreative sexuality came to a head in 1895 in a moral panic sparked by a Regina court case dealing with same-sex sexuality involving a prominent local merchant, aged 32, and two younger males, aged 20 and 17. Caught in the act of engaging in various homosexual activities in the basement of the merchant's dry goods store, these three men were charged, tried, and convicted on charges of "gross indecency," fined, and then informally exiled from the North-West Territories. Coinciding with the simultaneous prosecution of Oscar Wilde in London for "gross indecency," the trial was covered by six territorial newspapers, including the *Qu'Appelle Progress*, one of the principal local newspapers for Abernethy readers, as well as the *Regina Standard* and the *Manitoba Free Press* in Winnipeg. The extensive coverage of this case simultaneously marked both the first acknowledgement of same-sex sexuality in the region's popular discourse, and its subsequent repression, as expressions of revulsion by prairie newspapers concerning this affair were succeeded by stony silence. Notwithstanding a petition by the town's citizens for leniency, the discourse of opprobrium and silence surrounding this case, accompanied by the exile of these men from the region, delivered a clear message that alternatives to heteronormative sexuality within the traditional nuclear family would not be tolerated in early Saskatchewan society.[41]

By far the greatest moral battle waged by Anglo-Canadian farmers was against the use and abuse of alcohol. From the 1890s the temperance movement was a preoccupation with W.R. Motherwell and his Presbyterian and Methodist neighbours. This was not surprising, since they perceived alcohol to threaten all they held dear – religion, the family, the work ethic, and private property. W.R. Motherwell and his colleagues in the movement were not mere advocates of temperance – they were proselytes of prohibition. They were opposed to any use of alcohol, whether for recreational or medicinal purposes. In 1899 Motherwell wrote to Prime Minister Wilfrid Laurier to express deep disappointment at Laurier's refusal to support the prohibitionist cause. He noted that he had been one of Laurier's strongest defenders in the past, but that he might now have to reconsider his personal support.[42]

The Temperance Movement also gave expression to that other distinctly Victorian trait, the sense of mission,[43] particularly in the Banish-the-Bar Crusade after 1913. Temperance fervour intensified after the outbreak of the First World War in 1914 when reformers equated prohibition with patriotism. Since Saskatchewan's eastern European population was often identified in the popular mind with the enemy Central Powers, prohibitionists saw the provincial referendum as a kind of loyalty test.[44] The emphasis placed on sobriety was exaggerated, but demonstrated the dogmatic rigidity of Victorian attitudes that still held sway on the prairies. Figure 50 shows the members of the Regina Lodge of the Royal Templars of Temperance around 1890. Similarly, the emerging Anglo-Canadian middle class on the prairies was determined to stamp out any perceived threat to its position. Its concern with self-preservation was manifested in attempts to assimilate Aboriginal peoples and other ethnocultural groups.[45] Apart from the previously discussed attempt to promote the work ethic and the nuclear family, Anglo-Canadians also used the temperance movement as a lever against the cultural traditions of these groups. Efforts to impose a uniformity of language instruction in the schools were also key components of the assimilationist drive.[46]

It is clear, then, that the prairie Anglo-Canadian worldview in the Victorian period reflected traditional social mores, but there was also some evidence of a changing outlook in the period. This was particularly true of the perceived role of women. Before 1900, a dual ideology of womanhood prevailed. In keeping with the family-centredness of the Victorian era, women were idealized as the inspirational force of the home. As a Moose Jaw woman wrote in 1889, "Womanliness is a combination of qualities which may be cultivated by women in all ranks of life, they are essentially heart qualities...." But feminine virtues formed only part of the woman's role; pioneer wives were required to perform hard domestic labour on the homestead: "We are placed in a world where there is work to be done, ... life is no butterfly kind of existence." Despite the obvious importance of women to the farm economy, their role was still conceived as a supportive one; they were adjuncts to their husbands.

> ... on the womanliness of this generation will depend very largely the manliness of the next. It has been said, "Boys are what their mothers make them, young men are what their sweethearts make them, and husbands are what their wives make them."[47]

Even in the context of contemporary women's groups, the statements of the members of these organizations evinced an acceptance of their subordinate status. At a meeting of the Qu'Appelle Women's Christian Temperance Union in 1897, women heard portions of a scripture

Fig. 50. Royal Templars of Temperance, Regina Lodge members, ca. 1890. From the earnest expressions on their faces and their assorted weaponry, including a mace, gavel, and spears, the Templars are clearly ready to do battle. Saskatchewan Archives Board, Regina, Photo no. R-B 637.

"bearing on Christian quietness, that women should be adorned with a meek and quiet spirit...."[48]

By the First World War era the ideology of womanhood had shifted to encompass a new role, separate from that of supporting men. The economic precondition of this change in emphasis was the development of the Anglo-Canadian settlements into prosperous farm communities. Farmers' wives, burdened with housework in the pioneering period, now employed female domestics to do most of the household labour. Demographically, many early settlers' wives had also reached middle age and had already raised their families to adulthood. Freed of "women's work" and the tasks of raising young children, Anglo-Canadian women could devote themselves to enhancing the domestic environment and to community service. Figure 51 shows Catherine Motherwell reading in the lobby of their home in 1918. Speaking on the expanded role of women in 1916, Catherine Motherwell asserted:

> Women's most important present day responsibilities are in the uplifting of the homes. The homes must be raised to meet the present day conditions, raised intellectually, spiritually, socially, and aesthetically.[49]
>
> ... We must have influence at work that will reach the homes where these standards are low. We have a great avenue of service open to us through our church organizations and our schools through the intelligent use of these, much can be accomplished. But not only are these old-time institutions – the church and the school – avenues through which we may work, there are many other open doors ....[50]

As W.R. Morrison, in his study of late nineteenth-century feminism in Ontario, has noted, the adoption of community service by women was an extension of their traditional role of "mothering and maintenance" that they had performed in the home.[51] Most of the late Victorian members of the Ontario women's movement were the wives of middle-class businessmen and professionals. Their voluntary work focused on the non-confrontational areas of family life and child care. It was directed principally at alleviating the superficial problems of working-class mothers and their children, but Morrison notes that in their reform efforts they blamed the victims for their own suffering. Since these middle-class feminists sought to change not society but the victims, their movement was essentially conservative in tone.[52]

An analysis of the program of the Women Grain Growers Associations in Saskatchewan reveals that this prairie movement was equally conservative in its assumptions. In 1917, Catherine Motherwell's sister, Janet Gillespie, reported to the Abernethy chapter on the provincial association's convention. Among the issues discussed were: the female suffrage, liquor laws, laws affecting the home and family, municipal hospitals, women police "to look after wayward girls," juvenile and

Fig. 51. Catherine Motherwell in the lobby at Lanark Place, ca. 1918. Saskatchewan Archives Board, Photograph no. R87-219, MM68.1-94. Yet it was not enough to concentrate on one's own home; these virtues must be proselytized.

women's courts, social diseases, co-operation among women, compulsory education, and the need for "more patriotism."[53] As in the case of Ontario, a strong degree of maternalistic mission permeated the women's program. The women's groups blamed the victim. In Catherine Motherwell's words, "the weak must be made strong, the ignoble transformed."[54] In the case of the mentally deficient, however, the women were not so optimistic. At another meeting of the Abernethy chapter a speaker proposed that society must prevent the mentally ill "from mingling with those who are sound in mind and body and that would prevent their marrying and giving birth to children." She also advocated the sending of specialists to five different communities "to determine how suspected cases should be dealt with and thus prevent the formation of colonies of degenerates."[55]

Both the terminology and concerns expressed in this last quotation suggest that the Eugenics movement, which acquired such prominence in the United States and in Alberta after 1900, also made inroads into rural Saskatchewan and Manitoba. In the same era, the Canadian National Committee on Mental Hygiene, then agitating for immigration restrictions, carried out a survey of feeblemindedness in Manitoba, in which it was asserted "that the feeble-minded, insane, and psychopathic ... were recruited out of all proportion from the immigrant class..."[56] With a similar ethnocentric bias, this committee subsequently carried out similar surveys in Saskatchewan and Alberta. Eugenics was a social philosophy of genetic manipulation to promote racial betterment. Kenneth Ludmerer has analyzed the prejudice of American eugenicists in terms of the socio-economic background of the movement's leaders. They were essentially of Protestant native-American stock which had experienced "trying times." Ludmerer draws on Richard Hofstadter's use of the concept of a "status revolution" to explain the conservative reaction of the eugenicists. Like the leaders of the progressive movement, who were alienated by the perception of losing their cultural influence and prestige in the context of industrialization, the eugenicists felt threatened by the prospect of their displacement. Their program represented, despite its reform-oriented rhetoric, an "effort by the middle class to maintain the values, virtues, and social structure of the old way of life in which that group held a vested position."[57]

Underlying the program of the Women Grain Growers Associations, and the world view of Abernethy settlers generally, was their belief in the cultural superiority of the Anglo-Saxon "races." In a study of the Ontarian settlement fragment in the West, J.E. Rea has identified the central social theme of the prairies as the "struggle for cultural dominance." Rea showed how Anglo-Canadian settlers in Manitoba endeavoured to assert their control by imposing Ontario political institutions, compulsory English language instruction, and Prohibition.[58] He argued that Manitoba served as the prototype for similar action farther west. For example, in the District of Assiniboia the appearance of European ethnocultural settlements in the 1880s and particularly the large-scale influx of continental Europeans after 1896 contributed to fears that the Anglo-Canadians' dominant position in the West might be threatened. Western Anglo-Canadians responded by advocating full-fledged assimilation of immigrants.

In the Abernethy area, the assimilationist drive was spearheaded by the local press. In numerous articles in the 1880s and 1890s the *Qu'Appelle Vidette* voiced its strong opposition to any departure from unilingual English practice. For example, when German settlers at Long Lake requested in 1888 that territorial ordinances be printed in German, the *Vidette* suggested that the "curse of French" was enough. The newspaper argued that a precedent in favour of the German language would open the door to translation of government documents into other languages as well.[59] By the turn of the century, the *Vidette's* ethnocentrism extended to opposition even to French language rights. Yet even moderate liberals like W.R. Motherwell promoted the active assimilation of groups other than English- and French-speaking residents. In 1922 he wrote to his daughter Alma:

> ... You asked – and properly asked – why we should Canadianize these non-English. In the first place, that is essential if they are going to become good citizens and take a real interest in Canadian affairs. It is only right that, if people are going to live in Canada and make their homes in Canada, they should take an interest in Canadian institutions, and it is impossible for them to do that unless they get our point of view.[60]

It is important to place the prairie Anglo-Canadian ethnocentrism in its proper perspective. Cultural imperialism was common to all English-speaking countries in the nineteenth century. In Canada most men of W.R. Motherwell's ethnocultural and socio-economic position would not have questioned that assimilation of non-Anglo-Saxons was the proper course to follow. Within this general context there was considerable variation in the positions taken by individual Anglo-Canadians. Conservatives such as P.G. Laurie, editor of the *Saskatchewan Herald*, showed little tolerance or respect for the divergent cultural traditions of non-Anglo-Saxons.[61] On the other hand, Motherwell was a Laurier Liberal. His defence of limited French language instruction in Saskatchewan schools reflected a more tolerant perspective in line with the national Liberal Party position, while his support of Prohibition and other positions also showed his basic agreement with the principles of assimilation.

By 1929, anti-immigrant sentiment on the prairies had hardened into outright opposition to continued admission of newcomers from central and eastern Europe. Among many submissions presented by English-speaking groups to a provincial Royal Commission on Immigration and Settlement was a brief from the Provincial Orange Lodge of Saskatchewan. Its brief was a clear statement of the Anglo-Canadians' desire to preserve their leading position in Saskatchewan Society:

> We believe in Anglo-Saxon predominance within this province; that it is an unwise policy to, at any time, bring in more people than can be properly assimilated. We must protect our economic, social, moral and educational principles, and so regulate immigration that these principles may be advanced rather than retarded. We grant that the incoming peoples may add to what we already have, but we recognize as our foundations, those British ideals which are fundamental to our national existence....[62]

But as this statement also showed, "Anglo-Saxon predominance" was not conceived narrowly in the political sense but was broadly intended to encompass economic, social, and moral life.

8

Agrarian Unrest in the Central Qu'Appelle Region

In 1901–2, farmers in the Central Qu'Appelle region formed the Territorial Grain Growers' Association (TGGA). This organization was the wellspring of provincial farmers' associations and has been identified in western Canadian folklore with a successful struggle of prairie farmers to further their own interests. The almost legendary role attributed to agrarian leaders such as W.R. Motherwell found its most romantic expression in Hopkins Moorehouse's *Deep Furrows*, published in 1918. Moorehouse, a personal friend of Motherwell, portrayed the role of the "man from Abernethy" in heroic terms. He described the rendezvous at which Motherwell and Peter Dayman signed the notices for the association's founding meeting, commenting:

> When Peter Dayman drove away from the Motherwell place that night perhaps he scarcely realized that he carried in his pocket the fate of the farmers of Canada. Neither he, W.R. Motherwell, nor any other man could have foretold the bitter struggles which those letters were destined to unleash – the stirring events that were impending.[1]

An article by D.J. Hall reiterates the thesis that collective action by farmers in the early 1900s contributed to a "revolution in the relationship between the Dominion Government and the farmers of Western Canada."

> ... The events described occurred in the context of, and contributed to, a growing Western regional consciousness. The rapid expansion of the prairie West and the increasing

importance of western grain in Canada's export economy lent a weight to farmers' demands which the government no longer could afford to ignore.²

Despite the great significance attributed by these observers to the early farmers' movements in the District of Assiniboia, no detailed analysis has yet assessed the influence of the early agrarian unrest in detail. To establish its actual significance, it would be desirable to examine the precise character of the reforms initiated by W.R. Motherwell and his colleagues and to trace their impacts over a period of time. Further, a structural analysis of the movement is required in determining whether the agrarian agitation reflected a true "revolution" in the relationship between the Dominion government and prairie farmers or rather represented a modification of the existing structure. To understand the meaning of the founding of the TGGA within the overall context of western Canadian history, it is important to examine closely the chronology and progression of events, beginning with the initial appearance of agrarian activism almost 20 years earlier.

The origins of the agrarian protest can readily be situated in Prime Minister John A. Macdonald's National Policy of immigration and western settlement, protective tariffs, and a railway monopoly.³ Historians have not failed to note that almost immediately after emigrating from eastern Canada to the West, settlers began to raise grievances against interests in the central provinces from which they had just departed. These often focussed on what prairie farmers perceived to be their exploitation at the hands of monopolistic railway, grain, and elevator companies. Particularly discomfiting was the tariff system, which obliged the farmers to sell their grain at frequently depressed world prices, while paying artificially high prices for eastern Canadian manufactured goods, such as farm implements. In this era, successive Dominion governments continually supported their railway allies by repeatedly passing disallowance legislation to nullify provincial attempts to break the CPR's monopoly. The failure of both major parties to respond effectively to the cries of prairie discontent ensured that the farmers would seek new avenues to express their frustrations.

In 1883 Manitoba farmers founded the Manitoba and North West Farmers' Union.⁴ Manitoba was then in the midst of an economic depression. Farmers were frustrated by falling wheat prices and declining western immigration, which portended that the prairies would remain underpopulated and settlement diffuse. The lack of adequate rail service to many localities compounded the discontent. In October 1883 a Brandon-area settler wrote to the *Brandon Sun*, complaining of high costs and minimal profits, and proposed holding a convention,

> to consider our position and to determine which steps shall be taken to secure this country from the impending ruin and to obtain for ourselves that independence which is the birth right of every British subject.[5]

As a result, on 26 November Manitoba farmers met at Brandon to found the Farmers' Union, resolving to agitate for political reforms to further their interests. The subsequent provincial convention on 19 December approved a "Declaration of Rights." This platform supported the right of the Manitoba government to issue charters to any railway companies in the province, "free from any interference." It also demanded the removal of tariffs on agricultural implements and building materials and modifications in tariffs on goods needed for daily consumption. It called for amendments to the provincial Municipal Act to enable local municipalities to assist in the building of grain storage facilities and mills, requested the appointment of grain inspectors, called for unhindered provincial power to charter railways, and endorsed the construction of the Hudson's Bay Railway. The convention sent delegates to Ottawa to present the Union's demands to Prime Minister John A. Macdonald, who rejected them out of hand.

A second provincial convention in March 1884 revealed, as the historian Brian McCutcheon has shown, the deep divisions and weaknesses in the movement. The convention passed a resolution permitting the Union to advertise in eastern and foreign newspapers advising prospective immigrants not to come until necessary reforms had been carried out. Provincial business interests and the press turned away from the Union and attacked it. Meanwhile, Premier John Norquay seized the initiative from the Union when he introduced a draft "Bill of Rights," similar to the Farmers' Union's program, in the legislative assembly. Its position undercut, the Union immediately faded as a force.

Concurrently, a similar farmers' organization, the Manitoba and North West Farmers' Co-operative and Protective Union, rose to prominence. The Protective Union had also been founded in December 1883, at Manitou, Manitoba, but it centred its appeal on the need to repeal the monopoly clause in the Dominion's CPR legislation, which prohibited the building of rival rail lines in the west. The new organization pledged itself to seek the following objectives:

i) To concentrate the efforts of the agriculturalists of Manitoba and the Northwest in securing the repealing of laws that militate against their interest.
ii) The removal by agitation and other lawful means of the railway and all other monopolies that prevent the securing of a free market for the products of the soil.
iii) The securing of the cheapest freights possible to the markets of the world.
iv) The removal of unjust restrictions upon trade, and generally, to guard the interests of the people against unjust aggression from any quarter whatsoever.
v) The formation of subordinate unions in every portion of the province.[6]

Another important aspect of the Union's program was its promotion of the farmers' participation in grain marketing. In December 1883 it began to purchase wheat from farmers to ship to Ontario, and it paid prices of 10 to 15 cents more than elevator companies per bushel to participating members. In the spring of 1884 the Union obtained a provincial charter allowing it to acquire assets up to $100,000 in value to finance these activities. Branch Unions were formed to promote the Union's economic ventures, although it was unable to raise sufficient money to carry out its co-operative plans.

Despite the lack of working capital, the Protective Union negotiated a deal with Mitchell & Mitchell, a Montreal grain company, making this company the exclusive agent for marketing the Union's wheat. Not surprisingly, Winnipeg businessmen resisted these efforts to direct trade away from their companies. The Union also incurred the opposition of the Winnipeg grain merchants when it decided to purchase binder twine co-operatively for its members and awarded a contract to an Ontario firm which had submitted the lowest tender.[7]

At the same time, the Protective Union addressed the problem of monopolistic practices in the grain marketing system. Since the CPR required that all elevators on its rail lines possess a 25,000-bushel capacity, it was able effectively to prevent local businessmen, who lacked the necessary capital, from building their own elevators. Thus the existing line elevator companies enjoyed a virtual monopoly position. In some localities Branch Unions were able to establish their own companies, but success depended on a large pool of patrons, which was impossible to achieve in recently settled districts that were sparsely populated. The Union, accordingly, pressed the CPR to modify its regulations to permit farmers to load their wheat directly into boxcars from loading platforms. Alternatively, it asked for the right to build flat warehouses at sidings so that farmers could store their grain while awaiting the arrival of a grain car. This provision would enable farmers to obtain "track" wheat prices, thereby bypassing the elevators and concomitant storage charges. The CPR refused to grant these changes to its regulations. Continued problems with yields and low prices in 1884 impelled the Protective Union to a more political orientation. When Premier Norquay dropped the demands for crown lands and elimination of the CPR monopoly clause in exchange for an increased provincial subsidy, disappointed farmers perceived the need for renewed political action. The Union's executive called a convention for 4 March 1885. At this meeting several prominent Liberals persuaded the convention to pass a resolution promoting free trade. Another participant called for the adoption of the principle of representation by population for the Manitoba Legislative Assembly.

Yet, few concrete agrarian issues were addressed. At the same time the Protective Union's credibility was seriously compromised by its association in the public mind with radical groups during the North-West Rebellion. The Prince Albert Settler's Union, representing Anglo-Canadian settlers on the Saskatchewan River, made common cause with the Métis in the early stages of the revolt. Since the Prince Albert Union had been modelled after the Protective Union, the parent organization was open to the charge that it was a front for secessionists. Aggravating the situation, Charles Stewart, the original instigator of the Manitoba Union, called a secessionist meeting in Brandon in March 1885. Coinciding with several local Protective Union meetings throughout the province that month, the March meeting contributed to the perception that the Union's members were disloyal.

In May 1885, the Protective Union sealed its fate with the publication of an appeal to Queen Victoria. While most of this lengthy tome was a moderate and judicious treatment of the farmers' position in western Canada, it concluded with a defence of the North-West Rebellion as the result of long-neglected grievances. The appeal warned that if Manitoba and the Northwest

> are doomed much longer to their present anomalous position – in name only, the colony of a colony, denied all the rights that belong to the other colonies in the Confederation, under the "British North America Act," – then, indeed, it will only be a question of time as to when the people will become tired of their equivocal position and slip the yoke of servitude.[8]

Stirring as those sentiments might be to its audiences, the appeal could not have been more poorly timed. Associated in the public mind with the secessionists, the Protective Union was almost immediately politically discredited. Prominent Liberal Party politicians, who had promoted the Union in its infancy, now abandoned it. Moreover, the Union's co-operative grain-buying ventures were effectively sabotaged by a series of legal suits launched by grain merchants. The final deathblow to the movement occurred in March 1886, when Brandon farmers formed the Farmers' Alliance, a radical splinter group with a more political orientation. This further division in the farmers' ranks effectively destroyed the Union, and both organizations folded later that year.

By the end of the decade continued economic difficulties and unalleviated grievances among farmers contributed to a revival of agrarian radicalism. The appearance of the Patrons of Industry in 1887 heralded the first major agrarian challenge to the traditional political framework. Originating in the United States, the Patrons entered Canada in 1889 when a Michigan organizer spoke to a receptive audience of Lambton County, Ontario farmers, who formed a chapter the following year. The movement grew rapidly. By March 1891 the Patrons had formed more than 300 lodges in Ontario.[9]

The Patrons' platform, which they adopted at a meeting in London, Ontario, in September 1893, addressed a broad range of issues, centring on the Dominion's transportation and tariff policies. The platform called for an end to the system of providing government grants to private railways. With respect to trade it favoured the elimination of tariffs except for purposes of revenue, and stated that these tariffs should fall principally on luxury goods. It also called for reciprocity in trade between Canada and the rest of the world, when equitable terms could be arranged. Other planks included the maintenance of the British connection, reservation of crown lands for settlers, economy in government, simplification of laws, and the abolition of the Canadian Senate.[10]

The Patrons soon appeared in western Canada. In the fall of 1891 Manitoba farmers formed a provincial Grand Association of the Patrons of Industry. Their adopted slogan, "Manitoba for Manitobans," reflected the regional alienation that had hardened after a decade of Dominion Legislation disallowing provincial railway charters. Manitoba farmers were also angered by federal inaction in surrendering control of public lands to the province and by the recent boundaries case, which had awarded disputed territory to Ontario. The fourth clause of their platform, however, was noteworthy for its call upon farmers and waged labourers to unite in opposition to monopolistic interests:

> That we mutually agree as farmers and employees to band ourselves together for self-protection and for the purpose of obtaining a portion of the advantages that are now almost exclusively enjoyed by the financial, commercial and manufacturing classes, who by a system of combines and monopolies are exacting from us an undue proportion of the fruits of our toil and that we may have more time to devote to education and secure for our selves and [sic] equitable share of the profits of our industry. That our endeavor be to place the farmers and laborers of Manitoba in unison with the manufacturing laborers of the east to the exclusion of the middle men.[11]

For reasons outlined earlier, farmers, as a propertied class, possessed interests separate from those of labourers. As Brian McCutcheon has noted, the Manitoba Patrons soon abandoned their rhetorical statements of common cause with the working classes. By February 1892 the Patrons had formed more than 100 chapters in the West, including several in the North-West Territories. Between 1892 and 1895 the number of Patron sub-associations tripled to more than 300 in western Canada. In this period the Western patrons endorsed policies that reflected a distinctly western standpoint. They were, for example, more inclined to demand the elimination of tariff barriers; they also favoured the building of a government-controlled rail outlet to Hudson Bay. The Western Order also attached the Dominion policy of land grants to railways, and in 1894 it endorsed the principle of the enfranchisement of women.

It remained for the Patrons to test their policies in the political arena. Having elevated a sizable contingent to the Ontario legislature in 1894, they were also successful in a provincial by-election in Manitoba the same year. By the time of the Manitoba provincial election in January 1896, however, the Patrons' provincial rights plank had effectively been appropriated by Premier Thomas Greenway's Liberals. Only two Patrons were elected. Nationally, the Order's support was similarly undercut by the Liberal Party, as a paltry three candidates won election to the House of Commons in the Dominion election of June 1896. The Patrons' overwhelming defeat sounded the death knell of the movement.

Despite the Patrons' defeat, the farmers' grievances continued unabated in the late 1890s. It was true that their premature foray into the political arena had somewhat discredited the concept of direct political action. The principal sources of discontent, monopolistic transportation and grain handling practices, however, remained. In September 1897 news broke of an attempt by four major grain companies to combine to lower prices to farmers. At the same time a number of elevators refused to store grain they had not already purchase until their existing stocks had been shipped. Since the CPR granted a monopoly in grain handling to the elevator companies, farmers were denied the opportunity to sell their grain to a competitor. During the 1898 Parliamentary session, J.M. Douglas, Patron-Liberal MP for East Assiniboia, and R.L. Richardson, MP for the Manitoba constituency for Lisgar, each introduced private members' bills in the House of Commons. The bills, if passed, would require the railways to allow farmers to load their grain directly onto the grain cars from their wagons, on platforms or from flat warehouses. Richardson and Douglas also sought changes in the grain inspection system.[15] D.G. Hall has noted that lobbying by grain interests caused Douglas to dilute the provisions of his Bill. The amended Bill dropped reference to platform loading and permitted the farmers only two hours in which to load their grain cars. A penalty of 50 cents per hour, or $5 per 12-hour period was to be charged for demurrage or delays in loading. These amendments were passed by the Railway Committee of the Commons. When they learned of the new amendments, farmers reacted angrily and protested that they could not possibly load their wheat within the specified two-hour period. Western MPs approached D'Alton McCarthy, an Ontario member, to draft a new set of amendments. McCarthy's proposals provided for: the building of flat warehouses when requested by a person, the issuance of grain cars on demand to farmers to permit direct loading from wagons, and the reduction or elimination of demurrage charges.[16] McCarthy died before these amendments could be considered during the 1898 session. As a result of the bill's introduction, the CPR did provide some additional grain cars for platform loading that year.

When Douglas reintroduced his own bill in 1899, it incorporated the amendments drafted by McCarthy. This time, considerations of partisan politics prompted Clifford Sifton, Minister of the Interior, to attempt to sidetrack the bill by referring it again to a committee. An open split developed between Sifton and Douglas, whose efforts were winning demonstrable support in the West. To solve the impasse, on 7 October 1899 the Liberal government appointed a five-member Royal Commission on the Shipment and Transportation of Grain.[17] The Commission's terms of reference included the investigation of charges that elevator companies had unfairly docked farmers' grain and had given short weight. It was also instructed to investigate the allegation that the elevator companies enjoyed a monopoly position through thwarting attempts to build flat warehouses where their elevators were located, and that they conspired to profit from this situation by keeping prices lower than the true market value of western crops.

After holding a series of hearings at 22 localities throughout the West, the commissioners reported that the principal concerns expressed by western farmers pertained to restrictions on grain loading. The commissioners affirmed that car shortages had resulted in depressed prices for farm produce. They observed additionally that little actual competition existed among elevators with respect to prices offered to farmers. In the years that immediately preceded the investigation, mergers among elevator companies had progressively reduced competition. According to evidence presented before the Commission, 447 elevators then existed in the Manitoba inspection district, comprising Manitoba and the North-West Territories. Of these, 301, or 67 per cent, were owned by three line elevator companies and two milling companies. A further 120, or 27 per cent, were owned by individual millers and grain buyers. The farmers' own joint stock elevators accounted for only 26, or about 6 per cent of the total. The Grain Commission accordingly recommended that loading platforms be constructed by the railroad and that an officer be appointed to regulate the grain trade on the prairie.[12] During the 1900 session Charles N. Bell, Secretary of the Royal Commission, worked with Clifford Sifton to draft a new bill.[13]

In 1901 farmers in southwestern Manitoba and the District of Assiniboia reaped the largest harvest in their history, estimated at 62,820,000 bushels.[14] Harvesting was delayed by wet weather, and by the end of the Great Lakes navigation season, more than half the farmers' crop was still on their farms. Compounding the problems, the CPR lacked sufficient grain cars to handle the unprecedented harvest. When farmers hauled their loaded wagons to elevators on the railway, they discovered they were unable to ship their grain. At the same time the elevator companies, anticipating the need to cover storage costs for the winter, lowered their street prices for grain. Since farmers could not obtain cars for themselves they were obliged to sell to the elevators at the reduced price. The situation deteriorated as farmers' notes on implement purchases or land mortgage payments became due, and their anger and frustration mounted.

The first stirrings of the impending farmers' "revolt" occurred in November 1901. Two Indian Head farmers, John Sibbold and John A. Millar, the secretary of the local agricultural society, called an "indignation meeting" to permit an airing of grievances. About 50 farmers attended and demonstrated a high degree of unity in "opposition to the corporations which they stigmatized as their oppressors."[15] Indian Head was then one of the principal grain shipping centres in the West but only one-fourth of the huge 1901 harvest had been shipped by rail.[16] Hostility towards the grain "blockade" (Fig. 52) was therefore particularly acute among farmers in the Indian Head area.

In December 1901 two farmers in the Abernethy district, W.R. Motherwell and Peter Dayman, determined to exploit the momentum created by the Indian Head meeting. It was not surprising that the impetus for action should come from the settlements north of the Qu'Appelle River. Like their Indian Head counterparts, farmers from the northern districts had been thwarted in their attempt to market their wheat, but they had experienced the additional aggravation and

Fig. 52. Farmers lined up behind Indian Head elevators during the grain blockade. Saskatchewan Archives Board, Regina, Photo no. R-B 2809.

expense of hauling grain distances of 20 to 30 miles. Dayman and Motherwell met in the lobby of Motherwell's stone house and agreed to call a meeting at Indian Head for the purpose of initiating an organized response to the "wheat blockade." They sent letters to prominent farmers throughout the Central Qu'Appelle region, to Wolseley, Sintaluta, Indian Head, Qu'Appelle, Wide Awake, and other localities. To ensure that the meeting might not be construed as a partisan political rally, both farmers – Dayman, the Conservative supporter, and Motherwell, a Liberal – signed the meeting notices.[17]

The 18 December meeting was scheduled to coincide with the date of a political debate at Indian Head on the question of provincial status for the North-West Territories. Among the luminaries who were to participate were the premiers of the North-West Territories, F.W.G. Haultain, and of Manitoba, Rodmond P. Roblin. Motherwell and Dayman were thus assured of a large crowd for their own meeting, which was held in the afternoon.[18]

Motherwell and Dayman both addressed the meeting and spoke in favour of immediate organization to redress the farmers' grievances. Feelings evidently were running high; Motherwell later related that at the earlier indignation meeting angry farmers were virtually incited to violence.

> When this matter was broached many would jump to their feet in protest. "Now the time for organization is gone," they declared; "It's bullets we want and guns that are needed!"[19]

A more reasoned course prevailed, as the speakers persuaded the audience that the farmers' ingrained individualism must be overcome if they were to combat the monopoly interests successfully. Farmers agreed to found a "Territorial Grain Growers' Association" and to meet the following month to draft a constitution and to plan strategy. Angus Mackay, the Superintendent of the Indian Head Dominion Experimental Farm, nominated his long-time colleague W.R. Motherwell as the provisional president of the association. The assembled farmers also elected John Millar as provisional secretary along with a board of directors.[20]

On 6 January 1902 farmers met to draft a constitution for the TGGA. The turnout was disappointing, but the organizers, including W.R. Motherwell and Matthew Snow of Wolseley, embarked on a speaking tour to enlist wider support for the new organization.[21] They were well received at numerous localities, and by the time of the annual convention of the TGGA on 1 February 1902, 38 local grain growers' associations had formed and sent delegates. The convention confirmed the provisional officers in their positions and approved a plan of action. Specifically, the TGGA endorsed three specific recommendations by W.R. Motherwell for changes to the Manitoba Grain Act:

> That section 42 of the Manitoba Grain Act be amended to empower the Warehouse Commissioner to compel all railway companies to erect every loading platform approved by the said Commissioner within thirty days after said approval is given and in default the Commissioner shall have power to impose penalties on such defaulting railway, and

collect same through the courts, and that this amendment come into force on May 1, 1902.

That railway companies be compelled to provide farmers with cars to be loaded direct from vehicles, at all stations, irrespective of there being an elevator, warehouse or loading platform at such station or not.

That the Grain Act be amended making it the duty of the railway agent, when there is a shortage of cars, to apportion the available cars in the order in which they are applied for, and that in case such cars are misappropriated by applicants not entitled to them, the penalties of the act be enforced against such parties.[22]

All recommendations were sent to Parliament, where they were drafted almost verbatim into amendments to the Manitoba Grain Act. The amendments received a full debate on 17 March, were passed and received assent on 19 May 1902.[23] Among additional amendments passed at the same time was a provision removing limits on the construction of flat warehouses. Anyone living within 40 miles of a siding could now apply to build one. Moreover, where new warehouses were built, the CPR was required to underwrite the cost of land and sidings. At sidings where no station yard existed, the railways were required to build loading platforms, provided that 10 farmers made formal application.[24]

The legislation in itself did not ensure that the farmers' grievances would be redressed. In 1902 farmers reaped an even larger crop than the 1901 harvest, and grain cars continued to be in short supply. The CPR, moreover, evidently continued to give priority to the elevator companies in apportioning its cars. W.R. Motherwell wrote, "the plain provisions of the car distribution clause are disregarded at every shipping point, I believe, in the West ... Of 67 'spotted' cars at Sintaluta only 7 have been assigned to farmers."[25] On behalf of the TGGA, W.R. Motherwell and Peter Dayman travelled to Winnipeg to lodge a complaint with the CPR. They suggested to railway officials that unless the CPR abided by the new car distribution amendments of the Manitoba Grain Act, they would initiate action to force their compliance. William Whyte, second vice-president of the CPR, responded that the car shortage was the result of an inability on the part of the railway to keep pace with the rapidly expanding wheat production on the prairies.[26]

No noticeable change in the situation occurred, and farmers remained convinced of the CPR's continued abuse of the car distribution regulations of the Act. Several months later the TGGA laid a formal complaint before the Warehouse Commissioner, charging that the CPR agent at Sintaluta had violated the Act's provisions. On 28 November the Commissioner investigated the complaint at Sintaluta and initiated court proceedings against the agent. The case was tried before a tribunal of justices of the peace headed by Magistrate H.O. Partridge of Sintaluta.[27]

Evidence was adduced to show that two elevators at Sintaluta had been given priority in receiving cars ahead of A.W. Annis, a local farmer, despite the fact that his application had been submitted before theirs. Further evidence showed that another farmer had waited seven weeks for his car; in the meantime 80 cars had been allotted. The magistrates unanimously found the defendant guilty and imposed a fine of $50 and costs, or one month in jail. The CPR promptly appealed the judgment, but the verdict was upheld by Mr. Justice Hugh Richardson of the Supreme Court of the North-West Territories.[28]

Farmers viewed the test case as a vindication of their efforts. Historian Lewis Aubrey Wood wrote that, where farmers had received only seven cars in the first two months of the 1902–3 shipping season, the CPR dispatched "scores" of cars after the trial.[29] A problem remained, however, in the original wording of the 1902 shipping amendments to the Manitoba Grain Act, which had been hastily drafted and were still open to possible challenge in the courts.[30] In 1903 W.R. Motherwell and J.B. Gillespie of the TGGA and two officers of the new Manitoba Grain Growers' Association were sent to Ottawa to meet with representatives of grain and railway companies to develop a more satisfactory wording for the legislation.[31] Clifford Sifton, Minister of the Interior, introduced the new text in Parliament as a further amendment to the Manitoba Grain Act, which was subsequently passed and enacted during the 1903 session.[32]

The new amendments outlined the procedures for allocating cars and penalties for non-compliance. Railway agents were now required to keep a logbook of cars according to the format specified by the Warehouse Commissioner. Applicants were given numbers in the logbook according to the order in which they made application, and they were required to make two applications if they wished to have two cars. Agents were to allocate cars strictly according to the order of application, and no applicant could receive his second car until all other applicants had received their first. Farmers were prevented from signing their cars over to elevator companies by a provision requiring unused cars to be assigned to the next applicant on the list. The Act provided heavy penalties for selling the right to the use of a car. It further required railway agents to post a daily record of all the applicants who had been allocated cars in the preceding 24 hours. Beyond the procedural aspects of the car distribution clauses, the amended Act provided for larger loading platforms to be built at rail centres to handle more cars.[33]

The 1903 debate on amendments to the Manitoba Grain Act was noteworthy in that it revealed a dichotomy of interest between different socio-economic strata of prairie wheat farmers in terms of the legislation's effects. During this debate, T.O. Davis, MP for the constituency representing the District of Saskatchewan, comprising many of the more recently settled areas, objected that the government's amendments would create inequities in the distribution of cars among farmers. Davis stated that many new settlers could not produce the requisite 1,000 or 1,500 bushels of wheat to qualify for a grain car. These settlers would be forced to sell their grain to the elevator companies and thus would be subject to the abuses that the Manitoba Grain Act was intended to eliminate. Davis strongly recommended that the car distribution clauses be amended to permit two or three poor settlers, who had produced perhaps 500 bushels each, to apply collectively for a car.

James Douglas, Member for Assiniboia East, argued against Davis's suggestion, saying that under the proposed amendments, farmers could still combine to fill a car at a loading platform. Davis countered that many poorer settlers, who lived 25 miles or more from a railway station, were not in a position to take advantage of the platforms. Other western MPs, including Walter Scott, opposed Davis's proposal on the ground that it would open the door to possible fraud. Scott stated that if a farmer were permitted to order a car through the elevator, the elevator manager could conspire with other persons who possessed small quantities of grain (or none at all) to order more cars than the number to which they were entitled. For this reason, the grain growers' representatives, including W.R. Motherwell, had rejected the idea. Yet, as Davis pointed out, if the objective were to eliminate fraud, the provisions for the supply of cars to individuals contained no more safeguards than if they had been extended to encompass two or three joint applicants, as opposed to being reserved solely for an individual applicant.[34]

To appreciate the significance of Davis' objections, it is necessary to recall the distinctions between "street" and "track" wheat. Grain sold "on the street" was sold directly to grain dealers or elevator companies at the railhead. Since the buyers or elevator companies often enjoyed a position of monopoly in street sales, they could do several things to reduce the revenues accruing to the farmer. An elevator operator could insist that his bins of higher grades of wheat were already filled to capacity and thus offer the farmer the equivalent of a lower-grade price on a "take it or leave it" basis. Alternatively, the street sales were subject to possible fraudulent practices of short weight and excessive dockage for impurities.

The farmer could be certain that his wheat would bring the optimal price only if he could bypass the elevator and be assured of a car to ship his grain directly to Fort William. Wheat shipped in this way was known as "track wheat."[35] We have noted that only large-scale farmers produced enough grain to participate in track shipments. The more prosperous farmers usually also possessed sufficient capital reserves to retain their wheat until the price went up. The poorer settlers, who needed immediate cash to buy necessities or pay due "notes" on implements, often were obliged to sell "on the street" while the price was depressed. On both counts, the wealthier farmer competed much more effectively than his poorer counterpart. Figure 53 shows sleigh wagons loaded with bags of wheat lined up at an elevator in Kennedy, Saskatchewan in 1907.

Moreover, since the amended Manitoba Grain Act increased the allotment of cars to individual farmers, it reduced the number available to the elevators. In his treatment of the early grain growers' movements, Harald S. Patton noted that the car distribution provisions of the Grain Act actually

> benefit the carload grain shipper somewhat at the expense of the small grower who sells his grain on the street. The fact that the elevator can apply for only one car at a time, regardless of the amount of street grain it has to ship,

Fig. 53. Sleigh wagons loaded with bags of wheat lined up at an elevator in Kennedy, Saskatchewan in 1907. Saskatchewan Archives Board, Regina, Photo no. R-B 3663.

has a tendency to widen the spread between street and track prices, whenever car supply is restricted, or the close of navigation approaches.[36]

In other words, the reduction of cars available to ship street wheat forced the small producers into the position of having to pay higher storage charges. An unintended effect of the grain handling reforms, therefore, was that they reduced the poorer settlers' access to grain cars, further contributing to their competitively disadvantageous position.

The economic historian V.C. Fowke, in his treatment of the elevator issue in *The National Policy and the Wheat Economy*, drew attention to the difficulties caused by the indivisibility of the boxcars to new settlers. Since the basic homestead consisted of 160 acres of unimproved land, and most settlers broke their land with ox teams and mouldboard breaking ploughs, few were able to produce large quantities of grain for many years after settlement.³⁷ Fowke's statements respecting the slow rate of breaking on most farms were confirmed by the quantitative data presented in Chapter 2. In the fifth year after entry, Abernethy settlers were cropping an average of only 41 acres each; Neudorf settlers were cropping only 36 acres each. Indeed, by 1899, the average area cultivated per farmer in Territorial Crop District No. 2 in Eastern Assiniboia (incorporating Abernethy and other communities) was 88 acres.³⁸ Most farmers sowed oats on one-fourth to one-third of their crop lands to provide feed for their horses.³⁹ Assuming an average yield of 16 bushels per acre,⁴⁰ if the average farmer sowed wheat on three-fourths of his cultivated acreage he would reap 1,056 bushels, just over the 1,000-bushel limit for a small car. Yet the practice of summer fallowing further reduced the land available for crop.

Inequalities in the original disposition of free-grant land also greatly affected some settlers' capacity to exploit the car distribution provisions of the *Manitoba Grain Act*. Earlier, it was noted that in the 1880s Abernethy settlers usually were positioned to take advantage of generous *Dominion Lands Act* provisions for additional quarter-sections of land. They often acquired two or three quarters of open prairie land, many of which were highly cultivable. High quality soils in the Abernethy area also facilitated these farmers' capacity to produce large quantities of wheat. Only a few years later, Eastern European settlers at Neudorf arrived to discover that they were restricted to single quarter-section entries after 1889. Their lands were also of lesser productivity than those at Abernethy. More importantly, only about half of the Germans' lands were located on the open prairie, the remainder being consumed by intermittent poplar woods. The clearing of these wooded lands did not transpire until the Germans could afford to use more sophisticated clearing technology in the 1920s. Limited in production by smaller acreages and lesser productivity of their farmlands, their necessary reliance on street sales continued to place them at a disadvantage in grain marketing. As late as the mid-1920s, more than 50 per cent of all prairie wheat was still being sold in less-than-carload lots. Earlier, the proportion of street sales had been larger.⁴¹

Abernethy settlers like W.R. Motherwell and J.R. Gillespie, two of the four farmers' representatives who co-authored the 1903 Grain Act Amendments, would have known from personal experience that settlers typically spent many years developing their farms before producing enough wheat to fill a grain car. In his diary, Motherwell's neighbour Samuel Chipperfield reported that in 1892, 10 years after his arrival on his homestead, he had threshed only 900 bushels.[42] This harvest was 100 bushels short of the capacity of a small grain car. Chipperfield had expanded his yield by 1895, but his harvest of 1,542 bushels of wheat in that year was still only barely enough to fill a large car.[43] By the turn of the century, however, most Abernethy-area settlers were in a position to take advantage of the new car distribution provisions, for which they had lobbied.

Comparatively few loading platforms were built after 1900. Their chief value lay in their function as a competitive alternative to monopolistic elevator companies. The farmers believed that the presence of platforms kept the elevator companies "honest." Paradoxically, the emphasis on platform loading was a technological throwback. Direct loading was an outmoded and inefficient technique, out of step with modern technological innovations for lifting and distributing grain.[44] Much time and effort were wasted as farmers customarily used scoop shovels to throw the wheat manually from the grain wagons into the boxcars (Fig. 54). One farmer recognized the drawbacks when he wrote to the *Regina Standard* to suggest that farmers should continue to take advantage of the more advanced elevator technology to load their wheat. He proposed that special bins should be set aside in elevators for the specific storage of farmers' grain, as a prelude to loading on their own boxcars.[45] Yet none of the Manitoba Grain Act amendments reflected this writer's standpoint, and loading platforms continued to be used well into the 1930s, and even after.[46]

With the passage of the Manitoba Grain Act amendments, the enthusiasm for continued activism among the original officers of the TGGA waned. A new group of activists, most notably E.A. Partridge of Sintaluta, had begun to press the TGGA to agitate for more stringent regulation of the grading system and inspection of elevators. For W.R. Motherwell and John Millar, however, the TGGA had accomplished its principal objectives and the time had arrived for a re-assessment of the farmers' role. They stated their views in an open letter to the *Qu'Appelle Progress*, which appeared in the 1 December 1904 issue:

> ... The principal difficulties that confronted us as a body at the time of our organization, and indeed, was the occasion of our coming together as an organized body, have

Fig. 54. Loading wheat from a platform siding, ca. 1920. Saskatchewan Archives Board, Regina, Photo no. R-A 8829.

been largely corrected or removed, and it would appear as if we are now at the parting of the ways where we must either branch off into new lines of work more intimately allied with actual successful grain growing, or else suffer the natural consequences of not being fully or profitably employed.[47]

The authors also announced that papers on agricultural topics would be read at the forthcoming annual meeting of the TGGA. Clearly the officers were moving away from an activist stance to a more conservative position emphasizing practical day-to-day aspects of farming. Another statement, to the effect that the grain growers' advancement "lies with the individual farmer,"[48] illustrated Millar's and Motherwell's rejection of continued collective action.

Thereafter the momentum in the farmers' movement shifted to activists promoting formal co-operation in grain handling and marketing. The story of E.A. Partridge's attempts to enlist support from within the ranks of the TGGA for his concept of a producers' co-operative trading association has often been told.[49] Partridge and a group of like-minded Sintaluta area farmers launched the Grain Growers' Grain Company (GGGC) in 1906. The founding of the Company inaugurated a protracted struggle with the established grain companies over the GGGC's attempt to secure and maintain a seat on the Winnipeg Grain Exchange. By paying patronage dividends to its member clients, the company precipitated its expulsion from the Exchange. Partridge then enlisted the support of Manitoba Premier Roblin for reinstatement. Pressured by the Manitoba Grain Growers' Association (MGGA) on behalf of the GGGC, Roblin's government influenced the Exchange to give the GGGC its seat. In the process, D.W. McCuaig, President of the MGGA, had brought suit against three member companies for combining to obstruct trade. Partridge, as a member of the board of the Saskatchewan Grain Grower's Association (SGGA – successor to the TGGA), assured the Manitoba association of the SGGA's support. In an open letter W.R. Motherwell, now a member of the Saskatchewan cabinet, and John Millar of the SGGA attacked Partridge for acting unilaterally. Their letter underlined a split that had been long developing in the ranks of the agrarian movement.[50]

Differences between the radical and conservative elements in the farmers' movement reappeared in sharp relief in the context of the debate over publicly owned elevators. In early 1908, Partridge persuaded the Saskatchewan Grain Growers' Association to endorse the principle of provincial ownership of inland elevators and Dominion ownership of terminal elevators. At the same time, conventions of the provincial grain growers' associations in Manitoba and Alberta passed resolutions in support of the SGGA proposal.

The details of the "elevator issue" have been discussed at length in studies by V.C. Fowke and D.S. Spafford.[51] Briefly summarized, at the behest of Premier Roblin of Manitoba, the premiers of the prairie provinces met in May 1908 to discuss the Partridge Plan. They did not act on the proposal for publicly owned elevators, but agreed instead to recommend that the railways be pressed to implement some reforms.

The Interprovincial Council of Grain Growers rejected the premiers' suggestion as inadequate, and a second meeting of the premiers with agrarian leaders was convened in November 1908. The western premiers subsequently announced their decision not to become involved in publicly owned elevators in the absence of a monopoly in the elevator field. Such a restriction on trade would require an amendment to the British North America Act. Again, the Interprovincial Council rejected the notion that a monopoly was needed to ensure the success of the proposed public elevators. Also in 1908, an Interprovincial Council delegation, including Partridge, made overtures to Dominion officials for amendments to the Manitoba Grain and Grain Inspection acts. As part of their proposal, they urged the federal government to acquire the terminal elevators at Fort William and Port Arthur.[52]

In late 1909, the Saskatchewan Grain Growers' Association formally requested that the provincial Liberal government initiate a public elevator system. The Saskatchewan government responded by appointing a Royal Commission to investigate these proposals. The Commission was well aware of Premier Walter Scott's expressed opposition to public elevators. In its report, released 31 October 1910, the Commission rejected public ownership and recommended the creation of a system of farmers' cooperative elevators, with provincial financial support. Despite Partridge's continued opposition, the SGGA approved the Commission's plan at its annual meeting in February 1911. The Saskatchewan government had received the recommendation it desired; it quickly introduced a bill in the legislature to incorporate the "Saskatchewan Co-operative Elevator Company." The bill was passed by the provincial legislative assembly in March 1911.[53]

Partridge felt betrayed by the SGGA's shift in policy. He blamed W.R. Motherwell for having sabotaged his plan for publicly owned provincial elevators.

> There is no doubt in my mind that with Mr. Motherwell out of the way, Premier Scott would have lent a more sympathetic ear to the request of the Grain Growers' made through their representatives last spring.[54]

Inasmuch as Scott had already expressed his opposition to the concept of provincial elevators, Partridge's allegation was tenuous. He was, however, correct in identifying Motherwell as a major opponent of government enterprise. Motherwell's attitudes towards government ownership were well documented in a letter he wrote to J.H. Sheppard, Member of the Legislative Assembly for Moose Jaw, in 1908. Sheppard had written earlier to draw Motherwell's attention to the excessive distances that some of his constituents had been obliged to travel between their settlement and the town of Markinch. Sheppard suggested that the provincial government build a shelter to function as a "half-way house" for the settlers. In rejecting the proposal Motherwell stated that such an undertaking was best left to the private sector.

> ... I may as well candidly say that in all probability that before this condition of things is likely to exist for any length of time, that private enterprise will step in and provide the accommodation. In a new country such as this there is always a certain period of time during which the peoples' requirements are not fully provided for until business assumes such a proportion that some enterprising party is induced to go in and take the matter up. If the government was to step in every instance and provide every little want where are they going to get off at?[55]

Motherwell stated that when he "was a pioneer we never dreamed of having such a provision being made for us" and that he believed that new settlers were "quite as capable of looking after themselves as we were."[56]

Philosophically, Motherwell and Partridge represented diametrically opposing points of view *vis-à-vis* the role of government in the economy. Vernon Fowke has noted that until well after the First World War the dominant Canadian collectivist principle of using government funds for developmental purposes was qualified with the "proviso that public ownership and management should be avoided at all costs."[57] Motherwell and Premier Scott subscribed to this approach. Believing farmers to be effective competitors in a free market, they conceived a minimal role for government. Partridge believed that competition among farmers only served to erode their economic position. He envisaged government's more dynamic participation in the economy and the formal entrenchment of the co-operative principle. At the Farmers' Union convention in 1925, he stated in his address:

> Cooperation, to an agricultural population, is the primary step towards that true cooperation which will embrace the various elements of society. The type of cooperation that I ultimately hope to see is the cooperation of all those who live and desire to live by useful labor, whether of hand or brain – the cooperative commonwealth. If you content yourselves with mere vocational cooperation, it seems to me that you stop short of the ideal that will really be effective in changing human relationships and making them satisfactory.[58]

To appreciate why these divisions in the TGGA occurred it is useful to reflect on the underlying structural basis of the movement. Notwithstanding his rather monolithic treatment of farmers, C.B. Macpherson's *Democracy in Alberta* remains the most incisive structural analysis of Canadian prairie agrarian politics ever written. Macpherson believed that the political behaviour of prairie farmers was reducible to two main factors: the class basis of the western Canadian farm proprietor group, i.e., a broadly based group of independent commodity producers, and its "quasi-colonial" status in relation to central Canadian business. Having identified the structural basis for a degree of class action by a relatively homogeneous economic group of independent commodity producers, he elaborated the prairie farmers' collective relationship to the larger economic context. Wheat farmers, Macpherson observed, have occupied a precarious position between the spheres of capital and labour. They take pride in their "independence," that is, the nominal ownership of property and control over questions of production – specifically, what crops are to be raised and in what quantities. This element of independence within the confines of their farms has contributed to farmers' perceptions that they are masters of their own fates, that is, that they are entrepreneurs competing successfully in the overall free-enterprise economy. But farmers do not employ labour on a scale that would permit them to compete on an equal footing with industry, commerce, and finance. They depend on large markets to purchase and transport their grain and provide equipment and working materials for purchase, and on borrowed capital. For these reasons, the farmers' "independence" has been largely illusory.

Since the farmers' economic position is defined by market forces largely beyond their control, they occupy an inherently insecure position. In times of economic duress, their insecurity manifests itself in a recurrent radicalism. But since their identity is based on illusory notions of independence accompanying farm proprietorship, a tendency towards conservatism reasserts itself once periodic economic stresses subside. This "oscillation between conservatism and radicalism," to use Macpherson's words, has characterized the prairie farmers' response to economic and political questions that affect them. Macpherson writes of farmers:

> ... They cannot entirely identify themselves, or make permanent common cause, with either of the two classes. Yet, they are repeatedly driven by insecurity to find a solid basis somewhere. So they veer between attachment to one class and to the other; or rather, different sections of the whole class veer at different rates of speed and it may be in different directions at different times, depending on changes in their own position and on changes in the political outlook and action of the other classes.[59]

While a somewhat monolithic construct, Macpherson's thesis appears to hold considerable explanatory power when applied to various chapters in the history of prairie agrarianism. In his study of early farmers' movements in Manitoba in the 1880s, Brian McCutcheon showed that the outbreak of agrarian radicalism was directly related to immediate economic conditions. The convergence of a frosted crop and a marked drop in the price of grain, compounding long-term discontent over inadequate rail service, precipitated the founding of the original Manitoba and North West Farmers' Union. McCutcheon observed that both the Union and its successor, the Manitoba and North West Farmers Co-operative and Protective Union, failed as a result of membership of disparate and ultimately incompatible interest groups.[60]

An analysis of specific factors contributing to the outbreak of agrarian unrest in the Central Qu'Appelle area after 1900 indicates a repetition of the classic pattern. Here the precipitating issues were the grain bottleneck and reductions in the price of grain stemming from the blockade. Confronted with a situation of economic duress, Qu'Appelle area farmers resorted to ephemeral radicalism. After achieving its limited objectives, the propertied agrarian middle class quickly returned to a conservatism orientation occasioned by its members' perception of their role as entrepreneurs.

At the same time, the post-1900 agitation did achieve some momentary successes. Several conditions had changed. Unlike the earlier farmers' movements, the TGGA was not weakened by the membership of non-farmers within its ranks, a factor that had proved the undoing of the earlier Protective Union. It was also strengthened by the non-partisan nature of its leadership. W.R. Motherwell, a Liberal, and Peter Dayman, a Conservative, had taken pains to ensure that both persons signed the original notices that spawned the founding meeting of the TGGA. In this way they averted the charge that the association was politically motivated. The TGGA also addressed itself to a limited range of practical reforms that could feasibly be supported and implemented by the government.

Additional factors assisted the development of a broader basis for united class action by farmers in the post-1900 movement. First, a high level of communication among farmers facilitated a general awareness of issues. Communication links in the pre–telephone and radio era were ordinarily severely restricted, but the grain bottleneck in 1901 had obliged farmers to congregate at market centres along the CPR line while they waited to ship their grain. While aggravating frustrations, the blockade occasioned the airing of grievances and the exchange of information. It was, therefore, fairly easy for agrarian activists to recruit members, organize meetings, and mobilize a broad cross-section of farmers to their cause. In this way, John Sibbold and John Millar were able to organize the "indignation meeting" of October 1901 simply by approaching farmers in the long queues around Indian Head's elevators.

Another ingredient in the agrarian success after 1901 was the development of a sufficient class consciousness to permit a unified front, however temporary. Farmers had witnessed how divisions in their ranks contributed to the Patrons' defeat in 1896. At that time many had turned to the Liberal Party, whose electoral policy of reduced tariffs undermined the Patrons' platform.[61]

After the failure of Canadian negotiators at the Canadian-American Joint High Commission of 1898–99 to win any important trade concessions, western farmers felt betrayed. They had expected a more positive redress of their grievances from the Dominion Liberals than the Conservatives. Farmers north of the Qu'Appelle River, who for 12 years had petitioned the Dominion Conservative government unsuccessfully for local rail service, found after 1896 that the Liberals were equally unresponsive to their demands. The petition that Motherwell and his neighbours sent to Prime Minister Laurier in 1898 indicated the depth of frustration the settlers felt.[62] Despite the CPR's assumption of the Great North West Central Railway Company's charter that year, no construction on the Kirkella branch line had been started at the time of the agrarian outbreak three years later. This long-festering wound was reopened by the grain blockade. Not only were farmers obliged to haul their wheat 20 to 30 miles to the rail centre, on arrival they discovered they could not market it. A shared perception of exploitation by Central Canadian financial, commercial, and transportation interests had crystallized the farmers' formerly inchoate resentment into a consciousness of the need for collective action.

The post-1900 TGGA leaders were in a position to tap deep-rooted agrarian hostility towards corporate interests. In 1886 farmers met at Balcarres to consider the possibility of building a co-operative grist mill. John Teece argued that millers in the North-West Territories were "robbing" the farmers, by giving them only 25 to 30 pounds of flour per bushel of wheat, whereas farmers in Ontario or England received 40 to 45 pounds for the same weight.[63] In 1899 Abernethy farmers met to protest the practices of the Manitoba Farmers' Mutual Hail Insurance Company, which they believed had swindled them. A three-member committee, which included W.R. Motherwell, drafted a statement of grievance. One of their assertions was "that from the word 'go' this Mutual Hail Co. has in view the comfortable fleece of the farmers."[64] The farmers' response was particularly hostile towards implement companies, which could seize their tools if they defaulted on a note payment. In an editorial the *Regina Leader* claimed that no calamity could be compared to the spectacle of settlers "with the blood suckers of the lawyers at their veins and the fangs of cold and terrible corporations in their flesh."[65] These hyperbolic statements are indicative of the depth of the farmers' antagonism towards business. Well before the 1901 grain blockade, this anti-corporate sentiment provided fertile ground for agrarian agitation.

Further, the agrarian unrest of 1901 may also have been fuelled by the farmers' repeated experience of hope and disappointment, contributing to a shared psychological sense of victimization. In his study of modern homesteaders in the United States, Evon Vogt emphasized a characteristically optimistic orientation in the outlook of American farmers, which he described as as "hopeful mastery over nature."[66] A similar dynamic was evident among Abernethy's settlers, characterized by some second-generation farmers as "gamblers."[67] Banking everything on the next year's crop, their optimism was expressed in such ways as taking out short-term notes for their implements – effectively mortgaging their means of livelihood to moneylenders. Such notes typically came due at harvest time, when their hopes – and debts – were often again deferred until "next year." By 1900, Abernethy's farmers had improved substantial acreages and had begun to invest in farm buildings, machinery, and new land. As noted in Chapter 3, many went into debt to finance these improvements as expansion in their cultivable acreage appeared to position them well to capitalize on expanding markets. At the very moment when it appeared that prosperity had finally arrived – the bumper crop of 1901 – farmers were frustrated by the CPR's inability to move it. Momentarily buoyed by the real prospect of at last attaining success and then feeling yet again thwarted and betrayed by the railroad and elevator companies, their accumulated resentments as built up over years of disappointment erupted into widespread anger and active engagement in grassroots political action.

Despite a somewhat more highly developed degree of agrarian consciousness in the farmers' movements of the early 1900s, differences in economic interest ultimately drove a wedge between different socio-economic strata. An analysis of the composition of the first Board of Directors of the Territorial Grain Growers' Association shows that the prime movers of the association were from the upper stratum of the farming population.[68] W.R. Motherwell, the first president, possessed a farm of 800 acres, an elegant Victorian house, and an impressive farmstead campus. In addition to operating large farms at Indian head, two directors other directors – Walter Govan and John Millar, the latter the Association's secretary – had purchased lands other ands in the Abernethy district.[69] Elmer Shaw and Peter Dayman, two other directors, ranked among the most prosperous of Abernethy farmers at the turn of the century.[70] They were among the most prosperous of the successful proprietors who settled in the Indian Head and Abernethy districts on tracts with first-class clay soils, comprising some of the most productive farmland in the province.[71] As was shown in Chapter

3, by the early 1930s farmers in these areas had amassed an average capital more than double the provincial average. Far from being a representative group, the early leaders of the TGGA represented a comparatively privileged elite. Once this group had secured its immediate objectives, it turned away from radical action and even began to resist it.

The more radical approach of E.A. Partridge and the Sintaluta group may well have partly related to their occupancy of a somewhat lower economic stratum than their Abernethy counterparts. Twelve of the original 15 members of the Preliminary Organization Committee of Sintaluta Farmers can be identified on a 1906 map of land tenure in the central Qu'Appelle area.[72] Their land holdings ranged widely between one and eight quarter-sections, although some of the larger farms of this group included considerable tracts of non-productive land. As well, four of E.A. Partridge's five quarter-sections were partly broken up by coulee or creek-beds. At least two quarters of his neighbour David Railton's seven quarter-sections were similarly suitable only for grazing. Overall the average farm size among the 12 Sintaluta farmers was 640 acres, but it should also be noted that most of these farms possessed second-class loam soils. Further evidence of their somewhat more modest economic position vis-à-vis the Abernethy TGGA leaders may be discerned through the comparisons in material culture, specifically by investigating the houses in which members of the two respective groups lived. On the whole, Abernethy's leading farmers, such as W.R. Motherwell, James Morrison, and W.H. Ismond, occupied larger, more elegant residences than those of Partridge and his colleagues, indicative of greater accumulated wealth among these more conservative farm leaders. This interpretation is given greater credence through the study of the recently discovered minute book of the Abernethy chapter of the Territorial Grain Growers' Association, and after 1905, the Saskatchewan Grain Growers' Association. In a key early meeting of this local chapter of the association, convened on 10 December 1904, Motherwell and Elmer Shaw, another of the district's largest landowners, were clearly were the dominant actors behind several major resolutions pertaining to the grain trade.[73] At the time, both men were also prominent members of the executive of the central TGGA.

While the foregoing analysis may help explain why W.R. and other early agrarian leaders initially embraced collective action and then abandoned it, it does not account for the Dominion government's accession to the TGGA demands. In approaching this question it is fruitful to draw on V.C. Fowke's well-documented studies of the formation of federal agricultural policy in the period.[74] Fowke's analysis showed that at the early agricultural royal commissions, which were created in the context of the agrarian unrest around 1900, were dominated by farm members. The Commission of 1899–1900 comprised three Manitoba farmers under the chairmanship of an Ontario judge, Mr. Justice Senklar. Senklar resigned for health reasons but his successor, A.E. Richards, was from Winnipeg. Fowke notes that the records of this commission show that C.C. Castle, one of the farm members, dominated its deliberations. Similarly, the Royal Commission of 1906 consisted of three western farmers, including John Millar, the chairman. Millar, it should be remembered, was an organizer and the first secretary of the Territorial Grain Growers' Association. Hence, the composition of the first agricultural royal commissions was strongly weighted in the farmers' favour.

Fowke attributed the Dominion government's appointment of commission members sympathetic to the farmers' cause to the as yet unrealized goals of the National Policy. In the context of its promotion of western settlement, Laurier's government perceived that reforms in grain handling were essential to the continued expansion of western Canada:

> ... The federal government supported the western protest against monopoly in the grain trade as soon as it became convinced that such monopoly would stifle rather than promote, western expansion and would thus imperil the national policy. From 1899 the federal government used royal commissions ... chiefly for the purpose of getting this protest on the record.... So sure was the Dominion government of what it wanted to be forced to do that it would entrust to no one but farmers that task of manning its early agricultural commissions.[75]

Despite the weighting of the early agricultural royal commissions with members oriented towards responding positively to the agrarian agitation, the Dominion government was also careful to choose members who could be counted on not to recommend the option of government ownership. For example, John Millar, the chairman of the 1906 Commission, was recruited from the wealthier, and hence, more conservative elements of the farming population, who were less inclined to demand state intervention in the grain trade.

In light of these facts, the package of grain handling reforms initiated by prairie farmers in the late 1890s and early 1900s appears in rather a different light than the idealistic picture presented by Hopkins Moorehouse and D.G. Hall. W.R. Motherwell and other agrarian activists were catalysts in pressing the Dominion government to remove some of the worst abuses of the system, but their successes were largely attributable to the larger aims of the still unrealized National Policy. Not wishing to risk a curtailment of immigration that would surely result from continued agitation, the Laurier government was anxious to demonstrate its responsiveness to farmers at least at this minimum level.

The amended Manitoba Grain Act did little to alter the structural anomalies that ensured the prairie farmers' continued subordinate status. The principal features of the National Policy, that is, the tariff system and western regional dependency on the matrix of central Canadian financial, manufacturing, and transportation interests, were continued intact. V.C. Fowke has demonstrated that the agrarian leaders' capacity to influence public agricultural policy tailed off significantly after the high water mark of agrarian success in the grain trade regulation in the decade 1910 to 1920.[76] Moreover, the disproportionate weighting of benefits from the Grain Act amendments ensured that at least the upper stratum of the class of farm proprietors was sufficiently appeased to limit its agitation. As the earlier discussion has shown, the competitive position of poorer farmers was further weakened by the grain handling amendments; the early agrarian organizations were dominated by more prosperous Anglo-Canadian farmers who sponsored reforms reflecting their own economic interests as a group. While in this era western Canadian farmers overall experienced an inordinate transfer of their surplus product to corporate, and principally central Canadian interests, the revenues they did receive were divided unequally. The farmers' illusory view of themselves as independent entrepreneurs competing effectively in the Canadian economy, coupled with divisions of economic interest within their own ranks, ultimately limited the effectiveness of the agrarian agitation.

9

Conclusions

Between 1880 and 1920 the early Ontarian and British settlers at Abernethy established social structures reflecting their political and economic dominance in prairie society. By virtue of hard work, perseverance, and the central advantage of being the first homesteaders to arrive, they transformed the Pheasant Plains into a prosperous, settled farming community. Many of these early settlers began farming with little capital. Farming in the initial period was often a part-time proposition and did not assume the dimensions of full-fledged market activity until the settler had accumulated a sufficient base of improved land, buildings, and implements. By 1900 the settlers who had persisted as farmers had begun to display the signs of affluence marking them as a successful entrepreneurial class. Their success contrasted sharply with the contrary experience of many settlers who were obliged to cancel their homestead applications or those patentees who were unable to convert their homesteads into viable farming operations. While the established Abernethy farmers benefited from increased wheat prices after 1900, their economic success was largely attributable to the capital gains that accrued as land prices jumped rapidly in this period. Freed to some extent of the economic insecurity that plagued subsequent settlement groups, the Anglo-Canadian settlers pursued their political and social goals. To a significant degree, they succeeded in replicating the social structures and relationships that had prevailed in eastern Canada on the Saskatchewan prairie.

While the emerging social structure established by these settlers was ultimately stratified and hierarchical, significant social differences were not apparent in the early settlement era. Several factors served initially to mitigate social stratification within this group. Before 1914 the idealization of the yeoman farmer conferred upon most independent producers of the desired ethnicity a privileged social status.

One's status could also be enhanced by noteworthy farming success or diminished by poor management and profligate habits. In the early homesteading period the community was too sparsely populated and dispersed to permit the establishment of rigidly differentiated social levels. Social interaction tended to be tied to work or community service activities. Out of necessity neighbours often worked together during the busy seeding and harvest seasons in the spring and late summer. One's neighbour might be of a somewhat lesser status, but the shared experience of working together in the fields temporarily lessened social barriers.

This is not to suggest that Abernethy society was ultimately egalitarian. Following the period of farm consolidation, within the dominant class of independent farm proprietors, individual members became differentiated by significant variations in farm size and accumulated capital. Once the settlers established themselves successfully, they often also built impressive Victorian homes serving as status symbols. The spatial layouts of their dwellings indicated a propensity on the part of the transplanted Ontarians to recreate formal and stratified notions of social intercourse that had prevailed in the parent province. Other physical icons of status, including the Motherwells' silver tea service and elegant formal dress, indicated how far some settlers chose to go to distance themselves from their humble homestead beginnings.

Social stratification showed up most clearly in relations between the class of farm proprietors and the wage labourer group. Farm labourers possessed little status or even recognized membership in Anglo-Canadian prairie society. Their non-status was particularly evident in the case of "hired girls," who were seldom included in family activities and whose long and arduous working days were generally poorly compensated. In most cases, hired girls from continental European immigrant settlements or First Nations reserves had little alternative but to work in such low-status occupations until married.

Hired men, who often hoped to become farm proprietors at some point, occupied an equally insecure and transient position. Often they worked for a farmer for a year or two, during which time they acquired the money and knowledge to attempt a homestead entry themselves. Since most of the arable homestead lands in settled areas had long since been claimed, and lands offered for purchase were too expensive, these men were obliged to seek land on the periphery of settlement. Therefore, they did not often remain in the community long enough to establish a presence. If they did remain, the hired men were constrained by a fairly rigid stratification between the propertied and the propertyless.

A further dichotomy existed in terms of the relative social, economic, and political positions of the province's various ethnocultural groups. Settlers of non-Anglo-Canadian or British origin experienced considerable difficulty in breaking into the power structure established by these earlier homesteaders, in part perhaps because they were not in an economic position to be able to devote time or resources to political involvement. At the same time, the European immigrants were at a comparative disadvantage in that they arrived too late to benefit from more generous Dominion Lands provisions of the 1880s, and also lacked access to lands of good soil quality on the open prairie. Preoccupied with eking out a marginal living, the first two generations of eastern European farmers lived on the periphery of the emerging social structure, which placed Anglo-Canadians in the dominant position. Not until well after the First World War were significant numbers of non–Anglo-Canadians elected to the provincial legislative assembly.

First Nations and Métis residents occupied a more marginal position. These groups were effectively shunted aside by treaty and federal Indian administration, scrip grants, or fraud as a prelude to sustained Euro-Canadian or European settlement. In the Abernethy area, as Treaty Indians the Cree at the nearby File Hills reserves were at least able to fall back on their reserve lands as a basis for subsistence and were fairly successful in pursuing self-sufficient agriculture. No such reserve existed for the Métis, who, since displaced from their lands, lived in continual economic insecurity.

The Anglo-Canadian settlement group possessed its own cultural dynamics and placed its stamp on rural prairie society in a number of distinctive ways. Ontarian social institutions, including churches, fraternal orders, agricultural societies, mechanics' institutes, temperance groups, and agrarian political organizations, to name but a few, were transplanted in rather complete form to the prairie. Through the medium of these institutions, the Ontarians laid the groundwork for the establishment of a new prairie society according to their values.

The irony was that the particular Ontarian consciousness that envisioned a prosperous transplanted eastern Canadian society on the prairies was turned against the parent province when this dream was not fully realized. For farmers who had been raised in the heyday of Ontarian expansionist zeal and confidence, the myth of independent yeomanry died hard. Farm protest movements, such as W.R. Motherwell's Territorial Grain Growers' Association, attempted to stem the tide towards increasing concentration of power in the East. To some extent farmers were able to mitigate the worst effects of monopoly in the grain handling and transportation industries. Lacking sufficient economic and political power to challenge the existing structure, they opted for pragmatic solutions that sought merely to regulate these industries. While their political activism included sponsoring reforms to grain handling and distribution legislation, thereby enhancing their access to markets, the reforms they sponsored simultaneously undermined grain marketing opportunities of poorer farmers. Whether intended or not, the grain handling legislation had the effect of reinforcing differences in economic and political power even within the dominant Anglo-Canadian group.

While it is possible to define the social structure that emerged in Abernethy at the turn of the century, it is important to recognize the essentially ephemeral nature of its respective groupings. As sociologist Rodolfo Stavenhagen has noted, social class is above all a historical category. Classes are found in "specific historical formations," that is, they develop from the structural conditions in society.[1] Abernethy's structures were markedly short-lived due to particular circumstances in its evolution. The community developed in the twilight of the Victorian era, when the contradictions of the Victorians' assumptions regarding society were already becoming increasingly untenable in the context of rather pronounced stratification and reduced social mobility in eastern Canada. In a sense, the Canadian West offered the last "safety valve" for the continuation of nineteenth-century ideals of individual enterprise and unfettered exploitation of resources. So long as an open frontier of settlement persisted, Canadians could avoid reconciling the contradiction between the "myth of the self-made man" and an increasingly closed social framework.

Figure 55 schematically represents Abernethy's social and economic structure for the period 1880–1920. It suggests Abernethy's more successful farmers were not really the "upper class" since under the National Policy all prairie farmers were subordinate to the central Canadian business domination of the wheat economy. Rather, they were

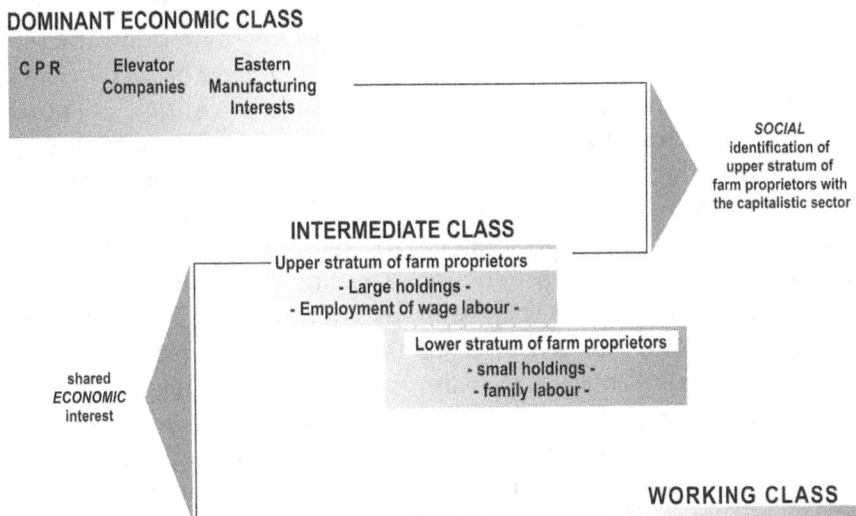

Fig. 55. Schematic representation of Abernethy's social and economic structure, ca. 1920.

part of the upper stratum of the farm proprietor group. By employing hired labour and farming extensive holdings, Abernethy farmers were relatively prosperous. Despite periodic outbursts against perceived corporate oppressors, they endeavoured to emulate the material culture of the business class, suggesting they identified socially with this group. Abernethy farmers' "conspicuous consumption" was a cheaper version of the lifestyle of the dominant class. It announced that they, too, had "made good." Actually, in the long run their economic interests were probably closer to those of the poorer farmers and farm labourers. Following the two-class model introduced in Chapter 5, these farm proprietor groups may be viewed collectively as an intermediate, transitional class. This class was essential to the capital accumulation envisioned in the National Policy, but under relentless assault of the price system it was short-lived.

The First World War represented a turning point in the social and economic structure of the prairie provinces. Many of the contradictory elements of Edwardian prairie society could no longer be reconciled as the old Victorian social order gave way to modern realities. Economically the war brought about a more complete integration of prairie agriculture into the national economy than had previously been the case. With the carrot of rapidly increasing agricultural prices before them, western farmers eagerly responded to the Dominion government's call for increased wheat production. Despite W.R. Motherwell's exhortations not to deplete their soils by overcropping, many farmers continuously cultivated wheat crops throughout the war period. Soil exhaustion was reflected in ever-diminishing yields,[2] but the farmers' revenues briefly maintained their high levels only because higher agricultural prices offset the reduced production. Many farmers who had purchased land and implements at inflated prices during the war found themselves burdened with excessive mortgage and loan payments by the early 1920s. Continuing low prices in that decade obliged many debt-ridden farmers to default on their financial commitments and to quit farming altogether. The well-established farmers were thus able to capitalize on their neighbours' failures and add to their holdings cheaply. By the early 1930s the effects of this consolidation of farmlands were manifest in the dramatic increase in average farm size in Abernethy and other farm districts.

While prairie farmers were slow to mechanize their operations during the First World War, a rapid switch to gasoline tractors occurred in the succeeding decade. This change was partly attributable to improvements in tractor design that translated into demonstrably improved efficiency and effectiveness. Generally, the switch to tractors both permitted and resulted from the consolidation of farms into larger units, enabling and then necessitating the capacity to work larger acreages more economically. Accordingly, the process of mechanization reduced the need for farm labourers, whose ranks became increasingly thin after 1930.

The passing of the era of small farms in which the 160-acre homestead was the basic unit represented a partial resolution of the contradiction between two themes that had accompanied prairie agriculture from the outset. The myth of the independent yeoman farmer, self-sufficient in providing a lifestyle and livelihood for himself and his family, was in direct opposition to the reality of accelerating consolidation of farming into large market-oriented units. Richard Hofstadter has observed of American farmers of the same era:

> The American farmer was unusual in the agricultural world in the sense that he was running a mechanized and commercialized agricultural unit of a size far greater than the small proprietary holdings common elsewhere, and yet he was running it as a family enterprise on the assumption that the family could supply not only the necessary capital and managerial talent but also most of the labour.[3]

While the more marginal farm operators were continually being forced out in the process of modernization, the social structure was sufficiently unstable that even members of the upper stratum were similarly vulnerable. The Motherwell family is a case in point. Motherwell's son Talmage moved off his father's farm in 1913. The recipient of two quarter-sections from his father, Talmage was promptly written out of his will. He had married Marian Diehl, the daughter of German immigrant parents from Minnesota and a future Christian Scientist. Thereafter, Talmage's family seldom visited Lanark Place. It seems probable that Motherwell's vision of ethnic and religious tolerance was put to a severe test by this marriage.

Talmage, for his part, was not an entrepreneur in the manner of his father. By all accounts he was a good family man, well liked by his neighbours. He did not expand his half-section inheritance, but concentrated on providing dairy milk for the village population. His nondescript stucco farmhouse, surrounded by a modicum of rudimentary landscaping, stood in sharp contrast to his father's elegant farmstead a mile down the road.

Talmage became a farmer in a period of decreasing opportunities in settled agricultural areas. Land prices had skyrocketed in the previous decade, and few persons, unless they homesteaded in the fringe areas of settlement or inherited land from their parents, could afford to start farming. Talmage's operation of his half-section farm represented a holding action, common among second-generation farmers, that persisted for some time. Yet Talmage's more modest ambitions also signified a fundamental shift in values in rural prairie society. Dick Harrison, in a comparative study of the fictional literature of homesteading on the American and Canadian plains, has noted the failures of domestic life occasioned by the pioneer experience. The settlers in the fiction of Ole Rolvaag and Frederick Philip Grove are "proud, stubborn, ambitious men, whose will to conquer is excited by the space and openness of the prairie."[4] Obsessed with their visions of prosperity and success, both Abe Spalding and Per Hansen were somewhat incapable of dealing with domestic affairs and of

> ... creating the homes that should have been the purpose of those visionary mansions. Abe and Per have brought to the prairies reluctant wives, temperamentally, who deteriorate physically or mentally as their husbands forge the material parts of their visions.[5]

To an extent these fictionalized accounts of the settlement experience, particularly Grove's, represent forces at work at the Motherwell farm and thousands of other homesteads in western Canada. Motherwell's first wife, Adeline, never strong physically and worn out by two decades of arduous labour on the homestead, succumbed to asthma in 1905. Motherwell's remarriage to Catherine Gillespie in 1908 brought to the farm a strong personality who was much better suited to the role of a cabinet minister's wife. But Talmage's departure in 1913 dashed Motherwell's hopes of leaving his farm to a male heir who would preserve his lineage. Like Abe Spalding, Motherwell must have experienced the profound disappointment of an Ontarian settler who had come west to rise in the social structure, only to see that his success would die with him. As Ian Clarke has noted, his son's marriage to a woman of non-Anglo-Canadian background presumably was another signal that the anticipated Anglo-Canadian hegemony in Saskatchewan was already in decline.[6]

Motherwell's Lanark Place in 1912 was an excellent example of a prosperous Ontarian prairie settler's farmstead at its peak. Its six quarter-sections comprised one of the largest farms in Abernethy, a rich wheat-growing district. The farm was crowned by a handsome stone house and impressively landscaped grounds, replete with lawn tennis court, ornamental flower beds and cropped hedges. To complete the picture of bucolic harmony, Motherwell possessed a family of considerable promise and achievement. His wife Catherine's achievements as a teacher, missionary, and promoter of women's rights were outstanding in their own right. His son Talmage had studied agriculture at the colleges in both Saskatoon and Guelph, and his daughter Alma had completed one year of her Normal School training. The farm also boasted two hired men and two hired girls, who, supplemented by seasonal staff, performed the bulk of manual labour on the farm. It was all the young settler from Ontario in 1882 could have hoped to achieve.

The farm in fact embodied all of the strengths and weaknesses of the Victorian Ontario mindset that Motherwell had brought with him to the North-West. The obvious conspicuous consumption of Lanark Place reflected a preoccupation with status that seemed inappropriate even for a prominent member of the upper stratum of this Edwardian farming community. The elaborate grounds belied the considerable expense entailed in their upkeep. It seems probable that after his ascension to political office, Motherwell's farming operations never fully underwrote the expense of maintaining the showpiece farm and that he dipped into his ministerial salary to maintain it.[7]

The year 1912, then, was a snapshot in time. It was a brief interlude between the early development of Anglo-Canadian society on the prairies and its subsequent erosion by forces greater than the settlers themselves could envisage. In restoring his homestead to its appearance in that year, Parks Canada is not only commemorating W.R. Motherwell's life and career, but also interpreting the historical role of the central Canadian settlement group he represented. When, in 1966, the Historic Sites and Monuments Board of Canada recommended that Motherwell be recognized as a figure of national historic significance, it also proposed that his farm be preserved to illustrate the home of a "typical prairie settler." This book has modified that initial perception. Motherwell was not typical of all settlers, but in important respects was representative of the upper stratum of independent farm proprietors within the dominant Anglo-Canadian settlement group. In this regard, he represents a group that has had an enormous impact in making western Canadian society what it is today.

Appendix A

Research Design for the Quantitative Analysis of Abernethy Settlement History

by David Greenwood, Parks Canada, Prairie and Northern Region

The quantitative analysis presented in Chapter I was composed of the following steps:

Data capture and verification
Data analysis
Data interpretation.

The *data capture and verification* consisted of recording on computer coding sheets information from: microfilmed Saskatchewan Department of the Interior Homestead Files; Surveyor's Township maps; a 1938 Soil Survey; and topographical maps which were used to measure distances from homesteads to railways and supply centres. Overall information was obtained from 461 entries in the Homestead Files at the Saskatchewan Archives Board and consisted of recording a total of 104 variables, according to a code book that was prepared to ensure consistent coding. The information on the sheets was then checked against the source documents to ensure accuracy, and to determine the interpretation used by the coder during situations not identified in the code book or in the notes accompanying the coding sheets. After this coding verification, the data were keypunched on computer cards and then two computer programs were run. The first printed a copy of the raw data from the cards and the second printed a frequency table for each of the 104 variables. These frequency tables provided a first level of analysis and were also used to check for extraneously coded values. For example, in considering the variable Range, it could only be assigned a value of 8 or 11. Therefore if we obtained a value of 6, we knew it was incorrect.

The printed data were then examined. A random sample of records was drawn and these were compared to the appropriate coding sheets. Any incorrectly coded values were identified and corrected on the cards. The two programs were then rerun. The first provided a reference copy of the raw data, and the second created nine new variables:

i) number of years between entry and application/cancellation for patent;
ii) number of years between when homesteader started to break land and application/cancellation;
iii) average number of acres broken per year;
iv) average number of acres cropped per year;
v) average number of cows;
vi) average number of horses;
vii) average number of sheep;
viii) average number of pigs;
ix) a variable, calculated from the Ethnicity and Geographic Range variables, which could take the four possible values of British Range 8, Non-British Range 8, British Range 11 and Non-British Range 11. The output from [his second program consisted of frequency fables for the 113 variables and can also be used as a reference document.

The remainder of the *data analysis* consisted of running computer programs using the Statistical Analysis System (S.A.S.) computer package. The S.A.S. procedures used include:

Frequencies provided the frequency tables as described previously and was used to generate various *cross-tabulations* which provide insights into the similarities or differences between specific sub-groups of the data set such as British in Range 11 versus Non-British in Range 8.

Means calculated the means of specified variables, such as year of entry, year of application and average number of children per family and was run for the entire data set and for specific groups of files in the data set. For example, the average year of entry for the Abernethy area was 1890 and for the Neudorf area was 1896.

Sort was used to sort the data set alphabetically according to the variables specified, for example LAST NAME. This SORT procedure facilitated the remainder of the analysis as the more complex S.A.S. procedures require the data set to be arranged in some type of sequence in order to analyze subgroups of the data set.

The remaining procedures attempted to find a regression model which could explain some of the variation in the year of entry for homestead application in the study areas. The model was used to determine the amount of variability in the year of entry, which is accounted for by the different values of the other variables included in the model and is measured by the statistic R^2, the square of the multiple correlation coefficient. An example of a simple regression is to consider the change in the year of entry in relation to the change in another variable such as distance to the railway.

R square was used to provide information on variables which could be included in a multiple regression model. Numerous variables were specified (usually eight or nine at a time) and combinations of these variables were used as input to this procedure which calculates the portion of the variability in the dependent variable, in this case the year of entry explained by the model specified, and allows to specify numerous possible models at one time. The most appropriate model or models were then chosen for further investigation. In this case the model chosen contained the variables *range, township, distance to railway at entry, distance to railway at application and distance to supply centres.*

GLM (General Linear Models) was used to estimate the coefficients in the model which was deemed most appropriate in the R square procedure. A test of significance was run in order to determine if the estimates of the coefficients were significantly different from zero, and as expected from the R square procedure we found that they were indeed significantly different from zero.

NOTES

Preface

1. Nathaniel Benson, *None of It Came Easy: The Story of James Garfield Gardiner* (Toronto: Burns, 1955).
2. Annie I. Yule, *Grit and Growth: The Story of Grenfell* (Grenfell, SK: United Church Women's Committee, 1967).
3. Ivor J. Mills, *Stout Hearts Stand Tall* (Vancouver: author published, 1971).
4. See Allan R. Turner, "W.R. Motherwell and Agricultural Development in Saskatchewan, 1905–1918," M.A. thesis, University of Saskatchewan, Saskatoon, 1958; and Leonard J. Edwards, "W.R. Motherwell and the Crisis of Federal Liberalism in Saskatchewan, 1917–1926." M.A. thesis, University of Saskatchewan, Saskatoon, 1961.
5. David Spector, "Agriculture on the Prairies, 1870–1940," in History and Archaeology/Histoire et archaeologie, No. 65 (Ottawa: Parks Canada, 1983).

Introduction to the Revised Edition

1. The statistics are compiled from population figures given in Robert England, *The Colonization of Western Canada* (London: P.S. King and Son, Ltd., 1936), Appendix B, pp. 310–11.
2. W.A. Mackintosh, *Prairie Settlement: The Geographical Setting* (Toronto: Macmillan of Canada, 1934); C.A. Dawson, *Settlement of the Peace River Country: A Study of a Pioneer Area* (Toronto: Macmillan of Canada, 1934); W.A. Mackintosh, *Economic Problems of the Prairie Provinces* (Toronto: Macmillan of Canada, 1935); C.A. Dawson, *Group Settlement: Ethnic Communities in Western Canada* (Toronto: Macmillan of Canada, 1936); R.W. Murchie, *Agricultural Progress on the Prairie Frontier* (Toronto: Macmillan of Canada, 1936); Arthur S. Morton, *History of Prairie Settlement*, and Chester Martin, *Dominion Lands Policy* (Toronto: Macmillan of Canada, 1938).
3. John A. Irving, *Social Credit Movement in Canada* (Toronto: University of Toronto Press, 1959); C.B. Macpherson, *Democracy in Alberta: Social Credit and the Party System* (Toronto: University of Toronto Press, 1953); Vernon C. Fowke, *The National Policy and the Wheat Economy* (Toronto: University of Toronto Press, 1951); Jean Burnet, *Next-Year Country: A Study of Rural Social Organization in Alberta* (Toronto: University of Toronto Press, 1951).
4. Allan Bogue, "Social Theory and the Pioneer," *Agricultural History*, Vol. 34 (January 1960), pp. 21–34; Robert P. Swierenga, "Towards the 'New Rural History': A Review Essay," *History Methods Newsletter*, Vol. 6, No. 3 (June 1973), pp. 111–21, and Robert P. Swierenga, "The New Rural History: Defining Parameters," *Great Plains Quarterly*, Vol. 1 (Fall 1981), pp. 211–23; John Gjerde, *From Peasants to Farmers: the Migration from Balestrand, Norway, to the Upper Middle West* (Cambridge, U.K./New York: Cambridge University Press, 1985); Robert Clifford Ostergren, *A Community Transplanted: the Trans-Atlantic Experience of a Swedish Immigrant Settlement in the Upper Middle West, 1835–1915* (Madison: University of Wisconsin Press, 1980); Merle Curti, *The Making of an American Community: A Case Study of Democracy in a Frontier County* (Stanford, California: Stanford University Press, 1959); Michael P. Conzen, *Frontier Farming in an Urban Shadow: The Influence of Madison's Proximity on the Agricultural Development of Blooming Grove, Wisconsin* (Madison: State University of Wisconsin, 1971).

5. Allan G. Bogue, "Farming in the North American Grasslands: A Survey of Publications, 1947-80," *The Great Plains Quarterly*, Vol. 1 (Spring 1981), pp. 105-131.

6. Marc Bloch, *French Rural History: An Essay on its Basic Characteristics* (trans. Janet Sondheimer) (London: Routledge and Kegan Paul, 1966); Emmanuel Le Roy Ladurie, *Montaillou: The Promised Land of Error* (trans. Barbara Bray) (New York: Vintage Books, 1979); Fernand Braudel, *The Mediterranean and the Mediterranean World in the Age of Philip II*, Vols. I and II (trans. Sîan Reynolds) (New York: Harper Torchbooks, 1975).

7. John W. Bennett, *Northern Plainsmen: Adaptive Strategy and Agrarian Life* (Chicago, Illinois: Aldine Pub. Co., 1969).

8. Giovanni Levi, 'On Microhistory,' in Peter Burke, ed., *New Perspectives on Historical Writing*, 2nd ed. (Cambridge, UK.: Polity Press, 2001), pp. 97-119; and Carlo Ginzburg, "Microhistory: two or three things that I know about it," *Critical Inquiry*, Vol. 20 (Autumn 1993), pp. 10-35.

9. See, for example, István Szijártó, "Four Arguments for Microhistory," *Rethinking History*, Vol. 6, issue 2 (June 2002), pp. 209-15; Alf Lüdtke, ed., *The History of Everyday Life. Reconstructing Historical Experiences and Ways of Life* (trans. William Templer) (Princeton, NJ: Princeton University Press, 1995); Brad S. Gregory, "Is Small Beautiful? Microhistory and the History of Everyday Life," *History and Theory*, Vol. 38, No. 1 (February 1999), pp. 100-110.

10. Richard D. Brown, "Microhistory and the Postmodern Challenge," *Journal of the Early Republic*, Vol. 23 (Spring 2003), p. 17.

11. Lyle Dick, *Farmers "Making Good": The Development of Abernethy District, 1880-1920* (Ottawa: Canadian Parks Service, 1989); Paul Voisey, *Vulcan: The Making of a Prairie Community* (Toronto: University of Toronto Press, 1988); Royden Loewen, *Family, Church, and Market: A Mennonite Community in the Old and New Worlds* (Toronto: University of Toronto Press, 1993); and Ken Sylvester, *The Limits of Rural Capitalism: Family, Culture, and Markets in Montcalm, Manitoba, 1870-1940* (Toronto: University of Toronto Press, 2001).

12. Lyle Dick, "Motherwell Historic Site: The Social and Economic History of the Abernethy District, Saskatchewan, 1880-1920: Bibliography, Historiography, and Methodology," Parks Canada, Manuscript Report Series No. 320, Ottawa, 1979, pp. 1-70.

13. See, for example, Kenneth Norrie, "Dry Farming and the Economics of Risk Bearing: The Canadian Prairies, 1870-1930," in Thomas R. Wessell, ed., *Agriculture in the Great Plains*, and "The National Policy and the Rate of Prairie Settlement: A Review," in R. Douglas Francis and Howard Palmer, eds., *The Prairie West: Historical Readings* (Edmonton: Pica Pica Press, 1995), pp. 243-63; William Marr and Michael Percy, "The Government and the Rate of Canadian Prairie Settlement," *Canadian Journal of Economics / Revue canadienne d'Economique*, Vol. 11, No. 4 (November 1978), pp. 757-767; Frank Lewis, "Farm Settlement on the Canadian Prairies, 1898 to 1911," *Journal of Economic History*, Vol. 41 (September 1981), pp. 517-35.

14. Minute Book of the Territorial Grain Growers Association of Abernethy, 1904-05 and the Abernethy Grain Growers Association, 1905-1925. Copy in the possession of Mr. Bert Garratt of Abernethy, Saskatchewan.

15. Lyle Dick and Jean-Claude LeBoeuf, "Social History in Architecture: The Stone House of W.R. Motherwell," *Research Bulletin/Bulletin de recherches*, No. 122 (Ottawa: Environment Canada, Canadian Parks Service, 1980).

16. Randy William Widdis, *With Scarcely a Ripple: Anglo-Canadian Migration into the United States and Western Canada, 1850-1920* (Montreal and Kingston: McGill-Queen's University Press, 1998), p. xxii.

17 While definitions abound for this term, I will rely on a current straightforward statement of meaning in the *Merriam-Webster On-Line Dictionary*: "the theory or study of the role of public policy in influencing the economic and social welfare of a political unit." <http://www.m-w.com/dictionary/political+economy>

18 Jeremy Adelman, *Frontier Development: Land, Labour and Capital on the Wheatlands of Argentina and Canada, 1890–1914* (Oxford and New York: Clarendon Press, 1994).

19 Beth LaDow, *The Medicine Line: Life and Death on a North American Borderland* (New York and London: Routledge, 2001); Stirling Evans, ed., *The Borderlands of the American and Canadian Wests: Essays on Regional History of the Forty-ninth Parallel* (Lincoln: University of Nebraska Press, 2006); and Sheila McManus, *The Line Which Separates: Race, Gender, and the Making of the Alberta-Montana Borderlands* (Lincoln: University of Nebraska Press, 2005), and "'Their Own Country': Race, Gender, Landscape, and Colonization around the Forty-ninth Parallel, 1862–1900," in Stirling Evans, ed., *The Borderlands of the American and Canadian Wests*, pp. 117–30.

20 Warren M. Elofson, *Cowboys, Gentlemen, and Cattle Thieves: Ranching on the Western Frontier* (Montreal and Kingston: McGill-Queen's University Press, 2000).

21 Sarah Carter, Lesley Erickson, Patricia Roome, and Char Smith, eds., *Unsettled Pasts: Reconceiving the West Through Women's History* (Calgary: University of Calgary Press, 2005). See, in particular, the essay by Carter, "'Complicated and Clouded': The Federal Administration of Marriage and Divorce Among the First Nations of Western Canada, 1887–1906," pp. 151–78.

22 Sheila McManus, *The Line Which Separates: Race, Gender, and the Making of the Alberta-Montana Borderlands* (Lincoln: University of Nebraska Press, 2005), and "'Their Own Country': Race, Gender, Landscape, and Colonization around the Forty-ninth Parallel, 1862–1900," in Stirling Evans, ed., *The Borderlands of the American and Canadian Wests*, pp. 117–30.

23 Thematically driven studies tend to exhibit deductive approaches to data collection and synthesis, that is, approaches in which the conclusions are largely known at the outset and necessitated from previously known facts. Conversely, microhistorical studies tend to proceed according to a combination of induction, or the development of possible explanatory frameworks based on initial empirical observations, and abduction, a process of adopting, discarding, and refining hypotheses in the course of selecting suitable interpretive frameworks. Each selected heuristic framework then guides further data collection, as the abductive approach is repeated throughout the research and even the writing phase. See Lyle Dick, "Microhistory: Does it Work?", Paper presented to the Canadian Historical Association Conference, London, Ontario, 1 June 2005.

24 For a very good exposition on the value of microhistorical approaches for rural settlement histories in Canada, see R.W. Sandwell, *Contesting Rural Space: Land Policy and Practices of Resettlement on Saltspring Island, 1859–1891* (Montreal and Kingston: McGill-Queen's University Press, 2005), especially "Introduction," pp. 3–14.

25 On the continuing gap between the reality of diversity of family models, and the continuing conservatism of approach in mainstream academic discourse, especially in interdisciplinary family studies, see Katherine R. Allen, "A Conscious and Inclusive Family Studies," *Journal of Marriage and the Family*, Vol. 62 (February 2000), pp. 4–17. For a critique of traditional models in Canadian family historiography, see the essays edited by Nancy Christie and Michel Gauvreau, *Mapping the Margins: The Family and Social Discipline in Canada, 1700–1975* (Montreal and

Kingston: McGill-Queen's University Press, 2004).

26 Sarah Carter, "'Complicated and Clouded': The Federal Administration of Marriage and Divorce Among the First Nations of Western Canada, 1887–1906," in Sarah Carter, Lesley Erickson, Patricia Roome, and Char Smith, eds., *Unsettled Pasts* (Calgary: University of Calgary Press, 2005), pp. 151–78.

27 See, for example, Lyle Dick, "Male Homosexuality in Saskatchewan's Settlement Era: The 1895 Case of Regina's Oscar Wilde," Paper presented to the Annual Conference of the Canadian Historical Association, York University, Toronto, May 2006.

28 Walter Prescott Webb, *The Great Plains: A Study in Institution and Environment* (Boston: Ginn and Co., 1931); James C. Malin, *The Grassland of North America: Prolegomena to its History* (Gloucester, MA: P. Smith, 1967).

29 John W. Bennett and Seena Kohl, *Settling the Canadian-American West, 1890–1915: Pioneer Adaptation and Community Building: An Anthropological History* (Lincoln: University of Nebraska Press, 1995).

30 Barry Potyondi, *In Palliser's Triangle: Living in the Grasslands, 1850–1930* (Saskatoon: Purich Publishing, 1995).

31 David C. Jones, *Empire of Dust: Settling and Abandoning the Prairie Dry Belt*, 2nd ed. (Calgary: University of Calgary Press, 2002).

32 Rod Bantjes, *Improved Earth: Prairie Space as Modern Artefact, 1869–1944* (Toronto: University of Toronto Press, 2005).

33 Lyle Dick, "A History of Prairie Settlement Patterns, 1870–1930," Parks Canada, Microfiche Report No. 307 (Ottawa: Environment Canada, Parks, Prairie and Northern Region, 1987), especially chapter 1, pp. 1–12.

34 Carl Tracie, *"Toil and Peaceful Life": Doukhobor Village Settlement in Saskatchewan, 1899–1918* (Regina: Canadian Plains Research Centre, 1996); John Warkentin, "The Mennonite Settlements of Southern Manitoba," PhD Dissertation, University of Toronto, 1960; and John C. Lehr, "Mormon Settlements in Southern Alberta," M.A. thesis, University of Alberta, Edmonton, 1971.

35 Ian Clarke, "Motherwell Historic Park: Structural and Use History of the Landscape and Outbuildings," in Ian Clarke, Lyle Dick, and Sarah Carter, *Motherwell Historic Park*, History and Archaeology Series no. 66 (Ottawa: Parks Canada, 1983), pp. 1–212; and Lyle Dick, "A History of Prairie Settlement Patterns," Chapter 4, "Individual Dispersed Settlement Patterns on the Prairies, 1870–1930."

36 See Lyle Dick, "A History of Prairie Settlement Patterns," Chapter 4, 'Individual Dispersed Settlement Patterns on the Prairies,' pp. 77–127; and Lyle Dick and Jean-Claude LeBoeuf, "Social History in Architecture: The Stone House of W.R. Motherwell," *Research Bulletin/Bulletin de recherches*, No. 122 (Ottawa: Environment Canada, Canadian Parks Service, 1980).

37 Sarah Carter, "A Materials History of the Motherwell Home," in Ian Clarke, Lyle Dick, and Sarah Carter, *Motherwell Historic Park*, History and Archaeology Series no. 66 (Ottawa: Parks Canada, 1983), pp. 265–353.

38 Some of Glassie's seminal work appeared in his *Folk Housing in Middle Virginia: A Structural Analysis of Historic Artifacts* (Knoxville: University of Tennessee Press, 1975); *Patterns in the Material Folk Culture of the Western United States* (Philadelphia: University of Philadelphia Press, 1969); and *Vernacular Architecture* (Philadelphia: Material Culture/Bloomington: Indiana University Press, 2000). See also Michael P. Conzen, "Ethnicity on the Land," in Michael P. Conzen, ed., *The Making of the American Landscape* (Boston: Unwin Hyman, 1990), pp. 221–48; Terry G. Jordan, Jon T. Kilpinen, and Charles F. Gritzner, *The Mountain West: Interpreting the Folk Landscape* (Baltimore and London: Johns Hopkins University Press, 1997); and Terry G. Jordan, and Matti

Kaups, *The American Backwoods Frontier: An Ethnic and Ecological Interpretation* (Baltimore: Johns Hopkins University Press, 1989).

39   See, for example, Edward M. Ledohowski and David K. Butterfield, *Architectural Heritage: The Eastern Interlake Planning District* (Winnipeg: Manitoba Department of Cultural Affairs and Historic Resources, 1983); and *Architectural Heritage: The MSTW Planning District* (Winnipeg: Manitoba Department of Cultural Affairs and Historic Resources, 1984); John C. Lehr, *Ukrainian Vernacular Architecture in Alberta*. Occasional Paper No. 1 (Edmonton: Alberta Culture, Historic Sites Service, 1976); Sonia Maryn, *The Chernochan Machine Shed: A Land Use and Structural History*. Occasional Paper No. 12 (Edmonton: Alberta Culture, Historic Sites Service, 1985).

1   THE SETTLEMENT OF THE ABERNETHY DISTRICT

1   M.C. Wilson and I.J. Dijks, "Introductory essay – Land of no quarter, the Palliser Triangle as an environmental-cultural pump," in R.W. Barendrgt, M.C. Wildon, and F.J. Jakunis, eds., *The Palliser Triangle, A Region in Space and Time* (Lethbridge, AB: University of Lethbridge, 1993); "Fescue Prairie," Environment Canada, <http://www.pnr.ec.gc.ca/nature/whp/prgrass/df03s34.en.html>

2   Randy William Widdis, *With Scarcely a Ripple: Anglo-Canadian Migration into the United States and Western Canada, 1880–1920* (Montreal and Kingston: McGill-Queen's University Press, 1998), pp. 179–289.

3   Saskatchewan Archives Board, Regina [hereafter cited as SABR], Surveyors' Correspondence Series. Report of C.F. Miles, Toronto, to the Surveyor General, Ottawa, 17 January 1882.

4   Canada, *Dominion Lands Surveyors' Reports* (Ottawa: Queen's Printer, 1886), p. 111.

5   Chester Martin, "*Dominion Lands*" *Policy*, Canadian Frontiers of Settlement Series, Vol. 2 (Toronto: Macmillan, 1938), p. 395.

6   *Statutes of Canada*, "An Act Respecting the Canadian Pacific Railway," assented to 14 June 1872 (Ottawa: Queen's Printer, 1873) [hereafter cited as SC], 35 Victoria, Cap. 71.

7   Chester Martin, "*Dominion Lands*" *Policy*, p. 415.

8   Vernon C. Fowke, *The National Policy and the Wheat Economy* (Toronto: University of Toronto Press, 1957), p. 48.

9   James M. Richtik, "The Policy Framework for Settling the Canadian West, 1870–1880," *Agricultural History*, Vol. 49, No. 4 (October 1975), p. 617.

10   Ibid., p. 620.

11   "Homestead Regulations," *Qu'Appelle Vidette*, 16 April 1897, p. 2.

12   Arthur J. Ray, *Indians in the Fur Trade: Their Role as Trappers, Hunters, and Middlemen in the Lands Southeast of Hudson Bay, 1660–1870* (Toronto: University of Toronto Press, 1974), 3–23.

13   Liz Bryan, *The Buffalo People: Prehistoric Archaeology on the Canadian Plains* Edmonton: University of Alberta Press, 1991).

14   Alexander Morris, *The Treaties of Canada With the Indians* (Toronto: Belfords, Clarke & Co., 1880), pp. 77–125.

15   Archives of Manitoba [hereafter cited as AM], Alexander Morris Papers, passim.

16   Articles of the Qu'Appelle Treaty, No. 4, in Morris, *The Treaties of Canada With the Indians*, pp. 330–35.

17   J.L. Tobias, "Canada's Subjugation of the Plains Cree, 1879–1885," *Canadian Historical Review*, Vol. 50, No. 4 (December 1983), p. 547.

18   Arthur R. Ray, Jim Miller, and Frank Tough, *Bounty and Benevolence: A History of Saskatchewan Treaties* (Montreal and Kingston: McGill-Queen's University Press, 2000), pp. 105–20.

19. See Jacqueline J. Kennedy, "Qu'Appelle Industrial School: White Rites for the Indians of the Old North-West." M.A. thesis, Carleton University, Ottawa, 1970. The larger story of efforts by the Cree of the Qu'Appelle region to pursue agriculture, and federal government's efforts to discourage it, is extensively covered in Sarah Carter's *Lost Harvests: Prairie Indian Reserve Farmers and Government Policy* (Montreal: McGill-Queen's University Press, 1990).

20. See Jacqueline J. Kennedy, "Qu'Appelle Industrial School: White Rites for the Indians of the Old North-West." M.A. thesis, Carleton University, Ottawa, 1970.

21. Saskatchewan Archives Board, Saskatoon [hereafter cited as SABS], Department of the Interior, Homestead File Nos. 406350 and 351–17 (files relating to the homestead of Culbert St. Denis).

22. Canada, Privy Council, Order-in-Council, P.C. 135-1885. *Report of a Committee of the Privy Council*, 28 January 1885; and Canada, Privy Council, Order-in-Council, P.C. 821-1885. *Report of a Committee of the Privy Council*, 18 April 1885 (Ottawa: King's Printer, 1886).

23. D.N. Sprague, "Government Lawlessness in the Administration of Manitoba Land Claims, 1870–1887," *Manitoba Law Journal*, Volume 10, Number 4 (1980), 415–41; Gerhard Ens, "Métis Lands in Manitoba," *Manitoba History*, No. 5 (Spring 1983), 2–11.

24. SABS, Homestead File No. 97132, "Scrip issue by the North West Half Breed Commission in 1885 at Fort Qu'Appelle to holders of Water Fronts."

25. *Qu'Appelle Vidette* 2, 1 January 1886, p. 4.

26. Library and Archives Canada [hereafter cited as LAC], RG 15, Department of the Interior Records, Vol. 292, File No. 58906, Petition of Qu'Appelle Métis to the Hon. Edgar Dewdney, Lieutenant-Governor, N.W.T.

27. Ibid., Vol. 292, File 58906, The Hon. Edgar Dewdney to Sir John A. Macdonald, Minister of the Interior, 29 August 1882.

28. LAC, RG15, Department of the Interior Records, Vol. 292, File No. 58906, John R. Hall, Acting Secretary, Dept. of the Interior, to Hon. Edgar Dewdney, N.W.T., 6 July 1883.

29. Ibid.

30. LAC, RG15, Department of the Interior Records, Vol. 229, File No.63883.

31. SABS, Homestead File No. 118650, Norbert Welsh, Lebret, to the Hon. Edgar Dewdney, 5 June 1886, Transcription.

32. Ibid., A.E. Forget, Clerk of Council, N.W.T. to the Deputy Minister of the Interior, 24 June 1886.

33. Ibid., Report of W.A. Clarke, Fort Qu'Appelle to the Hon. Edgar Dewdney, 3 December 1886.

34. Canada, Department of Agriculture, Indian Head Experimental Research Station. "Map of Indian Head and Adjacent District," ca. 1906.

35. Interview with Hugh and Wanda Stueck, by Lyle Dick, Abernethy, Sask., 18 April 1979. Parks Canada, Western and Northern Service Centre [hereafter cited as Parks Canada, WNSC]

36. Georgina Binnie-Clark, *Wheat and Women* (Toronto: Bell and Cockburn, 1914), p. 166.

37. Angelina H. Campbell, *Man! Man! Just Look at That Land!* (Saskatoon: n.p., 1966), p. 11.

38. See the Journal of John Allen, one of the original settlers on the Primitive Methodist Colony northeast of Abernethy, which describes the trek west in detail. SABR. M. Film 2.171.

39. For western Canada, a general starting-off point is W.L. Morton's article, "The Significance of Site in the Settlement of the American and Canadian Wests," *Agricultural History*, Vol. 25 (1951), pp. 97–104.

40. Michael Conzen, *Frontier Farming in an Urban Shadow: The Influence of Madison's Proximity on Agriculture Development of Blooming Grove,*

Wisconsin (Madison: The State Historical Society of Wisconsin, 1971).

41 Data on the dates of entry for the study townships were obtained from the Township Registers of the Saskatchewan Department of Agriculture, Lands Branch, Regina.

42 See André N. Lalonde, "The North-West Rebellion and Its Effects on Settlers and Settlement in the Canadian West," *Saskatchewan History*, Vol. 27, No. 3 (Autumn 1974), pp. 95–102.

43 See Lottie Meek's unpublished note on the early history of the Blackwood District, Saskatchewan, 1955, pp. 3, 4. Copy held by Environment Canada, Canadian Parks Service, Prairie Region, Library.

44 *Melville Advance*, 29 June 1955, p. 1.

45 Don C. McGowan, *Grassland Settlers, The Swift Current Region During the Era of the Ranching Frontier* (Regina: Canada Plains Research Centre, 1975), p. 49.

46 LAC, RG17, Department of Agriculture Records, Vol. 466, File No. 51015, Annual Report of A.J. Baker, Immigration Agent, Qu'Appelle to the Minister of Agriculture, 31 December 1885.

47 A.S. Morton, *History of Prairie Settlement*. Canadian Frontiers of Settlement Series (Toronto: Macmillan, 1938), Vol. 2, p. 84.

48 *Census of Population and Agriculture of the North West Provinces, Manitoba ..., 1885* and *Census of Canada, 1891* (Ottawa: Queen's Printer, 1886 and 1892).

49 Report of William Pearce, Dominion Lands Inspector, 31 October 1882, in Canada, Parliament, *Sessional Papers*, 46 Victoria, 1883, Cap. 23, p. 7.

50 LAC, RG15, Department of the Interior Records, Vol. 326, File No. 80391 (2), Report of Rufus Stevenson, Inspector to A. Walsh, Dominion Lands Commissioner, 4 December 1884.

51 Ibid., File 80391 (1) A, Rufus Stevenson to Smith, Dominion Lands Commissioner, 11 January 1886.

52 "Pheasant Plains," *Qu'Appelle Progress*, 1 January 1885, p. 3.

53 LAC, RG15, Department of the Interior Records, Vol. 247, File No. 25149-2.

54 Ibid., "To His Excellency the Governor-General of Canada in Council," Received, Dept. of Interior, 10 July 1885.

55 SABS, Department of the Interior, Orders-in-Council, 1885. Microfilm No. 6.1, p. 307.

56 LAC, RG15, Department of the Interior Records, Vol. 247, File 25149-4, "The Petition of the President and the Directors representing the present Shareholders in the Great North West Central Railway Company to His Excellency the Governor General of Canada in Council."

57 LAC, RG15, Department of the Interior Records, Vol. 247, File 25149-4, "The Petition of the President and the Directors representing the present Shareholders in the Great North West Central Railway Company to His Excellency the Governor General of Canada in Council."

58 Ibid.

59 Ibid.

60 Gustavus Myers, *A History of Canadian Wealth*, Vol. 1 (Toronto: James Lewis and Samuel, 1972), p. 286.

61 AM, MG4, B4, Sir Wilfrid Laurier Papers, Vol. 224, fols. 63029- 63032, Petition of Abernethy Settlers to Sir Wilfrid Laurier, 2 February 1902.

62 *The Abernethan* (Abernethy), 23 August 1905, p. 1.

63 Ibid., 30 August 1905, p.1.

64 SABS, Homestead File No. 36741, Jacob Popp to the Dominion Lands Commissioner respecting his homestead (S.W. 114 of Section 32, Township 20, Range 8, West of the 2nd Meridian), 30 October 1894.

65 Ibid., File No. 510548, B.P. Richardson, D.L.S. Officer, Grenfell, to the Dominion Lands Commissioner respecting the application of Philip Tempel and John Pipher, 27 December 1892.

66 Ibid., File No. 455760, Karl Adolf, Pheasant Forks, to the Dominion Lands Commissioner, 8 January 1898.

See also the Declaration of Abandonment of Ludwig Hollinger and Frank Wirth in SABS, Homestead File Nos. 698266 and 795174.

67 Interview with Richard A. Acton by Lyle Dick, 24 June 1979, Parks Canada, WNSC. Mr. Acton, born in 1894, is the son of Samuel Acton who homesteaded in Township 19, Range 9, West of the 2nd Meridian.

68 Ibid.

69 Interview with John Mann, born 1896. Cited in Janice Acton, comp., *Lemberg Local History* (Lemberg, SK: n.p., 1972), pp. 111–15.

70 Ibid.

71 LAC, RG15, Department of the Interior Records, Vol. 724, File No. 394 194, Extract of letter from Senator W.D. Perley, Wolseley, N.W.T., to the Department of the Interior dated 29 November 1895.

72 *Regina Leader*, 13 September 1894, p. 8.

73 SABR, *Cummins Rural Directories*, 1915–1918.

74 Interview with Richard A. Acton, in Lemberg, Saskatchewan, by Lyle Dick, Parks Canada, WNSC, 27 June 1979.

75 Canada, Department of the Interior, *Annual Report for 1896*. Part 4, "Immigration" or see Canada, Parliament, *Sessional Papers*, 60 Victoria, 1897, Vol. 13.

76 Ibid.

77 SC, "An Act to Amend the Dominion Lands Act," assented to 19 April 1884, 47 Victoria, Cap. 25, S. 2, p. 104.

78 See Canada, Parliament, *House of Commons Debates*, 25 April 1889 (Ottawa: Queen's Printer, 1890), pp. 1537–38.

79 SABS, Homestead File No. 227164. Memorial of the Legislative Assembly of the North-West Territories to the Honourable Minister of the Interior, Regina, 21 November 1889.

80 SC, "An Act Further to Amend the Dominion Lands Act," assented to 9 July 1892. 56–7 Victoria, Cap. 15, S. 4, pp. 75, 76.

81 SABS, Homestead File No. 461413, Synopsis of contents of letter from Karl Hopp, Neudorf, Assiniboia, to the Commissioner of Dominion Lands, 27 December 1898.

82 Ibid., L. Pereira, Assistant Secretary, Dominion Lands Branch, to Karl Hopp, 13 January 1899.

83 Ibid., Homestead File No. 3677410, Jacob Popp, Neudorf, Assiniboia, to the Dominion Lands Commissioner, 30 October 1894; and File No. 461413, Report of Neil G. McCallum, Homestead Inspector, 2 August 1899.

84 Ibid., File No. 461413, Report of Neil G. McCallum, Homestead Inspector, 2 August 1899.

85 Ibid., File No. 511449.

86 Ibid., File No. 555277, L. Pereira to J. Adolf, 6 April 1895.

87 Ibid., File No. 463816, G. Balfour to the Minister of the Interior, 28 February 1889.

88 Ibid., File No. 126610.

2 Estimates of Homsteading Costs

1 Robert E. Ankli and Robert M. Litt, "The Growth of Prairie Agriculture: Economic Considerations," in Donald H. Akenson (ed.), *Canadian Papers in Rural History* (Gananoque, ON: Langdale Press, 1978), Vol. 1, pp. 35–64.

2 David Gagan, "Land Population and Social Change: The 'Critical Years' in Rural Canada West," *Canadian Historical Review*, Vol. 59, No. 3 (1978), pp. 295–318; and "The Indivisibility of Land in Nineteenth-Century Ontario," *Journal of Economic History*, Vol. 36 (1976), pp. 126–241. Gagan has noted that cash inheritances averaged $450 for sons who were not the principal heirs of Ontario farmers. One child in five received nothing at all. Gagan, "The Indivisibility of Land," p. 136.

3 J.F. Snell, "The Cost of Living in Canada in 1870," *Social History*, Vol. 12, No. 23 (May 1979), p. 189. Wage rates for Ontario farm labourers for four localities ranged between $12

and $15 per month, with board, and averaged $13.50. One Qu'Appelle area settler from Toronto described a year's earnings in Ontario at $200. See the testimonial of Samuel Copithorn, *Qu'Appelle Vidette*, 2 March 1893. Even the more arduous tasks of threshing by cradle brought only about $1.50 per day, or about $40 per month during harvest season, and summer wage rates generally ranged between $1 and $1.50 in the early 1870s in Ontario. Richard Pomfret, "The Mechanization of Reaping in Nineteenth Century Ontario: A Case Study of the Pace and Causes of the Diffusion of Embodied Technical Change," *Journal of Economic History*, Vol. 36, No. 2 (June 1976), Appendix C, p. 414.

4   SABR, Local Histories File "Qu'Appelle," Annual Report of A.J. Baker, Immigration Agent, Qu'Appelle to the Dominion Minister of Agriculture, 31 December 1888.

5   John Thompson, "Bringing in the Sheaves: The Harvest Excursionists, 1890–1929," *Canadian Historical Review*, Vol. 59, No. 4 (Summer 1978), p. 488.

6   John Burton, Letter to the Editor, *Qu'Appelle Vidette*, 2 March 1889.

7   "Started on $200," *Qu'Appelle Vidette*, 3 March 1889.

8   See, for example, "Experience and Opinions of Settlers," *Manitoba Official Handbook* (Liverpool: 1892), pp. 40–44.

9   *Regina Leader*, Supplement, 9 April 1889.

10 Ibid., 19 May 1898.

11 Ibid., "Successful Colonization in the German Colony of New Toulcha," 4 December 1888, p. 8.

12 *Western World*, Winnipeg, March 1891, p. 63.

13 SC, "An Act Respecting Public Lands," 49 Victoria, 1886, Cap. 54, S. 52, p. 832.

14 The average proving-up period at Abernethy was more than five years.

15 James B. Hedges, *Building the Canadian West* (New York: Russell & Russell, 1939), pp. 67–68.

16 James M. Minifie, *Homesteader* (Toronto: Macmillan, 1973), p. 140.

17 Annie I. Yule, *Grit and Growth: The Story of Grenfell*, p. 27.

18 Ibid., p. 76.

19 *Letters from a Young Emigrant in Manitoba* (London: Kegan Paul, 1883), p. 97.

20 W.M. Elkington, *Five Years in Canada* (London: Whittaker & Co., 1895), p. 135.

21 *Western Canada: Manitoba, Alberta, Assiniboia, Saskatchewan, and New Ontario* (immigration pamphlet), 1902, p. 68. Copy in Manitoba Legislative Library.

22 Herman Ganzevoort, translator and editor, *A Dutch Homesteader on the Prairies* (Toronto: University of Toronto Press, 1973), pp. 27–29.

23 *Revised Statutes of Canada*, "An Act Respecting Public Lands." 49 Victoria, Cap. 54, S. 38, s.s. 6 & 7, pp. 830–31, 1886.

24 *Qu'Appelle Vidette*, 5 November 1891, p. 4.

25 Isaac Cowie, *The Agricultural and Mineral Resources of the Edmonton Country* [hereafter cited as *Edmonton Country*] (Edmonton: Western Canada Immigrant Association), 1901, p. 51.

26 Georgina Binnie-Clark, *Wheat and Women*, p. 60.

27 Isaac Cowie, *Edmonton Country*, p. 51.

28 *The Star Almanac* 1893 (Montreal: Hugh Graham, 1893), p. 255, "Hints to Intending Settlers"; *Letters from a Young Emigrant ...*, pp. 96–98; John Macoun, *Manitoba and the Great North-West* (Guelph: World Publishing Co., 1882), "Advice to Immigrants," pp. 637–39.

29 Isaac Cowie, *Edmonton Country*, p. 51.

30 John Macoun, *Manitoba and the Great North-West*, pp. 637–39.

31 *Letters from a Young Emigrant ...*, p. 130.
32 *Western Canada: Manitoba, Alberta, Assiniboia, Saskatchewan, and New Ontario*, p. 68.
33 Herman Ganzevoort, *A Dutch Homesteader on the Prairies*, p. 88.
34 Canada, Department of Agriculture, *The Province of Manitoba and North West Territories, Information for Intending Immigrants*, 5th ed. (Ottawa, 1880), pp. 14–19. Reprinted in Kevin H. Burley, ed., *The Development of Canada's Staples, 1867–1939* (Toronto: McClelland and Stewart, 1971), pp. 37, 38.
35 Georgina Binnie-Clark, *Wheat and Women*, p. 350.
36 Ibid., p. 166. Georgina Binnie-Clark has provided a cost breakdown for a two-strand barbed wire fence in 1907:

8 rolls wire	$24.00
660 pickets	19.80
labour	20.00
	$63.80 per mile

Since two miles of fencing were required to enclose a quarter-section, Binnie-Clark would have spent about $127.60 or $0.80 per acre.
37 Herman Ganzevoort, *A Dutch Homesteader on the Prairies*, p. 35.
38 Leo Thwaite, *Alberta: An Account of Its Wealth and Progress* (London: George Routledge and Sons, 1912), p. 88.
39 Herman Ganzevoort, *A Dutch Homesteader on the Prairies*, p. 40.
40 *Letters from a Young Emigrant ...*, pp. 86–88.
41 W.M. Elkington, *Five Years in Canada*, p. 135.
42 Herman Ganzevoort, *A Dutch Homesteader on the Prairies*, p. 72.
43 Henry J. Boam, comp., *The Prairie Provinces of Canada*, edited by Ashley G. Brown (London: Sells Ltd., 1914), pp. 388–89.
44 Janice Acton (comp.), *Lemberg Local History*, p. 164.
45 See, for example, John Macoun, *Manitoba and the Great North-West*, p. 637; *Letters from a Young Emigrant ...*, pp. 96–98; Isaac Cowie, *Edmonton Country*, pp. 18–19; "Hints to Settlers," quoted in Annie Yule, *Grit and Growth*, p. 67.
46 Herman Ganzevoort, *A Dutch Homesteader on the Prairies*, pp. 36–37.
47 *The Province of Manitoba and North West Territories, Information for Intending Immigrants*, in Kevin H. Burley, ed., *The Development of Canada's Staples* (1867–1939), p. 37.
48 All of the following sources cite $80 as the price of a wagon in the 1880s: Macoun, *Manitoba and the Great Northwest*, p. 637; *Farming and Ranching in the Canadian Northwest, the Guide Book for Settlers* (Ottawa: Queen's Printer, 1886); *The Province of Manitoba and North West Territories: Information for Intending Immigrants of Canada*, in Burley, *The Development of Canada's Staples* (1867–1939), p. 37; and R. Goodridge, *A Year in Manitoba* (London: W.R. Chambers, 1882), pp. 107–8. Cowie's *Edmonton Country*, p. 24, gives $75 as the price of a wagon in 1897 and Boam's *The Prairie Provinces*, published in 1914, states that farm wagons at that time could be purchased for $70 (pp. 388–89). Second-hand wagons, of course, could be purchased more cheaply. In 1907 Georgina Binnie-Clark purchased a used wagon for $45. See *Wheat and Women*, p. 37.
49 *Letters from a Young Emigrant ...*, p. 130.
50 See, for example, *Western Canada: Manitoba, Alberta, Assiniboia, Saskatchewan, and New Ontario*, p. 68; *The Manitoba Official Handbook*, pp. 44, 45.
51 SABS, S-X2, Pioneer Questionnaire of George A. Hartwell, Pheasant Forks, District of Assiniboia, 1882, p. 5.
52 Canada, Parliament, *Sessional Papers*, 48 Victoria, 1885, No. 8, Table B, p. 150. "Table showing the Price of Agricultural Implements as sold at different places in the United States and Canada,

during the season of 1884." Prices of implements at Brandon, Manitoba were selected for this book.

53 Canada, Department of Agriculture, Indian Head Dominion Experimental Farm, Ledger Book, 1888–89. The figures quoted in the ledger book represent actual expenditures for farm implements. Comparable prices for 1889 farm implements are revealed in an Oak Lake settler's letter to the *Manitoba Colonist*, No. 42, November 1889, p. 14.

54 Isaac Cowie, *Edmonton Country*, p. 51.

55 Saskatchewan, Department of Agriculture, Agricultural Machinery Administration, Series 11, *Catalogues*, Group A. 20 "Massey-Harris." Box 1, "No. 1 Retail Price List 1909. Manitoba Branch. Massey Harris Company Ltd., Winnipeg, Manitoba."

56 Saskatchewan, *Royal Commission on Farm Machinery, 1914,* Implement Catalogue File A(v), Cockshutt Plow Company, Box 1. "Retail Price List B," 1914.

57 Saskatchewan, Department of Agriculture, Agricultural Machinery Administration, Series 11, *Catalogues*. Group A. 20. "Massey-Harris." Box 1. "No 2 X Retail Price List, 1915–1916" (effective 1 May 1915). Regina Branch, Massey-Harris.

58 AM, Diary of Claude H. Manners, Moosomin, 1883.

59 Isaac Cowie, *Edmonton Country*, p. 19, gives $1.95 as the cost of backsetting in 1901.

60 See "Wheat: Cost of Production," *Canadian Thresherman* (Winnipeg), November 1905, p. 3. According to this article steam plowing entailed a cost of 57.5 cents per acre. "Using the Traction Engine on the Plow," *Nor'-West Farmer* (Winnipeg), 5 March 1902, p. 193.

61 W.M. Elkington, *Five Years in Canada*, p. 64.

62 William Compton, Opawaka, Manitoba, Letter to the Editor, 25 August 1890, printed in *The Western World* (Winnipeg), September 1890.

63 "A Farm Started in '82," *Nor'-West Farmer*, 20 November 1928.

64 SABR, Department of Agriculture, Statistics Branch, File re: General Publicity, 1914. John Teece to J. Cromie, 27 December 1913.

65 J.M. Bonnor, "Early History of the Blackwood District," unpublished local history manuscript (Blackwood, Saskatchewan: 1963). See also SABS, Homestead File No. 71859.

66 *Letters from a Young Emigrant*, pp. 85–86.

67 SABS, Homestead Files.

68 See, for example, *Letters from a Young Emigrant* (1883), and W.M. Elkington, *Five Years in Canada* (1895).

69 James Mavor, *Report to the Board of Trade on the North West of Canada, With Special Reference to Wheat Production for Export* (London: King's Printer, 1904), p. 26.

70 Statistical surveys of prairie farmers conducted in the 1920s and 1930s indicate that a large proportion of settlers began with relatively little capital. In 1930–31, the Saskatchewan College of Agriculture carried out a series of surveys in areas representative of the different soil groups in the province. Overall, these surveys showed that farmers who settled before 1900 had begun with an average capital of $357. Evidently, the costs of farm-making increased quickly after 1900, as farmers who arrived in the 1900–1905 period reported an average initial worth of $1,249. Starting capital increased rapidly thereafter to a high of $5,000 for those who commenced farming after the First World War. Further evidence for the low capital outlay thesis is found in a 1926 study of 360 Manitoba farmers. Fully 60.1 per cent of the surveyed group reported having started farming with less than $500. See R.W. Murchie, *Agricultural Progress on the Prairie Frontier*, Canadian Frontiers of Settlement Series, Vol. 5 (Toronto: Macmillan, 1936), pp. 72–73; and R.W. Murchie and H.C. Grant, *Unused Lands of Manitoba* (Winnipeg: Department of Agriculture and Immigration, 1926), pp. 71–72.

3 ECONOMIC DEVELOPMENT OF THE ABERNETHY DISTRICT, 1880–1920

1 *Regina Leader*, 10 December 1884.
2 Annie J. Yule, *Grit and Growth* ..., p. 23.
3 Vernon C. Fowke, *Canadian Agricultural Policy: The Historical Pattern* (Toronto: University of Toronto Press, 1946), p. 242.
4 The prices were taken from market quotations published in the *Qu'Appelle Progress*, *Qu'Appelle Vidette*, and *Regina Leader*. Qu'Appelle prices pertain to the principal market centres for Abernethy wheat in the period, including Indian Head, Fort Qu'Appelle and Sintaluta. Some market quotations provide a range of wheat prices for different grades and others indicate only a single price. It is assumed that these single prices pertain to the highest possible price, i.e. of No. 1 Hard Wheat. On this basis they were averaged with the highest quoted prices on individual dates in the same season, i.e., 1 September to 31 August of the following year.
5 *Qu'Appelle Vidette*, 31 December 1885.
6 *Qu'Appelle Progress*, 12 July 1894.
7 A tabulation of farm improvements recorded by 461 homestead applicants confirms the slow rate of land clearing. Abernethy settlers broke an average of only about 12 acres per year.
8 *Qu'Appelle Vidette*, 31 December 1885.
9 Ibid., 28 December 1893.
10 LAC, RG17, Department of Agriculture Records, Vol. 384, File No. 41291, Report of A.J. Baker, 1 October 1883.
11 Ibid., Vol. 427, File No. 46581, Annual report of A.J. Baker, 31 December 1884.
12 *Qu'Appelle Vidette*, 1 October 1885.
13 Letter of Samuel Chipperfield to the editor, *The Abernethan* (Abernethy), 4 October 1907, p. 5.
14 LAC, RG17, Department of Agriculture Records, Vol. 562, File 63154, Annual report of A.J. Baker, Qu'Appelle, 31 December 1887. See also *Qu'Appelle Progress*, 1 September 1887, p. 6.
15 SABS, Homestead File No. 133067.
16 Letter of Elizabeth Motherwell, Perth, Ontario, to W.R. Motherwell, December 1888. Copy in Parks-PNRO Library.
17 SABS, S-A32, A.S. Morton, Transcription of Department of Interior File No. 225330, Letter of A.G. Thorburn, MLA, 29 November 1889, appended to memo of Lieutenant-Governor E. Dewdney to the Privy Council, 11 January 1890. In his account, Thorburn included the following statistics vis-à-vis gopher destruction in 1889:

RETURNS DISTRICT	ACRES SEEDED	ACRES DESTROYED	PERCENTAGE DESTROYED
WALLACE DISTRICT	369	342	92.70
KINBRAE DISTRICT	404	214	53.00
WHITEWOOD DISTRICT	1508	827	54.80
BROADVIEW DISTRICT	1065	365	34.30
KENDRICKS DISTRICT	1634	776	47.50
YORKTON DISTRICT	957	780	81.50
GRENFELL DISTRICT	483	302	62.50
QU'APPELLE DISTRICT	419	231	55.13
QU'APPELLE DISTRICT	1377	583	43.34
QU'APPELLE DISTRICT	403	227	56.33
TOTALS OF 10 RETURNS	8619	4647	54 PER CENT

18 "Chickney Settlement," *The Western World* (Winnipeg), June 1890, p. 100.
19 LAC, RG18, R.C.M.P. Records, Vol. 254, 1890, "Crop and Weather Bulletins for 1890."

20 James Bonnor, "Early History of the Blackwood District," p. 1.
21 *Melville Advance*, 29 June 1955.
22 LAC, RG15, Department of the Interior Records, Vol. 702, File No. 351884, Petition to His Excellency the Governor General in Council of the Dominion of Canada, Received by the Department of Interior, 12 March 1894.
23 LAC, RG17, Department of Agriculture Records, Vol. 682, File No. 78020, "Canada as a Field for Emigration," Letter of a settler, Regina, N.W.T. to the *Belfast Newsletter*, March 1891.
24 "Hard Times at Lorlie," *Qu'Appelle Progress*, 3 January 1895, p. 1.
25 *Qu'Appelle Progress*, 2 March 1893.
26 *Qu'Appelle Progress*, 24 June 1897, p. 4.
27 The table of yields at the Motherwell and Indian Head Farms is taken from S.D. Clarke, "Settlement in Saskatchewan with Special Reference to the Influence of Dry Farming," M.A. thesis, University of Saskatchewan, Saskatoon, 1931, p. 79.
28 SABS, W.R. Motherwell Papers, fol. 23308, news clipping from *Canadian Farms and Homes*, Christmas–New Year's Issue, 1910.
29 "Review of Conditions in the West," *Regina Leader*, 5 January 1899, p. 1.
30 *Qu'Appelle Vidette*, 26 January 1898, p. 8.
31 "Actual Farming Results," *The Western World*, March 1897, p. 12.
32 Canada, Parliament, *Sessional Papers*, Annual Report of the North-West Mounted Police, 1895, Report of Superintendent, Depot Division, Regina, 1 December 1895, No. 15, Appendix C., p. 58.
33 A.S. Morton, *History of Prairie Settlement*, p. 125.
34 W.A. Macintosh, *Economic Problems of the Prairie Provinces*, Canadian Frontiers of Settlement Series (Toronto: Macmillan, 1935). Table 5, "Costs of Transporting Wheat," p. 284.
35 Canada, *Census of Population and Agriculture of the North West Provinces*, 1906.
36 The data of wheat yields were derived from *Annual Reports of the Department of Agriculture, North-West Territories, 1898–1904*, and from Saskatchewan, Department of Agriculture, *Annual Reports, 1905–1912*.
37 North-West Territories. *Annual Report of the Department of Agriculture, 1901*.
38 See Herman Ganzevoort, *A Dutch Homesteader on the Prairies*, p. 6.
39 Saskatchewan, Regina Land Titles Office, Certificate of Title to the North-East Quarter of Section 14, in Township 20, Range 11, West of the 2nd Principal Meridian. Mortgage registered by the Law Union and Crown Insurance Company for $5,000, dated 2 September 1907.
40 See for example the Certificate of Title to the North-West Quarter of Section 22 in Township 20, Range 11, West of the 2nd Meridian.
41 Saskatchewan, *Royal Commission on Agricultural Credit* (Regina: King's Printer, 1913), p. 65.
42 Saskatchewan, Department of Agriculture, *Annual Report, 1912* (Regina: King's Printer, 1913), p. 102.
43 Ibid., p. 103.
44 *The Abernethan* (Abernethy), 22 September 1908.
45 Saskatchewan, *Royal Commission on Agricultural Credit*, p. 63.
46 Georgina Binnie-Clark, *Wheat and Women*, p. 354.
47 All the details on Georgina Binnie-Clark's farming venture were obtained from her personal account in *Wheat and Women*.
48 Saskatchewan, Department of Agriculture, *Annual Report, 1918*, p. 14.
49 Ibid., p. 13.
50 Ibid.
51 Ibid.
52 Somewhat variable figures for the costs to the farmer for wheat cultivation are cited in two different sources for 1913

Notes 249

and 1914. The Saskatchewan Grain Markets Commission tabulation gives the cost of producing an acre of wheat in 1914 to be $10.13 for the province and $9.35 for the Southeastern Region (Saskatchewan, *Royal Commission on Grain Markets* [Regina: King's Printer, 1914, p. 16]). The *Canada Yearbook*, on the other hand, gives a figure of $12.53 for 1913. *Canada Yearbook, 1914* (Ottawa: King's Printer, 1915), p. 200.

53 John H. Thompson, *The Harvests of War: The Prairie West, 1914-1918* (Toronto: McClelland and Stewart, 1978), pp. 63-65.

54 See Figure 16, "Wholesale Price Indexes by Commodity Group, 1890-1924," compiled from M.C. Urquhart and K.A.H. Buckley, *Historical Statistics of Canada* (Toronto: Macmillan, 1965), p. 292.

55 William Allen, E.C. Hope, and F.C. Hitchcock, "Studies of Financial Indebtedness and Financial Progress of Saskatchewan Farmers," Report No. 3, University of Saskatchewan, *Agricultural Extension Bulletin No. 68*, 1935 [hereafter cited as W. Allen, et al., "Studies of Farm Indebtedness"].

56 William Allen, E.C. Hope, and F.C. Hitchcock, "Studies of Probable Net Farm Revenues for the Principal Soil Types of Saskatchewan," University of Saskatchewan, *Agricultural Extension Bulletin No. 64*, 1935.

57 Robert Ankli, "Farm Income on the Plains and the Prairies," *Agricultural History*, Vol. 51, No. 1 (January 1977), p. 99.

4 Work and Daily Life at the Motherwell Farm

1 Marc Bloch, *French Rural History: An Essay on its Basic Characteristics* (trans. Janet Sondheimer) (Berkeley and Los Angeles: University of California Press, 1966).

2 Interview with Nina Gow and Ben Noble, Abernethy, by Lyle Dick, Abernethy, Saskatchewan. Parks Canada, WNSC, January 1978. Mrs. Gow recalled that Motherwell hired her late husband George Gow to work solely on the farmstead landscape.

3 Alma Mackenzie, correspondence with H. Tatro, Calgary, 17 March 1968. Parks Canada, WNSC. Mrs. Mackenzie was W.R. Motherwell's daughter.

4 Ian Clarke, Motherwell Historic Park, "Landscape and Outbuildings- Structural and Use History," Manuscript Report Series, No. 219, Environment Canada, Canadian Parks Service, National Historic Parks and Sites Branch, Ottawa, 1977, p. 3.

5 E.A.W. Gill, *A Manitoba Choreboy: The Experiences of a Young Emigrant Told From His Letters* (London: The Religious Tract Society, 1912), p. 27.

6 Interview with Dan Gallant, Regina, by Lyle Dick, 6 March 1983. Parks Canada, WNSC.

7 Interview with Jack Bittner, Abernethy, by Lyle Dick, January 1978. Parks Canada, WNSC.

8 Interview with Dan Gallant by Lyle Dick, 6 March 1983. Parks Canada WNSC.

9 Herman Ganzevoort, *A Dutch Homesteader on the Prairies*, p. 4.

10 Interview with Dan Gallant by Lyle Dick, 6 March 1983. Parks Canada, WNSC.

11 Ibid.

12 Interview with Major McFadyen, Dan Gallant and Olive Gallant, Abernethy, by Ian Clarke, Parks Canada, WNSC, September 1976.

13 Interview with Elizabeth Large, Balcarres, by Lyle Dick, 7 July 1980. Parks Canada, WNSC,

14 Ibid.

15 Interview with Dan Gallant by Lyle Dick, 6 March 1983. Parks Canada, WNSC.

16 Ibid.

17 Telephone interview with Walter Brock, Abernethy, by Lyle Dick, February 1983. Parks Canada, WNSC.

18 Interview with Dan Gallant by Lyle Dick, 6 March 1983. Parks Canada, WNSC.

19 Interviews with Marie Bittner, Lemberg, 21 July 1980, and Elizabeth Large, Balcarres, 7 July 1980, and with Elizabeth Morris, Indian Head, by Lyle Dick, 22 March 1978. Parks Canada, WNSC.
20 Interview with Major McFadyen, Regina, by Ian Clarke, Parks Canada, WNSC, 25 June 1976.
21 Interview with Marie Bittner by Lyle Dick, 21 July 1980. Parks Canada, WNSC.
22 Interview with Margretta Evans Lindsay by Ian Clarke and Margie Lou Shaver, Regina, May 1977. Parks Canada, WNSC.
23 Interview with Elizabeth Large by Lyle Dick, Balcarres, Saskatchewan, 7 July 1980, Parks Canada, WNSC.
24 Ibid.
25 Ibid.
26 Ibid.; and interview with Marie Bittner by Lyle Dick, 21 July 1980, Parks Canada, WNSC.
27 Joann Vanek, "Work, Leisure and Family Roles: Farm Households in the United States, 1920–1955," *Journal of Family History*, Vol. 5, No. 4 (Winter 1980), pp. 422–31.
28 Interview with Laura Jensen, Sun City, Arizona, 7 December 1977. Parks Canada, WNSC.
29 Interview with Gertrude Barnsley, Abernethy, by Lyle Dick, 15 July 1980, Parks Canada, WNSC.
30 Interview with Margretta Evans Lindsay by Ian Clarke and Margie Lou Shaver, May 1977. Parks Canada, WNSC.
31 Alma Mackenzie, untitled written reminiscences, No. B10. Parks Canada, WNSC.
32 Nellie McClung, *Clearing in the West* ... (Toronto: Thomas Allen Ltd., 1935), p. 364.
33 See SABS, Philip Crampton Manuscript, p. 39.
34 Bruce B. Peel, "R.M. 45: The Social History of a Rural Municipality," M.A. thesis, University of Saskatchewan, Saskatoon, 1946, p. 224.
35 Georgina Binnie-Clark, *Wheat and Women*, p. 275.
36 Ibid.
37 Ibid.
38 See SABS, S-M12, W.R. Motherwell Papers, File No. 81, fols. 12041-44, "The Use and Abuse of the Common Drag Harrow." Address delivered at the 1915 International Dry Farming Congress by W.R. Motherwell in Denver, Colorado.
39 Interview with Major McFadyen by Ian Clarke, September 1976. Parks Canada, WNSC.
40 Interview with Dan Gallant by Lyle Dick, 6 March 1983. Parks Canada, WNSC.
41 W.R. Motherwell, "Saskatchewan as a Field for Dry Farming Operations." Address delivered before the Fifth Dry Farming Congress at Spokane, Washington, 5 October 1910. Issued as Saskatchewan, Department of Agriculture, *Bulletin No. 21*, "Methods of Soil Cultivation Underlying Successful Grain Growing in the Province of Saskatchewan" (Regina: King's Printer, 1910), p. 11.
42 Samuel Chipperfield, who farmed at Chickney about six miles southeast of Abernethy, reported that his hired man performed seeding operations up to the middle of June in the 1902 crop year. SABR, Samuel Chipperfield Diaries, Entries for June 1902.
43 Bruce B. Peel, "R.M. 45: The Social History of a Rural Municipality," pp. 128–43.
44 Advisory Board of Manitoba, *Prairie Agriculture* (Toronto: W.J. Gage, 1896), p. 103.
45 W.C. McKillican, *Experiments with Wheat at the Dominion Experimental Farm, Brandon, Manitoba, 1899–1923: A Summary*, Canada, Department of Agriculture, Bulletin No. 42, New Series (Ottawa: King's Printer, 1924), p. 31.
46 Ibid.
47 SABS, S-M12, W.R. Motherwell Papers, File No. 121, fols. 17185-6, W.R.

Motherwell to A.J. Quigley, Sintaluta, Saskatchewan, 19 December 1907.
48 See also SABS, S-M12, Motherwell Papers, File No. 81, fols. 12041–44, "The Use and Abuse of the Common Drag Harrow."
49 Ibid.
50 SABS, S-M12, W.R. Motherwell Papers, File No. 121, fol. 17272, W.R. Motherwell to J.W. DeStein, Regina, 17 May 1916.
51 W.R. Motherwell, "Oat Growing for the Qu'Appelle Valley," *Nor'-West Farmer*, 20 April 1901, p. 239.
52 SABS, S-M12, W.R. Motherwell Papers. File No. 121, fols. 17185–86, W.R. Motherwell to A.J. Quigley, Sintaluta, Saskatchewan, 19 December 1907.
53 SABS, W.R. Motherwell Papers, File No. 121, fol. 17272. W.R. Motherwell to J.W. DeStein, 17 May 1916.
54 Advisory Board of Manitoba, *Prairie Agriculture*, p. 88.
55 Ibid., pp. 89–90.
56 SABR, Samuel Chipperfield Diaries, 1900–1910.
57 Ibid.
58 E.A.W. Gill, *A Manitoba Choreboy ...*, p. 49.
59 Ibid.
60 Interview with Dan Gallant by Lyle Dick, 6 May 1983. Parks Canada, WNSC.
61 Herman Ganzevoort, *A Dutch Homesteader on the Prairies*, p. 7.
62 A.G. Street, *Farmer's Glory* (Toronto: Macmillan, 1934), p. 129.
63 Advisory Board of Manitoba, *Prairie Agriculture*, p. 103.
64 Interview with Walter Brock by Lyle Dick, 26 July 1980. Parks Canada WNSC.
65 A.G. Street, *Farmer's Glory*, p. 136.
66 Quoted in John H. Thompson, "Bringing in the Sheaves: The Harvest Excursionists, 1890–1929," p. 478.
67 A.G. Street, *Farmer's Glory*, p. 130.
68 Advisory Board of Manitoba, *Prairie Agriculture*, p. 103.
69 SABR, Samuel Chipperfield Diary, 1900–1910.
70 E.A.W. Gill, *A Manitoba Choreboy ...*, pp. 53–54.
71 Ibid. Walter Brock, a neighbour and friend of W.R. Motherwell, has related that Motherwell "stacked for two men." Interview with Walter Brock, Abernethy, by Lyle Dick, Parks Canada, WNSC, 18 July 1980.
72 John Bracken, *Crop Production in Western Canada* (Winnipeg: Grain Growers' Guide Ltd., 1920), pp. 120–21. Willem de Gelder also gives an account of stacking in Herman Ganzevoort, *A Dutch Homesteader on the Prairies*, p. 11.
73 James M. Minifie, *Homesteader*, p. 131.
74 Bruce B. Peel, "R.M. 45," pp. 224–26.
75 Interview with Alma MacKenzie by R. Dixon, 1969. Parks Canada, WNSC.
76 Herman Ganzevoort, *A Dutch Homesteader on the Prairies*, p. 12.
77 Philip Crampton in his reminiscences of the Carrot River District recalled that portable threshers had come into general use by 1910. SABS, Philip Crampton Manuscript. W.R. Motherwell owned a Case 50-inch separator in this period.
78 Interview with Jack Bittner by Lyle Dick, January 1978. Parks Canada, WNSC.
79 See the account of threshing operations in Herman Ganzevoort, *A Dutch Homesteader on the Prairies*, p. 12.
80 S.J. Ferns and H.S. Ferns, *Seventy Five Years in Canada* (Winnipeg: Queenston House, 1978), pp. 50–51.
81 Herman Ganzevoort, *A Dutch Homesteader on the Prairies*, p. 12.
82 James M. Minifie, *Homesteader*, p. 74.
83 A.G. Street, *Farmer's Glory*, p. 106.
84 Ibid., p. 107.
85 Ibid., p. 106.
86 Herman Ganzevoort, *A Dutch Homesteader on the Prairies*, p. 55.

87 J.F. McCorrell, "Threshing in the West," *Qu'Appelle Progress*, 2 February 1902, p. 2.
88 SABS, Pioneer Questionnaire of Mrs. Edith Stilborne.
89 *The Abernethan*, 21 September 1902, p. 4.
90 Ibid., 18 October 1907, p. 1.
91 James M. Minifie, *Homesteader*, p. 76.
92 See text, Chapter 3, above.
93 See James M. Bonnor, "Early History of the Blackwood District," 1963, and Annie I. Yule, *Grit and Growth*, p. 37.
94 Ibid.
95 *Qu'Appelle Progress*, 3 December 1896, p. 1.
96 A.G. Street, *Farmer's Glory*, p. 130.
97 Interview with Howard Dinnin, Abernethy, by Lyle Dick, 26 July 1980. Parks Canada, WNSC.
98 James M. Minifie, *Homesteader*, p. 79.
99 Bruce B. Peel, "R.M. 45," Chapters 6 and 7, pp. 124-128 and 194-234.
100 Ibid.
101 SABR, Samuel Chipperfield Diaries, 1902-03.
102 Ibid.
103 Catherine Motherwell's expertise in domestic bookkeeping prompted her husband to encourage her to give a lecture on this subject at the founding convention of Saskatchewan Homemakers' Club in 1911. See *Report of the First Annual Convention of the Homemakers' Clubs of Saskatchewan* (Regina: n.p., 1911), p. 56. See also Ann Oakley, *Housewife* (London: Allen Lane, 1974), p. 49.
104 Thorstein Veblen, *The Theory of the Leisure Class* (New York: Modern Library, 1934).
105 Ann Oakley, *Housewife*, p. 49.
106 Henry Hamilton, *History of the Homeland*, quoted in Ann Oakley, *Housewife*, p. 49.
107 Kate Caffrey, *The 1900s Lady* (London: Gordon and Cremonesi, 1976), p. 24.
108 Interview with Ralph Stueck, Abernethy, by Ian Clarke, 8 June 1976, p. 6. Parks Canada, WNSC.
109 Interview with Laura Jensen, Patricia Motherwell and Laura Murray, Calgary, by Lyle Dick and Sarah Carter, 2 July 1978. Parks Canada, WNSC.

5 ABERNETHY'S SOCIAL AND ECONOMIC STRUCTURE

1 Seymour M. Lipset, *Agrarian Socialism: The Co-operative Commonwealth Federation in Saskatchewan* (Berkeley: University of California Press, 1971), p. 212.
2 Ibid., p.51.
3 C.B. Macpherson, *Democracy in Alberta: Social Credit and the Party System* (Toronto: University of Toronto Press, 1953), pp. 10-20.
4 Morton Herbert Fried, *The Evolution of Political Society: An Essay in Political Anthropology* (New York: Random House, 1967).
5 Michael B. Katz, Michael J. Doucet, and Mark J. Stern, *The Social Organization of Early Industrial Capitalism* (Cambridge, MA: Harvard University Press, 1982).
6 Ibid., p. 42.
7 See Vernon C. Fowke, *The National Policy and The Wheat Economy*, pp. 281-97, and Harold A. Innis, "The Penetrative Powers of the Price System," in Mary Q. Innis (ed.), *Essays in Canadian Economic History* (Toronto: University of Toronto Press, 1956), pp. 252-72.
8 W. Allen et al., "Studies of Farm Indebtedness," Table 26, p. 46.
9 Ibid., Table 33, p. 43.
10 Ibid., Table 26, p. 46.
11 Ibid., Table 26, p. 46.
12 SABS, S-M12, W.R. Motherwell Papers, File No. 85, fol. 12462, W.R. Motherwell, Regina, to F. Riley, Outlook, Saskatchewan, 1 April 1909.
13 Ibid., fol. 12414.
14 Rural Municipality of Abernethy, Tax Assessment Roll, 1926.

15 Canada, *Census of Population and Agriculture of the North-West Provinces, Manitoba, Saskatchewan, Alberta, 1906*, p. 45. Between 1901 and 1906, for example, Township 20, Range 11, West of the 2nd Meridian, encompassing the Village of Abernethy, experienced a drop in population from 215 to 167.

16 W. Allen et al., "Studies of Farm Indebtedness," p. 11.

17 Robert Hunter, *Poverty* (1904), quoted in Michael Katz, *The People of Hamilton, Canada West* (Cambridge, MA: Harvard University Press, 1975), pp. 79–80.

18 Saskatchewan. Department of Agriculture, *Annual Report for 1913* (Regina: King's Printer, 1914).

19 Disputes over the non-payment of farm wages sometimes ended up in court. See, for example, SABS, Judicial District of Melville, Court of King's Bench, Civil Case No. 11 of 1919 (Scott vs Moonie). In this case a farm labourer brought suit against an Abernethy farmer who had failed to pay accumulated wages of $645 earned between December 1917 and October 1918.

20 See Herman Ganzevoort, *A Dutch Homesteader on the Prairies*, pp. 6–8.

21 John H. Blackburn, *Land of Promise* (Toronto: Macmillan 1970), p. 226.

22 See T.B. Bottomore, *Classes in Modern Society* (New York: Vintage Books, 1966), pp. 24–26.

23 Paul H. Landis, *Rural Life in Process* (New York: McGraw Hill, 1940), p. 73.

24 John Bennett, *Northern Plainsmen: Adaptive Strategy and Agrarian Life* (Arlington Heights, Illinois: A.H.M. Publishing Co., 1969), pp. 72–73.

25 Ibid.

26 Frederick P. Grove, *Fruits of the Earth* (Toronto: McClelland and Stewart, 1965), p. 72.

27 See Canada, Department of Agriculture, Indian Head Dominion Experimental Farm Collection, "Map of Indian Head and Adjacent District," ca. 1906.

28 Allan R. Turner, "W.R. Motherwell and Agricultural Development in Saskatchewan, 1905–1918," p. 6.

29 See interview with Andy Sproule by Thomas White, Balcarres, Saskatchewan, 1968. Parks Canada, WNSC.

30 Michael Katz, *The People of Hamilton, Canada West*, especially pp. 44–93.

31 Speech of John Morrison, quoted in L.D. Courville "The Conservatism of the Saskatchewan Progressives," Canadian Historical Association, *Historical Papers, 1974*, p. 169.

32 Herman Ganzevoort, *A Dutch Homesteader on the Prairies*, p. 6.

33 Canada, *Census for 1901* (Ottawa: King's Printer, 1902).

34 Ibid.

35 Maps of land tenure in the Abernethy district for 1906 and 1914 show only a handful of non-Anglo-Saxon names in the area bounded by the Rural Municipality of Abernethy.

36 Howard Palmer, *Land of No Second Chance: A History of Ethnic Groups in Southern Alberta* (Lethbridge: The Lethbridge Herald, 1972), p. 253.

37 *Qu'Appelle Vidette*, 25 January 1899.

38 SABS, S-M12, W.R. Motherwell Papers, File No. 85.

39 "The Conservatives Rally," *The Abernethan*, 16 October 1908, p. 1.

40 See L.L. Dobbin, "Mrs. Catherine Gillespie Motherwell, Pioneer Teacher and Missionary," *Saskatchewan History*, Vol. 14, No. 1 (Winter 1961), pp. 17–26.

41 Interview with Mrs. Eleanor Brass, Regina, by Lyle Dick, September 1978, p. 11. Parks Canada, WNSC.

42 *Report of the First Annual Convention of the Homemakers' Clubs of Saskatchewan*.

43 SABS, S-M12, W.R. Motherwell Papers, File No. 123, "Women," fols. 17471–17479.

44 St. Andrew's College Library, University of Saskatchewan, Saskatoon, "Knox

Presbyterian Church, Abernethy," Unpublished manuscript, n.d.

45 Allan G. Bogue, "Social Theory and the Pioneer," *Agricultural History*, Vol. 34 (January 1960), p. 21.

46 Canada, Department of Agriculture, Indian Head Experimental Research Station. "Map of Indian Head and Adjacent District," ca. 1906, and SABS, Department of the Attorney General, Justice of the Peace Files.

47 See SABR, Minutes of the Abernethy Women's Grain Growers' Association, 1917–25; St. Andrew's College Library, Minute Book of the Women's Missionary Society, Knox Presbyterian Church, Abernethy, 1915–25.

6 SOCIAL RELATIONSHIPS OF THE SETTLEMENT ERA

1 Canada, *Census of Canada 1890–91* (Ottawa: Queen's Printer 1893), Vol. II, General Table of Subjects, Table I: 'Ages of the People: The Territories,' 14–17. The raw numbers in these age brackets in Assiniboia West were: males, 3234, and females, 1746; Table III, 'Civil Condition: The Territories,' 220–21.

2 Allan G. Bogue, "Social Theory and the Pioneer," pp. 21–34.

3 Walter M. Elkington, *Five Years in Canada*, pp. 111–12.

4 *Letters from a Young Emigrant ...*, pp. 89–90.

5 Walter M. Elkington, *Five Years in Canada*, pp. 111–12.

6 SABR, Diocese of Qu'Appelle, Assiniboia, Occasional Paper No. 25 (April, 1891), *Some Impressions of Life in the North West*, pp. 17–18.

7 LAC. Census of the Prairie Provinces, 1906, District No. 14, Qu'Appelle, 203, Sub-district no. 35, p. 3, Lines 1 and 2.

8 LAC, Fourth Census of Canada, 1901, District 203, Sub-District C-3, Assiniboia East, Polling Sub-division No. 23, North Qu'Appelle, p. 2, lines 38–39, p. 3, lines 44–45, p. 8, lines 39–40; and Polling Sub-division No. 108, South Qu'Appelle, p. 3, lines 29–30, p. 4, lines 42–44, and p. 6, lines 5–6.

9 LAC, Fourth Census of Canada, 1901, District No. 203, East Assiniboia, Sub-district G-2, Polling Sub-division no. 92, Kenlis, p. 9, lines 38 and 39.

10 LAC, Fifth Census of Canada, 1911, District 215, Sub-District C-3, Assiniboia East, Enumeration District no. 8, Township 20, Range 11, West 2nd Meridian, p. 3, lines 24 and 25.

11 LAC, Fifth Census of Canada, 1911, District 215, Sub-District C-3, Assiniboia East, Enumeration District no. 8, p. 2, lines 11 and 12.

12 LAC, Fifth Census of Canada, 1911, District 215, Sub-District C-3, Assiniboia East, Enumeration District no. 8, Indian Head, p. 9, lines 35 and 36.

13 F.P. Grove, *Fruits of the Earth*, p. 37.

14 SABR, Ferdinand David Diary, 1893–1899.

15 *Letters from a Young Emigrant ...*, p. 108.

16 Walter M. Elkington, *Five Years in Canada*, p. 112.

17 SABS, Pioneer Questionnaire of Mrs. Florence Kenyon, Abernethy, p. 1.

18 Ibid., p. 3.

19 Walter M. Elkington, *Five Years in Canada*, pp. 111–12.

20 Ibid.

21 *Letters from a Young Emigrant ...*, pp. 89–90.

22 See Bruce Peel, "R.M. 45."

23 J. Burgon Bickersteth, *The Land of Open Doors: Being Letters from Western Canada 1911–13* (Toronto: University of Toronto Press, 1976), pp. 82–84.

24 Bachelor Billy, 'Social Jottings from Spencerville,' *Qu'Appelle Progress*, 6 June 1895, p. 1.

25 See E. Anthony Rotundo, "Romantic Friendship: Male Intimacy and Middle-Class Youth in the Northern United States, 1800–1900," *Journal of Social History*, Vol. 23, No. 1 (Autumn 1989), pp. 1–25.

26 J.K. McLean, an Ontarian visitor to the Abernethy district in 1895, used

the term "bachelor den" in expressing his disgust with the predominance of single men's habitations. *Qu'Appelle Vidette*, 25 July 1895, p. 4.
27 Walter M. Elkington, *Five Years in Canada*, p. 56.
28 Ibid., pp. 60–61.
29 SABS, Autobiography of Mrs. Frank Jordans, p. 47.
30 Canada, Parliament, *Sessional Papers, 1883*, Report of the Commissioner, North-West Mounted Police, Appendix D. "Cases tried before Lt.-Col. A.G. Irvine, S.M. and other Magistrates in the various Police Districts throughout the Territories, during the year 1882," pp. 44–46.
31 James Gray, *Red Lights on the Prairies* (Toronto: Macmillan, 1971).
32 *Qu'Appelle Vidette*, 18 June 1885, p. 3.
33 Ibid.
34 Ibid., 25 March 1886, p. 3.
35 Herman Ganzevoort, *A Dutch Homesteader on the Prairies*, p. 24.
36 SABS, Pioneer Questionnaire (No. 2) of W.H. Ismond, Abernethy, p. 5.
37 SABS, Pioneer Questionnaire (No. 1) of W.H. Ismond, Kenlis, Saskatchewan, p. 5.
38 *Qu'Appelle Progress*, 4 March 1897.
39 "Literary Society," *Lemberg Leader*, 13 December 1897.
40 *The Abernethan*, 26 July 1911, p. 8.
41 Ibid., 27 November 1908, p. 8.
42 Ibid., 19 July 1911, p. 8.
43 "Abernethy L.O.L. 1892," *The Abernethan*, p. 5.
44 Georgina Binnie-Clark, *Wheat and Women*, p. 314.
45 SABS, Pioneer Questionnaire of Lottie Meek.
46 Interview with Mrs. Marie Bittner, Lemberg, by Lyle Dick, 21 July 1980. Parks Canada, WNSC.
47 SABS, Pioneer Questionnaire of Edith Stilborne, Pheasant Forks, Saskatchewan.
48 *Qu' Appelle Vidette*, 12 January 1898.
49 The sociological analysis of Victorian Canadian prairie houses was earlier developed by Lyle Dick and Jean-Claude Lebeouf, "Social History in Architecture: The Stone House of W.R. Motherwell," *Research Bulletin/Bulletin de recherche*, No. 122 (Ottawa: Environment Canada, Canadian Parks Service, 1980). This paper was presented at the annual meeting of the Society for the Study of Architecture in Canada, in Saskatoon, 25 May 1979.
50 Kenneth L. Ames, "Meaning in Artifacts: Hall Furnishings in Victorian America," *Journal of Interdisciplinary History*, Vol. 9, No. 1 (Summer 1978), p. 30.
51 See Lyle Dick, "W.R. Motherwell's Stone House: A Structural History," (Ottawa: Environment Canada, Canadian Parks Service, National Historic Parks and Sites, Manuscript Report No. 267, 1978). Reprinted in *History and Archaeology/Histoire et archéologie*. No. 66 (Ottawa: Parks Canada, 1983), pp. 213–63.
52 Herman Ganzevoort, *A Dutch Homesteader on the Prairies*, pp. 12–13.
53 Letter of "Dabs" to the Editor, *The Abernethan*, 3 November 1907.
54 Nellie McClung, *Sowing Seeds in Danny*, quoted in Susan Jackel, "The House on the Prairies," *Canadian Literature*, No. 42 (Autumn 1969), p. 48.
55 F.P. Grove, *Fruits of the Earth*, pp. 147–53.
56 Genevieve Leslie, "Domestic Service in Canada 1880–1920," in Janice Acton, Pennie Goldsmith, and Bonnie Shepard, eds., *Women at Work, 1850–1930* (Toronto: Canadian Women's Educational Press, 1974), pp. 71–126.
57 Ibid., p. 86.
58 Ibid.
59 Marie Bittner and Lizzie Lutz were both 18 years of age when they came to work at the Motherwell farm.
60 Interview with Miss Elizabeth Morris (née Lutz), Indian Head, by Lyle Dick, Parks Canada, WNSC, 22 March 1978.

61 Interview with Mrs. Marie Bittner by Lyle Dick, 21 July 1980.
62 Interview with Mrs. Elizabeth Morris by Lyle Dick, 22 March 1978.
63 Saskatchewan, Department of Agriculture, *Practical Pointers for Farm Hands* (Regina: King's Printer, 1915), p. 3.
64 Ibid.
65 Ibid.
66 Interview with Major McFadyen by Ian Clarke, 25 June 1976.
67 SABS, Pioneer Questionnaire of George A. Hartwell, Primitive Methodist Colony, Pheasant Forks, District of Assiniboia.
68 SABS, Pioneer Questionnaire of Edith Stilborne.
69 Interview with Jack Bittner by Lyle Dick, January 1978.
70 James M. Minifie, *Homesteader*, pp. 43–44.
71 Ibid.
72 See Notice of Application for Letters Patent to form "The Pheasant Plains Steam Threshing Company, Ltd.," *Qu'Appelle Progress*, 24 December 1891, p. 4.
73 "Important Judgement," *Qu'Appelle Progress*, 17 October 1901.
74 North-West Territories, *Consolidated Ordinances of the Legislative Assembly 1898*, Cap. 80, "An Ordinance Respecting Estray Animals," and "An Ordinance Respecting the Herding of Animals" (Regina: Queen's Printer, 1899), pp. 786–91 and pp. 792–803.
75 *The Abernethan*, 8 November 1907, p. 4.
76 Ibid.
77 Nellie McClung, *Clearing in the West*, p. 369.
78 Ibid.

7 ABERNETHY'S SOCIAL CREED

1 *Qu'Appelle Progress*, 21 September 1893, p. 1.
2 See Fried's distinctions between egalitarian, rank, stratified, and state-based societies in Morton Herbert Fried, *The Evolution of Political Society: An Essay in Political Anthropology* (New York: Random House, 1967).
3 David Bidney, *Theoretical Anthropology*, 2nd ed. (New Brunswick, NJ: Transaction Publishers, 1996), p. 407.
4 Allen Smith, "Myth of the Self-Made Man ...," pp. 189–219.
5 Nellie McClung, *Clearing in the West*, p. 369.
6 Allan Bogue, "Social Theory and the Pioneer," p. 22.
7 See, for example, "Experience and Opinions of Settlers," *Manitoba Official Handbook* (immigration pamphlet), pp. 40–44.
8 Lyle Dick and Jean Claude Lebeouf, "Social History in Architecture: The Stone House of W.R. Motherwell," *Research Bulletin/Bulletin de recherches* (Ottawa: Environment Canada, Canadian Parks Service, No. 122, 1980).
9 George Britnell, *The Wheat Economy* (Toronto: University of Toronto Press, 1939), p. 173n.
10 See the treatment of Victorian hall furnishings in Kenneth L. Ames, "Meaning in Artifacts: Hall Furnishings in Victorian America." *Journal of Interdisciplinary History*, Vol. 9, No. 1 (Summer 1978), pp. 19–46.
11 Allen Smith, "Myth of the Self-Made Man ...," p. 207.
12 Saskatchewan, Department of Agriculture, *Practical Pointers for Farm Hands*.
13 W.L. Morton, "Victorian Canada" in W.L. Morton (ed.), *The Shield of Achilles: Images of Canada in the Victorian Age* (Toronto: McClelland and Stewart, 1968), p. 328.
14 Quoted in Paul Rutherford, "The New Nationality, 1884–1897: A Study of

the National Aims and Ideas of English Canada in the late 19th Century," Ph.D. thesis, University of Toronto, 1973, p. 28.

15 Taking Township 20, Range 11, West of the 2nd Meridian as example, in 1911 the census enumerator counted 30 Presbyterians, 10 Methodists, five members of the Anglican Church, two Roman Catholics, and one Baptist, meaning that more than 80 per cent of the population in the township was either Presbyterian or Methodist. LAC, Fifth Census of Canada, Saskatchewan District no. 215, Schedule I, "Population," Enumeration District No. 2, p. 1.

16 Ernst Troeltsch, *The Social Teachings of the Christian Churches*, Vol. 2 (New York: Harper Torchbooks, 1960), pp. 489, 576–602, and 721–24.

17 Ibid., p. 590.

18 Richard Allen, *The Social Passion: Religion and Social Reform in Canada 1914–28* (Toronto: University of Toronto Press, 1971), pp. 5–6.

19 "Presbytery of Minnedosa," *Canada Siftings*, Russell, Manitoba, 9 July 1898, p. 9. Printed supplement in *Saltcoats Siftings*.

20 See Jacqueline J. Kennedy, "Qu'Appelle Industrial School," pp. 141–42.

21 *The Abernathan*, 16 October 1908, p. 1.

22 See Barbara Schrodt, "Sabbatarianism and Sport," *Journal of Sport History*, Vol. 4, No. 1 (Spring 1977), pp. 22–23.

23 "An Ordinance to Prevent the Profanation of the Lord's Day," North-West Territories. *Consolidated Ordinances of the North-West Territories, 1898*, p, 877.

24 "Magistrate's Court," *Qu'Appelle Progress*, 25 December 1885, p. 2.

25 *Qu'Appelle Progress*, 9 January 1886.

26 Interview with Mrs. Laura Jensen, Sun City, Arizona, by Lyle Dick, 7-8 December 1978.

27 SABS, Pioneer Questionnaire of Mrs. Edith Stilborne and Mrs. Florence Kenyan.

28 Nellie McClung, *The Stream Runs Fast* (Toronto: Thomas Allen Ltd., 1945), pp. 59–60.

29 SABR, Ferdinand David Diary. Entry for 16 December 1894.

30 Interview with Mr. Howard Dinnin, Abernethy, by Lyle Dick, Parks Canada, WNSC, 26 July 1980.

31 Among numerous social purity tracts published in Canada in this era, see: W.J. Hunter, *Manhood Wrecked and Rescued* (Toronto: William Briggs, 1894), and C.S. Clark, *Of Toronto the Good: A Social Study. The Queen City of Canada as It Is* (Montreal: The Toronto Publishing Company, 1898). See also Aaron M. Powell, *The National Purity Congress: Its Papers, Addresses, Portraits* (First Published, 1896; New York: Arno Press Reprint, 1976).

32 Susan Jackel, "Introduction," in Georgina Binnie-Clark, *Wheat and Women* (Toronto: University of Toronto Press, 1979), xx–xxvii. Catherine Cavanagh has shown that a concerted campaign by Alberta women to achieve equality of ownership in the proceeds of homesteading ventures was ultimately unsuccessful in the settlement era. Catherine Cavanagh, "The Alberta Campaign for Homestead Dower, 1909–1925," in Catherine Cavanagh and Jeremy Mouat, eds., *Making Western Canada: Essays on European Colonization and Settlement* (Toronto: Garamond Press, 1996), pp. 186–214; Arthur S. Morton, *History of Prairie Settlement* (Toronto: Macmillan of Canada, 1938), Map: Manitoba and Northwest Territories, p. 97.

33 *The Progress* (Qu'Appelle, N.W.T.), 28 September 1893, p. l; "Proposition to Tax the Bachelors," *Qu'Appelle Progress* (Qu'Appelle, N.W.T.), 23 November 1893, p. 4.

34 "The Decline of Marriage: The Marriageable Age for Men has Advanced from Twenty-Five to Thirty-Two," *The Leader* (Regina), 25 April 1895, p. 6.

35 James G. Snell, "'The White Life for Two': The Defence of Sexual Morality in Canada, 1890–1914," *Histoire*

36   See, for example, "Matrimony," *The Leader*, 9 July 1889; and T. De Witt Talmage, "Thy Land Shall be Married," *The Standard*, 16 June 1899, 3 Among innumerable articles in the region linking Christian religion and the family, see "The Religion of Marriage," *The Standard*, 2 January 1896, 6; "A Revival is Needed: There is Too Much Neglect of Home Duties Now," *The Vidette* (Qu'Appelle and Indian Head), 31 December 1896, 2. *sociale/Social History*, Vol. 16, No. 31 (May 1983), pp. 111–28.

37   Saskatchewan Archives Board, Saskatoon (SABS), A391 GS168, United Church of Canada, Saskatchewan Conference Records, II. The Methodist Church in Canada, Saskatchewan Conference, A. Conference Records, 1884–1925. 2. Minutes, 1884–1918, *Minutes of the Manitoba and North-West Conference 1893* (Toronto: William Briggs, 1894), "Report of the Temperance Committee," p. 52; A.13. Miscellaneous, 1905–1925, "General Session, 29 May 1925: Report of the Committee on Evangelism and Social Service," 1925; A. Conference Records, 1884–1925, 2. Minutes, 1884–1918; and The Methodist Church, *Minutes of the Manitoba and North-West Conference 1893* (Toronto: William Briggs 1893), "Report of the Epworth League Committee," p. 63.

38   SABS, Pioneer Questionnaire (No. 1) of W.H. Ismond, Kenlis, Saskatchewan, p. 5.

39   "Social Purity," *The Standard*, 7 March 1895, p. 5; "Men's Meeting: Rev. H.T. Crossley Speaks Plainly to a Promiscuous Audience of the Men of Regina," *The Standard*, 8 May 1891, p. 5.

40   "Young Men Warned: Dr. Talmage on Bad Company," *The Saskatchewan Herald* (Battleford, N.W.T.), 9 October 1896, p. 3; and T. De Witt Talmage, *The Night Sides of City Life* (St. John, NB: J. & A. McMillan 1878), p. 80. In a column carried by the *Regina Standard in 1898*, Talmage again directed his wrath at urban immorality: "Then look at the impurities of these great cities. Ever and anon there are in the newspapers explosions of social life that make the story of Sodom quite respectable, 'for such things,' Christ says, 'were more tolerable for Sodom and Gomorrah than for the Chorazins and Bethsaidas of greater light.'" T. De Witt Talmage, "The Sins of American Cities," *The Standard*, 6 October 1898, p. 3. See also T. De Witt Talmage, "Dangers of the Cities," *The Vidette*, 12 January 1898, p. 7.

41   Lyle Dick, "Male Homosexuality in Saskatchewan's Settlement Era: The 1895 Case of Regina's Oscar Wilde," Paper presented to the Annual Conference of the Canadian Historical Association, York University, Toronto, 30 May 2006.

42   AM, MG4, B4, Sir Wilfrid Laurier Papers. Vol. 106, fol. 32068, Letter of W.R. Motherwell to Sir Wilfrid Laurier, 3 April 1899.

43   On the Victorians' sense of mission, see Walter E. Houghton, *The Victorian Frame of Mind, 1830–1870* (New Haven, CT: Yale University Press, 1957), pp. 244–50, and 349.

44   William Brennan, "A Political History of Saskatchewan," Ph.D. thesis, University of Alberta, 1976.

45   See the discussion in Marilyn Barber, "The Assimilation of Immigrants in the Canadian Prairie Provinces, 1896–1918, Canadian Perception and Canadian Policies," Ph.D. thesis, University of London, 1975.

46   See Erhard Pinno, "Temperance and Prohibition in Saskatchewan," M.A. thesis, University of Saskatchewan at Regina, 1969; and John Thompson, "The Prohibition Question in Manitoba, 1892–1928," M.A. thesis, University of Manitoba, 1969. Pinno points out that while a large majority throughout the province voted for prohibition in the 1916 plebiscite, Eastern European and some other ethnic groups voted heavily against its imposition.

47   "Womanliness," The Moose Jaw Times, 31 May 1889, p. 1.

48 SABR, Minute Book of the Qu'Appelle Chapter, Women's Christian Temperance Union, 1894–1903, Minutes of the 18 August 1897 Meeting.
49 SABS, S-M12, W.R. Motherwell Papers, File No. 123, "Women," fol. 17478, "Mrs. Motherwell's Address on Women's Present-Day Responsibilities" (1916).
50 SABR, Abernethy Women's Grain Growers' Association Minute Book, 1917–25, 22 February 1917.
51 W.R. Morrison, "'Their Proper Sphere': Feminism, the Family and Child Centered Social Reform in Ontario, 1875–1900," Part 11, *Ontario History*, Vol. 68, No. 2 (June 1976) pp. 71–72.
52 Ibid., p. 73.
53 SABR, Abernethy Women's Grain Growers' Association Minute Book, 1917–25, 22 February 1917.
54 SABS, W.R. Motherwell Papers, File No. 123, "Women," fol. 17478, "Mrs. Motherwell's Address on Women's Present-Day Responsibilities" (1916).
55 SABR, Minutes of the Abernethy Women's Grain Growers' Association, 13 June 1917.
56 Angus McLaren, *Our Own Master Race: Eugenics in Canada, 1885–1945* (Toronto: McClelland and Stewart, 1990), p. 60.
57 Kenneth M. Ludmerer, *Genetics and American Society: A Historical Appraisal* (Baltimore: Johns Hopkins University Press, 1972), p. 32. On the development of the eugenic movement in Canada, see Angus McLaren, *Our Own Master Race*.
58 J.E. Rea, "The Roots of Prairie Society," in David Gagan, ed., *Prairie Perspectives*, Vol. 1 (Toronto: Holt, Rinehart and Winston, 1970), pp. 46–55.
59 *Qu'Appelle Vidette*, 19 July 1888.
60 Alma (Motherwell) Mackenzie papers. Letter of W.R. Motherwell, Minister of Agriculture to Miss Alma Motherwell, Abernethy, 24 February 1922 (held in private collection of Donald and Dorothy Mackenzie, Montague, P.E.I.).
61 Walter Hildebrandt, "P.G. Laurie of Battleford: The Aspirations of a Western Conservative," *Prairie Forum*, Vol. 8, No. 2 (Fall 1983), pp. 157–78.
62 *Royal Commission on Immigration and Settlement 1930*, Appendix XII, p. 204.

8 Agrarian Unrest in the Central Qu'Appelle Region

1 Hopkins Moorehouse, *Deep Furrows* (Toronto and Winnipeg: George J. McLeod Ltd., 1918), p. 48.
2 D.J. Hall, "The Manitoba Grain Act: An Agrarian Magna Charta?" *Prairie Forum*, Vol. 4, No. 2 (Spring 1979), p. 118.
3 The classic study of the National Policy and its far-reaching impacts on the prairie farm economy is Vernon C. Fowke's *The National Policy and the Wheat Economy*.
4 Brian R. McCutcheon, "The Birth of Agrarianism in the Prairie West," *Prairie Forum*, Vol. 1, No. 2 (November 1976), pp. 81–83.
5 Ibid., p. 81.
6 *Manitoba Free Press*, 6 December 1883, p. 4, quoted in Brian R. McCutcheon, "The Patrons of Industry in Manitoba, 1890–1898," *Transactions of the Historical and Scientific Society of Manitoba*, Series 3, No. 22 (1965–66), p. 8.
7 B.R. McCutcheon, "Birth of Agrarianism," p. 87.
8 Ibid., p. 87.
9 Lewis Aubrey Wood, *A History of Farmers' Movements in Canada: The Origins and Development of Agrarian Protest, 1872–1924* [hereafter cited as *History of Farmers' Movements*] (Toronto: University of Toronto Press, 1975), p. 113.
10 Ibid., p. 166.
11 Ibid.
12 Ibid.
13 SC, 63–64 Victoria, 1900, Cap. 39.
14 W.A. Wilson, *A Century of Canadian Grain: Government Policy to 1951*

(Saskatoon: Western Producer Prairie Books, 1978), p. 32.
15 L.A. Wood, *History of Farmers' Movements*, p. 173.
16 Ibid. The grain "blockade" continued throughout the winter. By late February, only 800,000 bushels of a total crop of 2.2 million in the Indian Head district had been shipped. At Sintaluta, the proportion of the crop shipped was even less – only 150,000 of 1,230,000 bushels. *Regina Standard*, 26 February 1902, p. 1.
17 Hopkins Moorehouse, *Deep Furrows*, pp. 46, 47.
18 Ibid., p. 50.
19 "Story of the Early Days – Hon. W.R. Motherwell, Regina" (from an interview by Hopkins Moorehouse, 16 April 1916), *Saskatchewan History*, Vol. 8, No. 3 (Autumn 1955), p. 109.
20 Hopkins Moorehouse, *Deep Furrows*, p. 51.
21 L.A. Wood, *History of Farmers' Movements*, p. 175.
22 Harald S. Patton, *Grain Growers' Cooperation in Western Canada* (Cambridge, MA: Harvard University Press, 1928), p. 34.
23 SC, 2 Edward VII, 1902, Cap. 19, pp. 121–23.
24 Ibid., Section 5, p. 122.
25 Letter of W.R. Motherwell to the Territorial Grain Growers' Association, 1902, quoted in H.S. Patton, *Grain Growers' Cooperation in Western Canada*, p. 35n.
26 Hopkins Moorehouse, *Deep Furrows*, p. 54.
27 L.A. Wood, *History of Farmers' Movements*, p. 180; Lyle Dick, "Henry Oscar Partridge," in Ramsay Cook, ed., *Dictionary of Canadian Biography*, Vol. XIV: 1911–1920 (Toronto: University of Toronto Press, 1998), pp. 822–24.
28 H.S. Patton, *Grain Growers' Cooperation*, p. 36.
29 L.A. Wood, *History of Farmers' Movements*, p. 180.
30 H.S. Patton, *Grain Growers' Cooperation*, p. 38.
31 Ibid.
32 Canada, Parliament, *House of Commons Debates*, 1903, p. 9351.
33 SC, 3 Edward VII, 1903, Cap. 33, pp. 182–91.
34 The debate is recorded in Canada, Parliament, *House of Commons Debates*, 1903, pp. 7993–7995.
35 For a detailed discussion of the logistics of shipping "track" wheat, see V.C. Fowke, *The National Policy and the Wheat Economy*, pp. 110–14.
36 H.S. Patton, *Grain Growers' Cooperation*, p. 39n.
37 V.C. Fowke, *The National Policy and the Wheat Economy*, pp. 112–13.
38 North-West Territories, *Annual Report of the Department of Agriculture, 1899* (Regina: King's Printer, 1900).
39 See, for example, the testimonials of Qu'Appelle area farmers in the *Qu'Appelle Vidette*, 2 and 3 March 1889. See, for example, the testimonials of Qu'Appelle area farmers in the *Qu'Appelle Vidette*, 2 and 3 March 1889.
40 Between 1918 and 1932, the average annual wheat yield in Crop District No. 2 was 16 bushels per acre. See fn. 4., p. 328, W. Allen et. al., *Studies of Farm Indebtedness and Financial progress of Saskatchewan Farners*, Report No. 3 (1935), p. 37, Table 29.
41 V.C. Fowke, *The National Policy and the Wheat Economy*, p. 113.
42 SABR, Diary of Samuel Chipperfield, 1892.
43 Ibid., Diary, 1895.
44 V.C. Fowke, *The National Policy and the Wheat Economy*, p. 115.
45 "A Farmer's Suggestions," Letter of A. Ormiston to the Editor, dated 27 December 1902, printed in *Regina Standard*, 1 January 1903.
46 W.A. Wilson, *A Century of Canadian Grain*, p. 687.
47 *Qu'Appelle Progress*, 1 December 1904, p. 1.
48 Ibid.

49 See, for example, Hopkins Moorehouse, *Deep Furrows*, pp. 61–72; L.A. Wood, *History of Farmers' Movements*, pp. 183-90, and James B. Hamilton and Donald E. Reid, "E.A. Partridge: Prairie Radical," unpublished graduate paper, History Department, University of Saskatchewan, Saskatoon, 1971.

50 *Grenfell Sun*, 30 January 1907, p. 3, cited in J.B. Hamilton and D.E. Reid, "E.A. Partridge: Prairie Radical," p. 21.

51 V.C. Fowke, *The National Policy and the Wheat Economy*, pp. 127–52, and D.S. Spafford, "The Elevator Issue, the Organized Farmers and the Government, 1908–11," *Saskatchewan History*, Vol. 15, No. 3 (Autumn 1962), pp. 81–92.

52 J.B. Hamilton and D.E. Reid, "E.A. Partridge: Prairie Radical," pp. 31–32.

53 V.C. Fowke, The National Policy and the Wheat Economy, p. 144.

54 *Regina Standard*, 10 August 1908, p. 4.

55 SABS, S-M12, W.R. Motherwell Papers, File No. 80, fol. 11716, W.R. Motherwell, Commissioner of Agriculture, Regina, to J.H. Sheppard, Moose Jaw, 26 March 1908. The author would like to thank Sarah Carter for drawing his attention to this reference.

56 Ibid.

57 V.C. Fowke, *The National Policy and the Wheat Economy*, p. 143.

58 H.S. Patton, *Grain Growers' Cooperation in Western Canada*, p. 410.

59 C.B. Macpherson, *Democracy in Alberta*, p. 226.

60 B.R. McCutcheon, "Birth of Agrarianism," pp. 79–94.

61 That the Patrons and the Liberals shared affinities in terms of electoral policies was suggested by the election of James Douglas under the joint banner of both parties in Assiniboia East in the 1896 federal election.

62 AM, MG4, B4, Sir Wilfrid Laurier Papers, Series A, Vol. 224, fols. 63029-32, Petition of Abernethy Settlers to Sir Wilfrid Laurier, 2 February 1902.

63 "Balcarres," *Qu'Appelle Vidette*, 20 May 1886, p. 3.

64 "Abernethy: Scathing Censure Upon the Methods of the Manitoba Mutual Hail Insurance Company," *Regina Leader*, 23 November 1899, p. 4.

65 "Taking Out Patents – The Homestead Menaced," *Regina Leader*, 9 November 1893, p. 4.

66 Evon Z. Vogt, *Modern Homesteaders: The Life of a Twentieth Century Frontier Community* (Cambridge, MA: Harvard University Press, 1955).

67 Interview with Walter Brock, Abernethy, by Lyle Dick, 26 June 1981. Parks Canada, WNSC.

68 A list of the original Board of Directors of the central TGGA appears in Hopkins Moorehouse, *Deep Furrows*, Appendix, p. 295.

69 See the land titles for the SW quarter-section of Section 6 in Township 20, Range 11, West of the 2nd Meridian. Saskatchewan. Department of the Attorney-General. Property Management Branch. Regina District Land Titles Office, Regina.

70 According to a ca. 1906 map of land tenure in the Indian Head district, Elmer Shaw then owned six quarter-sections of land, and Peter Dayman, four quarters. Canada, Department of Agriculture, Indian Head Experimental Farm, "Map of Indian Head and Adjacent District."

71 See Figure 2 in text, above.

72 "Map of Indian Head and Adjacent District." Data on the composition of the original group of Sintaluta farmers was obtained from the Appendix in Hopkins Moorehouse, *Deep Furrows*, p. 296.

73 Minute Book, Territorial and Saskatchewan Grain Growers' Association, Abernethy Chapter, 1904–1924. This document is in the custody of Mr. Bert Garratt, Abernethy, Saskatchewan.

74 See V.C. Fowke, *Canadian Agricultural Policy: The Historical Pattern*, esp. pp. 249–50, and V.C. Fowke, "Political Economy and the Canadian Wheat Grower," in Norman Ward and Duff

Spafford, eds., *Politics in Saskatchewan* (Don Mills, ON: Longmans, 1968), pp. 207–20.

75 V.C. Fowke, "Royal Commissions and Canadian Agricultural Policy," *Canadian Journal of Economics and Political Science* (1948), p. 169.

76 Ibid.

9 CONCLUSIONS

1 Rodolfo Stavenhagen, *Social Classes in Agrarian Societies* (New York: Anchor Books, 1975), p. 26.

2 See John H. Thompson, "Permanently Wasteful but Immediately Profitable: Prairie Agriculture and the Great War," Canadian Historical Association, *Historical Papers* 1976, pp. 193–206.

3 Richard Hofstadter, *The Age of Reform* (New York: Random House, 1955), p. 46.

4 Dick Harrison, "Rolvaag, Grove, and Pioneering on the American and Canadian Plains," *Great Plains Quarterly*, Vol. 1, No. 4 (Fall 1981), p. 253.

5 Ibid.

6 Ian Clarke and Gregory Thomas, "The Garrison Mentality and the Canadian West: The British-Canadian Response to Two Landscapes: The Fur Trade Post and the Ontarian Prairie Farmstead," *Prairie Forum*, Vol. 4, No. 1 (1979), p. 101.

7 Ian Clarke, "Motherwell Historic Park: Structural and Use History of the Landscape and Outbuildings," in Ian Clarke, Lyle Dick, and Sarah Carter, *Motherwell Historic Park* (Ottawa: Environment Canada, Parks Canada: National Historic Parks and Sites Branch, 1983), p. 171.

BIBLIOGRAPHY

I. Manuscript and Other Unpublished Public Documents

Canada. Department of Agriculture. Indian Head Experimental Research Station.
Account Ledger Books, Indian Head Dominion Experimental Farm, 1888–1905.
Daily Journal of Activities, 1889–1914.
"Map of Indian Head and Adjacent District," ca. 1906.

Canada. Library and Archives Canada (LAC).
Government Archives Division
RG 17, Department of Agriculture Records, Vols. 384, 427, 466.
RG15, Department of the Interior Records, Vols. 229, 247, 292, 326, 702, 724.
RG18, R.C.M.P. Records, Vol. 254.
Manuscript Division
Fourth Census of Canada, 1901, manuscript schedules.
Census of the Prairie Provinces, 1906, manuscript schedules.
Fifth Census of Canada, 1911, manuscript schedules.

Manitoba. Archives (AM).
MG4, B4, Sir Wilfrid Laurier Papers (microfilmed copy of the originals at the National Archives of Canada).
M68, B4, Diary of Claude H. Manners, Moosomin, District of Assiniboia, 1883.
MG12, B 1, Alexander Morris Papers.

Rural Municipality of Abernethy
Tax Assessment Roll, 1926.

Saskatchewan. Archives Board, Regina (SABR).
M. Film 1.20 (1) Abernethy Women Grain Growers Association, Minutes, 1917–25.
M. Film 2.171 Journal of John Allen, 1883.
M. Film 2.864 Samuel Chipperfield Diaries, 1890–1910.
Coumans Saskatchewan Land Map Series, ca. 1917.
Cummins Rural Directories, 1915–1930.
Ferdinand David Diary
Local Histories File "Qu'Appelle," Annual Report of A.J. Baker, Immigration Agent, Qu'Appelle, to the Dominion Minister of Agriculture, 31 December 1888.
M.Film 2.180 Records of the Synod of the Diocese of Qu'Appelle, Anglican Church of Canada.

Saskatchewan, Department of Agriculture, Statistics Branch, File Re: General Publicity, 1914.
Women's Christian Temperance Union, Qu'Appelle District, Minutes, 1894- 1903.

Saskatchewan. Archives Board, Saskatoon (SABS).
Canada, Department of the Interior, "Homestead Files" for townships 19A, 19, 20 and 21 in Ranges 8, 10, 11 and 12, West of the Second Meridian.
Orders-in-Council, 1885. Microfilm No. 61.
S-A76 Frederick Charles Gilchrist Diary, 1859–96.
S-A93 William Hays Diary, 1883.
S-A32 A.S. Morton Papers.
S-M12 W.R. Motherwell Papers.
Philip Crampton Manuscript, n.d.
S-X2 Pioneer Questionnaires.
    J.H. Behrns, Abernethy, District of Assiniboia, date of settlement, 1899.
    Sidney Chipperfield, Chickney, 1883.
    Kenneth Foster, Abernethy, n.d.
    Alfred W. Garratt, Blackwood, 1884.
    George A. Hartwell, Primitive Methodist Colony, Pheasant Forks, 1882.
    W.H. Ismond, Abernethy, 1892.
    Florence Kenyon, Lemberg, 1895.
    Lottie Meek, Blackwood, 1884.
    Harriet Stueck, Abernethy, 1883.
    Edith Stilborne, Pheasant Forks, 1883.
    Mrs. Harry Teece, Lemberg, 1900.
    Rosanna Thompson, Abernethy, 1884.
    Elizabeth Webster, Balcarres, 1884.
A391 GS168, United Church of Canada, Saskatchewan Conference Records, II. The Methodist Church in Canada, Saskatchewan Conference, A. Conference Records, 1884–1925. 2. Minutes, 1884–1918, *Minutes of the Manitoba and North-West Conference 1893* (Toronto: William Briggs, Publisher, 1894), 'Report of the Temperance Committee,' 52; A.13. Miscellaneous, 1905–1925, 'General Session, 29 May 1925: Report of the Committee on Evangelism and Social Service, 1925; A. Conference Records, 1884–1925, 2. Minutes, 1884–1918; and The Methodist Church, *Minutes of the Manitoba and North-West Conference 1893* (Toronto: William Briggs 1893).

Saskatchewan. Department of the Attorney General.
Judicial District of Melville. Court of King's Bench and Supreme Court Docket Books, 1905–20.
Justice of the Peace Files.
Property Management Branch. Regina Land Titles Office, Regina.
Certified Land Titles for all quarter-sections in township 20, Range 11 and township 19 Range 8, West of the Second Meridian.

St. Andrew's College Library. University of Saskatchewan, Saskatoon.
Minutes of the Knox Presbyterian Church, Abernethy, 1915–1925.
Minutes of the Board of Controllers, Women's Missionary Society, Knox Presbyterian Church, Abernethy, 1905–1925.

II. Oral History Interviews

Parks Canada, Western and Northern Service Centre, Cultural Resource Services (WNSC).
These interviews represent the core of oral history research for the Motherwell project. With the exception of the Alma MacKenzie materials, these interviews were tape recorded. Typed transcripts are on file at the Western and Northern Service Centre office in Winnipeg.

    Richard A. Acton by Lyle Dick, 24 June 1979,
    Gertrude Barnsley by Lyle Dick, Abernethy, Saskatchewan, 15 July 1980.
    Jack Bittner by Lyle Dick, Abernethy, January 1978.
    Marie Bittner by Lyle Dick, Lemberg, Saskatchewan, 21 July 1980.
    Eleanor Brass, Regina, by Lyle Dick, September 1978.
    Walter Brock by Lyle Dick, Abernethy, 18 July 1980, 26 June 1981 and February 1983.
    Allen Burton by Lyle Dick, Abernethy, 21 July 1980.
    Lyle Dick, Balcarres, Saskatchewan, 22 July 1980.
    Howard Dinnin by Lyle Dick, Abernethy, 26 July 1980.
    Dan Gallant by Lyle Dick, Regina, 6 March 1983.
    Dan and Olive Gallant
        (i) and Major McFadyen by Ian Clarke, Lanark Place, September 1976.
        (ii) by Ian Clarke and Margie Lou Shaver, Regina, May 1977.
    Nina Gow and Ben Noble by Lyle Dick, Abernethy, January 1978.
    Laura Jensen by Lyle Dick, Sun City, Arizona, 7 and 8 December 1977.
    Laura Jensen, Laura Murray and Patricia Motherwell by Sarah Carter and Lyle Dick, Calgary, 2 July 1978.
    Dick and Elizabeth Large by Lyle Dick, Balcarres, 7 July 1980.
    Margretta Evans Lindsay
        (i) by Ian Clarke and Margie Lou Shaver, Regina, May 1977.
        (ii) by Lyle Dick, Regina, November 1977.
    Alma MacKenzie
        (i) Correspondence with H. Tatro, Calgary, 6 March 1968.
        (ii) Correspondence with H. Tatro, Calgary, 17 March 1968.
        (iii) Taylor Interview, Prince Edward Island, 17 April 1968.
        (iv) Response to questions from W. Naftel, 8 November 1968.
        (v) Interview by R. Dixon re: Naftel questions, 15 January 1969.
        (vi) "Recollections" by R. Dixon, n.d.
    Major McFadyen by Ian Clarke, Regina, 25 June 1976.
    Elizabeth Morris by Lyle Dick, Indian Head, 22 March 1978.
    Annie Morrison by Sarah Carter and Lyle Dick, Abernethy, July 1978.

Mr. and Mrs. Rick Penney by Lyle Dick, Abernethy, September 1978.
Harry Smith by Lyle Dick, Balcarres, July 1980.
Andy Sproule by Thomas White, Balcarres, 1968.
Hugh and Wanda Stueck by Lyle Dick, Abernethy, 18 April 1979.
Nelson Stueck by Lyle Dick, Abernethy, 21 July 1980.
Ralph Stueck by Ian Clarke, Abernethy, 8 June 1976.
Louis Wendell by Lyle Dick, Neudorf, 17 July 1980.

III. Manuscripts in Private Collections

Alma (Motherwell) Mackenzie Papers.
Private collection of Donald and Dorothy Mackenzie, Montague, P.E.I.

Minute Book of theTerritorial Grain Growers Association of Abernethy, 1904–05; and of the Abernethy Grain Growers' Association, 1905–1925.
Copy in the possession of Mr. Bert Garratt of Abernethy, Saskatchewan.

IV. Primary Printed Sources

Canada.
Census of Canada, 1891, 1901, 1911 and 1921 Population and Agriculture. Queen's Printer (King's Printer after 1901), Ottawa, 1892, 1902, 1912 and 1922.

Canada.
Census of Population and Agriculture of the North West Provinces, Manitoba, Saskatchewan and Alberta, 1885 and 1906. Queen's/King's Printer, Ottawa, 1886 and 1907.

Canada. Department of Agriculture.
Annual Reports of the Department of Agriculture, 1882–1900. Queen's Printer, Ottawa, 1883–1901.

Canada. Department of the Interior.
Annual Reports of the Department of the Interior for the Years 1882 to 1900. Queen's Printer, Ottawa, 1883–1901.

Canada.
Dominion Lands Surveyors' Reports. Queen's Printer, Ottawa, 1886.

Canada. North West Mounted Police.
Annual Reports of the Commissioner of the North-West Mounted Police Force, 1882–1900. Maclean Roger & Co., Ottawa, 1883–1901.

Canada. Parliament.
House of Commons Debates, 1882–1885, 1889–1893, 1902–1905. Queen's/King's Printer, Ottawa.
—. *Sessional Papers, 1883–1900.* Queen's Printer, Ottawa.

Canada. Privy Council.
Reports of a Committee of the Privy Council, 28 January 1885 and 18 April 1885, Queen's/King's Printer, Ottawa, 1886.

Canada. Laws, Statutes, etc.
Statutes, 1873, 1885, 1886, 1893, 1900, 1902, 1903, 1904. Queen's/King's Printer, Ottawa.
———. *Revised Statutes*, 49 Victoria. Queen's Printer, Ottawa, 1886.

Canada Yearbook, 1914.
King's Printer, Ottawa, 1915.

North-West Territories.
Annual Reports of the Department of Agriculture 1898–1904. Queen's/King's Printer, 1899–1905.
———. *Consolidated Ordinances of the Legislative Assembly, 1898.* Queen's Printer, Regina, 1899.

Saskatchewan. Department of Agriculture.
Annual Reports, 1905–20. King's Printer, Regina, 1906–21.
———. *Pamphlets, Bulletins and Circulars [and Catalogues]*, 1905–1920, King's Printer, Regina, 1906–21.
———. Government Publications Series, *Practical Pointers for Farm Hands.* King's Printer, Regina, 1916.

Saskatchewan.
Royal Commission on Agricultural Credit. King's Printer, Regina, 1913.
———. *Royal Commission on Farm Machinery.* Implement Catalogues. King's Printer, Regina, 1914.
———. *Royal Commission on Grain Markets.* King's Printer, Regina, 1914.
———. *Royal Commission on Immigration and Settlement, 1930.* King's Printer, Regina, 1931.

V. Immigration Pamphlets

Cowie, Isaac.
The Agricultural and Mineral Resources of Edmonton Country. Western Canada Immigrant Association, Edmonton, 1901.

Farming and Ranching in the Canadian Northwest, The Guide Book for Settlers. Queen's Printer, Ottawa. 1886.

Manitoba Official Handbook. Liverpool, England, 1892.

Western Canada: Manitoba, Alberta, Assiniboia, Saskatchewan, and New Ontario. N.p., 1902.

VI. Newspapers and Farm Journals

The Abernethan, Abernethy (intermittent issues), 1905–14.
Canadian Thresherman, Winnipeg, 1905–10.
Saltcoats Siftings, Saltcoats, 1897–8.
Lemberg Leader, Lemberg, 1897, 1905–10.
Manitoba Colonist, Winnipeg, 1888–90.
Melville Advance, Melville, 29 June 1955.
Moose Jaw Times, Moose Jaw, 1889.
Nor' West Farmer, Winnipeg, 1883–1910; 1928.
Qu'Appelle Progress, Qu' Appelle, 1885–1901.
Qu'Appelle Vidette, Fort Qu'Appelle, 1884–1897.
Regina Leader, Regina, 1883–1905.
Regina Standard, Regina, 1891–1905.
Star Almanac, Montreal, 1893.
The Western World, Winnipeg, 1889–97.

VII. Secondary Sources

Acton, Janice, comp.
Lemberg Local History. n.p., Lemberg, Sask., 1983.

Adelman, Jeremy
Frontier Development: Land, Labour and Capital on the Wheatlands of Argentina and Canada, 1890–1914. The Clarendon Press, Oxford and New York, 1994.

Advisory Board of Manitoba
Prairie Agriculture. W.J. Gage, Toronto, 1896.

Allen, Katherine R.
"A Conscious and Inclusive Family Studies," *Journal of Marriage and the Family*, Vol. 62 (February 2000), pp. 4–17.

Allen, Richard
The Social Passion: Religion and Social Reform in Canada 1914–28. University of Toronto Press, Toronto, 1971.

Allen, William, E.C. Hope, and F.C. Hitchcock
"Studies of Farm Indebtedness and Financial Progress of Saskatchewan Farmers." Report No. 3, Surveys made at Indian Head and Balcarres; Grenfell and Wolseley; and Neudorf and Lemberg. *Agricultural Extension Bulletin No. 68*, University of Saskatchewan, 1935.
——. "Studies of Probable Net Farm Revenues for the Principal Soil Types of Saskatchewan." *Agricultural Extension Bulletin No. 64*, University of Saskatchewan, 1935.

Ames, Kenneth L.
"Meaning in Artifacts: Hall Furnishings in Victorian America." *Journal of Interdisciplinary History*, Vol. 9, No. 1 (Summer 1978), pp. 19–46.

Ankli, Robert
"Farm Income on the Plains and the Prairies." *Agricultural History*, Vol. 51, No. 1 (January 1977). pp. 92–103.

Ankli, Robert E., and Robert M. Litt
"The Growth of Prairie Agriculture: Economic Considerations," in Donald H. Akenson, ed., *Canadian Papers in Rural History*. Vol. 1. Langdale Press, Gananoque, Ont., 1978, pp. 35–64.

Anonymous
Letters from a Young Emigrant in Manitoba. Kegan Paul, London, 1883.

Bantjes, Rod
Improved Earth: Prairie Space as Modern Artefact, 1869–1944. University of Toronto Press, Toronto, 2005.

Barber, Marilyn
"The Assimilation of Immigrants in the Canadian Prairie Provinces, 1896–1918: Canadian Perception and Canadian Policies." Ph.D. dissertation, University of London, 1975.

Bennett, John W.
Northern Plainsmen: Adaptive Strategy and Agrarian Life. Aldine Pub. Co., Chicago, 1969.

Bennett, John W., and Seena Kohl
Settling the Canadian-American West, 1890–1915: Pioneer Adaptation and Community Building: An Anthropological History. University of Nebraska Press, Lincoln, 1995.

Benson, Nathaniel
None of It Came Easy: The Story of James Garfield Gardiner. Burns Publishing Co., Toronto, 1955.

Berger, Carl
The Sense of Power: Studies in the Ideas of Canadian Imperialism, 1867–1914. University of Toronto Press, Toronto, 1970.

Bickersteth, J. Burgon
The Land of Open Doors: Being Letters from Western Canada, 1911–13. University of Toronto Press, Toronto, 1976.

Bidney, David
Theoretical Anthropology, Second Edition Transaction Publishers, New Brunswick, NJ, 1996.

Binnie-Clark, Georgina
Wheat and Women. Bell and Cockburn, Toronto, 1914.

Blackburn, John H.
Land of Promise. Macmillan, Toronto, 1970.

Bloch, Marc
French Rural History: An Essay on its Basic Characteristics. Translated by Janet Sondheimer. University of California Press, Berkeley and Los Angeles, 1966.

Boam, Henry J., comp.
The Prairie Provinces of Canada: Their History, People, Commerce, Industries and Resources. Edited by Ashley G. Brown. Sells Ltd., London, 1914.

Bogue, Allan C.
From Prairie to Cornbelt: Farming on the Illinois and Iowa Prairies in the Nineteenth Century. University of Chicago Press, Chicago, 1963.
——. "Social Theory and the Pioneer." *Agricultural History*, Vol. 34 (January 1960), pp. 21-34.
——. "Farming in the North American Grasslands: A Survey of Publications, 1947–80." *The Great Plains Quarterly*, Vol. 1 (Spring 1981), pp. 105–31.

Bonner, James. M.
"Early History of Blackwood District." Unpublished local history manuscript, Blackwood, Sask., 1963.

Bottomore, T.B.
Classes in Modern Society. Vintage Books, New York, 1966.

Bracken, John
Crop Production in Western Canada. Grain Growers' Guide Ltd., Winnipeg, 1920.

Braudel, Fernand
The Mediterranean and the Mediterranean World in the Age of Philip II, Vols. I and II (Trans. Sîan Reynolds). Harper Torchbooks, New York, 1975.

Brennan, William
"A Political History of Saskatchewan." Ph.D. thesis, University of Alberta, Edmonton, 1976.

Brown, Richard D.
"Microhistory and the Postmodern Challenge." *Journal of the Early Republic*, Vol. 23 (Spring 2003), pp. 1–20.

Bryan, Liz
The Buffalo People: Prehistoric Archaeology on the Canadian Plains. University of Alberta Press, Edmonton, 1991.

Burley, Kevin H., ed.
The Development of Canada's Staples, 1867–1939. McClelland and Stewart, Toronto, 1971.

Burnet, Jean
Next Year Country: A Study of Rural Social Organization in Alberta. University of Toronto Press, Toronto, 1951.

Bryan, Liz
The Buffalo People: Prehistoric Archaeology on the Canadian Plains. University of Alberta Press, Edmonton, 1991.

Caffrey, Kate
The 1900s Lady. Gordon and Cremonesi, London, 1976.

Campbell, Angelina H.
Man! Man! Just Look at that Land! n.p., Saskatoon, 1966.

Carter, Sarah
Lost Harvests: Prairie Indian Reserve Farmers and Government Policy. McGill-Queen's University Press, Montreal and Kingston 1990.
——. "A Materials History of the Motherwell Home," in Ian Clarke, Lyle Dick, and Sarah Carter, *Motherwell Historic Park*. History and Archaeology Series no. 66. Ottawa: Parks Canada, 1983., pp. 265–353.
——. "'Complicated and Clouded': The Federal Administration of Marriage and Divorce Among the First Nations of Western Canada, 1887–1906," in Sarah Carter, Lesley Erickson, Patricia

Roome, and Char Smith, eds., *Unsettled Pasts: Reconceiving the West Through Women's History*. University of Calgary Press, Calgary, 2005, pp. 151–78.

Cavanagh, Catherine
"The Alberta Campaign for Homestead Dower, 1909–1925," in Catherine Cavanagh and Jeremy Mouat, eds., *Making Western Canada: Essays on European Colonization and Settlement*. Garamond Press, Toronto, 1996, pp. 186–214.

Christie, Nancy and Michel Gauvreau
Mapping the Margins: The Family and Social Discipline in Canada, 1700–1975. McGill-Queen's University Press, Montreal and Kingston, 2004.

Clark, Clifford Jr.
"Domestic Architecture as an Index to Social History: The Romantic Revival and the Cult of Domesticity in America, 1840–1870." *Journal of Interdisciplinary History*, Vol. 7, No. 1 (Summer 1976), pp. 33–56.

Clark, C.S.
Of Toronto the Good: A Social Study. The Queen City of Canada as It Is. The Toronto Publishing Company, Toronto, 1898.

Clark, S.D.
"Settlement in Saskatchewan With Special Reference to the Influence of Dry Farming." M.A. thesis, University of Saskatchewan, Saskatoon, 1931.

Clarke, Ian
"Motherwell Historic Park, Landscape and Outbuildings, Structural and Use History." Manuscript Report Series No. 219, National Historic Parks and Sites, Canadian Parks Service, Environment Canada, Winnipeg, 1977. Published in *History and Archaeology/Histoire et archéologie*, No. 66, Ottawa, 1983, pp. 3–212.

Clarke, Ian, and Gregory Thomas
"The Garrison Mentality and the Canadian West: The British-Canadian Response to Two Landscapes: The Fur Trade Post and the Ontarian Prairie Farmstead." *Prairie Forum*, Vol. 4, No. 1 (Spring 1979), pp. 83–104. Regina.

Conzen, Michael
Frontier Farming in an Urban Shadow: The Influence of Madison's Proximity on Agricultural Development of Blooming Grove, Wisconsin. The State Historical Society of Wisconsin, Madison, 1971.
——. "Ethnicity on the Land," in Michael P. Conzen, ed., *The Making of the American Landscape*. Unwin Hyman, Boston, 1990, pp. 221–48.

Courville, Leo D.
"The Conservatism of the Saskatchewan Progressives." Canadian Historical Association, *Historical Papers* 1974 (1975), pp. 157–82. Ottawa.

Curti, Merle
The Making of an American Community. Stanford University Press, Stanford, CA., 1959.

Dawson, Carl A.
Group Settlement: Ethnic Communities in Western Canada. Canadian Frontiers of Settlement Series, Vol. 7. Macmillan, Toronto, 1936.

Dawson, C.A.
Settlement of the Peace River Country: A Study of a Pioneer Area. Canadian Frontiers of Settlement Series, Vol. 6. Macmillan of Canada, Toronto, 1934.

Dick, Lyle
"Profile of Settlement Community: Abernethy, Saskatchewan, 1880–1930," in *Canada, Confederation to Present* (CD-ROM). Edmonton: Chinook Micromedia Productions University of Alberta, 2001, pp. 1–24.
———. "Estimates of Farm-Making Costs in Saskatchewan, 1882–1914." *Prairie Forum*, Vol. 6, No. 2 (Fall 1981), pp. 182–202.
———. "Factors Affecting Prairie Settlement: A Case Study of Abernethy, Saskatchewan in the 1880s." Canadian Historical Association, *Historical Papers* 1985 (1986), pp. 11–28. Ottawa.
———. "A History of Period Farm Maintenance Techniques at the Motherwell Homestead, 1900–12." Microfiche Report No. 131, Canadian Parks Service, National Historic Parks and Sites 1984, pp. 1–270 & 56 illus.
———. "A Material Culture Study of Farm Implements at the Motherwell Homestead, 1882–1920." Microfiche Report Series No. 116, Environment Canada, Canadian Parks Service, National Historic Parks and Sites, Ottawa, 1984.
———. "W.R. Motherwell's Stone House: A Structural History." Manuscript Report Series No. 267, Environment Canada, Canadian Parks Service, National Historic Parks and Sites, Ottawa, 1978. Published in *History and Archaeology/Histoire et archéologie*, No. 66, Ottawa, 1983, pp. 213–63.
———. "The Social and Economic History of the Abernethy District, Saskatchewan, 1880–1920: Bibliography, Historiography, and Methodology." Manuscript Report Series No. 320, Environment Canada, Canadian Parks Service, National Historic Parks and Sites, Ottawa, 1979.
———. "A History of Prairie Settlement Patterns, 1870–1930." Parks Canada, Microfiche Report No. 307. Environment Canada, Parks, Prairie and Northern Region, Ottawa, 1987.
———. "Henry Oscar Partridge," in Ramsay Cook, ed., *Dictionary of Canadian Biography*, Vol. XIV: 1911–1920. University of Toronto Press, Toronto, 1998, pp. 822–24.
———. "James Moffat Douglas," in Ramsay Cook, ed., *The Dictionary of Canadian Biography* Vol. XIV: 1911–1920. University of Toronto Press, Toronto, 1998, pp. 305–308.
———. "Microhistory: Does it Work?" Paper presented to the Canadian Historical Association Conference, London, Ontario, 1 June 2005.
———. "Male Homosexuality in Saskatchewan's Settlement Era: The 1895 Case of Regina's Oscar Wilde." Paper presented to the Annual Conference of the Canadian Historical Association, York University, Toronto, 30 May 2006.

Dick, Lyle, and Jean Claude Lebeuf
"Social History in Architecture: The Stone House of W.R. Motherwell." *Research Bulletin/Bulletin de recherches*. Environment Canada, Canadian Parks Service, No. 122 (January 1980), Ottawa.

Dobbin, L.L.
"Mrs. Catherine Gillespie Motherwell, Pioneer Teacher and Missionary." *Saskatchewan History*, Vol. 14, No. 1 (Winter 1961), pp. 17–26. Saskatoon.

Edwards, Leonard J.
"W.R. Motherwell and the Crisis of Federal Liberalism in Saskatchewan, 1917–1926." M.A. thesis, University of Saskatchewan, 1961.

Elkington, Walter M.
Five Years in Canada. Whittaker & Co., London, 1895.

Elofson, Warren M.
Cowboys, Gentlemen, and Cattle Thieves: Ranching on the Western Frontier. McGill-Queen's University Press, Montreal and Kingston, 2000.

England, Robert
The Colonization of Western Canada. P.S. King, Publishers, London, 1936.

Ens, Gerhard
"Métis Lands in Manitoba." *Manitoba History,* No. 5 (Spring 1983), pp. 2–11.

Evans, Stirling, ed.
The Borderlands of the American and Canadian Wests: Essays on Regional History of the Forty-ninth Parallel. University of Nebraska Press, Lincoln, Nebraska, 2006).

Ferns, S.J. and H.S. Ferns
Seventy-Five Years in Canada. Queenston House, Winnipeg, 1978.

Fowke, Vernon C.
Canadian Agricultural Policy: The Historical Pattern. University of Toronto Press, Toronto, 1946.
——. *The National Policy and the Wheat Economy.* University of Toronto Press, Toronto, 1957.
——. "Political Economy and the Canadian Wheat Grower" in Norman Ward and Duff Spafford, eds., *Politics in Saskatchewan.* Longman's, Don Mills, Ont., 1968, pp. 207–20.
——. "Royal Commissions and Canadian Agricultural Policy." *Canadian Journal of Economics and Political Science.* Vol. 14 (1948). pp. 163–75. Toronto.

Fried, Morton Herbert
The Evolution of Political Society: An Essay in Political Anthropology. Random House, New York, 1967.

Gagan, David
"The Indivisibility of Land in Nineteenth Century Ontario." *Journal of Economic History,* Vol. 36 (1976), pp. 126–41. Atlanta, GA.
——. "Land Population and Social Change: The 'Critical Years' in Rural Canada West." *Canadian Historical Review,* Vol. 59, No. 3 (1978), pp. 295–318. Toronto.

Ganzevoort, Herman, trans. and ed.
A Dutch Homesteader on the Prairies. University of Toronto Press, Toronto, 1973.

Garratt Bert, ed.
Dance on the Bridge: A History of Abernethy and Area. Abernethy Heritage Association, Abernethy, Saskatchewan, 1983.

Gill, E.A.W.
A Manitoba Choreboy: The Experiences of a Young Emigrant Told From His Letters. The Religious Tract Society, London, 1912.

Ginzburg, Carlo
"Microhistory: Two or Three Things That I Know About It." *Critical Inquiry,* Vol. 20 (Autumn 1993), pp. 10–35.

Gjerde, John
From Peasants to Farmers: the Migration from Balestrand, Norway, to the Upper Middle West. Cambridge University Press, Cambridge, U.K./ New York, 1985.

Glassie, Henry H.
Folk Housing in Middle Virginia: A Structural Analysis of Historic Artifacts. University of Tennessee Press, Knoxville, 1975.

———. *Patterns in the Material Folk Culture of the Western United States.* Philadelphia: University of Philadelphia Press, 1969.

———. *Vernacular Architecture.* Material Culture / Indiana University Press, Philadelphia / Bloomington, 2000.

Goodridge, R.
A Year in Manitoba. W.R. Chambers, London, 1882.

Gracie, Bruce A.
"The Agrarian Response to Industrialization in Prairie Canada: 1900-1935." Ph.D. thesis, McMaster University, 1976.

Gray, James
Red Lights on the Prairies. Macmillan, Toronto, 1971.

Gregory, Brad S.
"Is Small Beautiful? Microhistory and the History of Everyday Life." *History and Theory*, Vol. 38, No. 1 (February 1999), pp. 100-110.

Grove, Frederick P.
Fruits of the Earth. McClelland and Stewart, Toronto, 1965.

Hall, D.J.
"The Manitoba Grain Act: An Agrarian Magna Charta?" *Prairie Forum*, Vol. 4, No. 1 (Spring 1979), pp. 105-20. Regina.

Hamilton, James B., and Donald E. Reid
"E.A. Partridge: Prairie Radical." Unpublished graduate paper, University of Saskatchewan, Saskatoon, 1971.

Harrison, Dick
"Rolvaag, Grove and Pioneering on the American and Canadian Plains." *Great Plains Quarterly*, Vol. 1, No. 4 (Fall 1981), pp. 252-62. Lincoln, NE.

Hedges, James B.
Building the Canadian West. Russell and Russell, New York, 1939.

———. *Federal Railway Land Subsidy Policy of Canada.* Harvard University Press, Cambridge, MA., 1934.

Hildebrandt, Walter
"P.G. Laurie of Battleford: The Aspirations of a Western Conservative." *Prairie Forum*, Vol. 8, No. 2 (Fall 1983), pp. 157-78.

Hofstadter, Richard
The Age of Reform. Vintage Books, New York, 1955.

Houghton, Walter E.
The Victorian Frame of Mind, 1830-1870. Yale University Press, New Haven, Connecticut, 1957.

Hunter, W.J.
Manhood Wrecked and Rescued. William Briggs, Toronto, 1894.

Irving, John A.
Social Credit Movement in Canada. University of Toronto Press, Toronto, 1959.

Jackel, Susan
"The House on the Prairies." *Canadian Literature*, No. 42 (Autumn 1969), pp. 46-55. Vancouver.

Jackel, Susan
"Introduction," in Georgina Binnie-Clark, *Wheat and Women.* University of Toronto Press, Toronto, 1979, xx-xvii.

Jordan, Terry G., Jon T. Kilpinen, and Charles F. Gritzner
The Mountain West: Interpreting the Folk Landscape. Johns Hopkins University Press, Baltimore and London, 1997.

Jordan, Terry G. and Matti Kaups
The American Backwoods Frontier: An Ethnic and Ecological Interpretation. Johns Hopkins University Press, Baltimore, 1989.

Jones, David C.
Empire of Dust: Settling and Abandoning the Prairie Dry Belt, 2nd Edition. University of Calgary Press, Calgary, 2002.

Katz, Michael
The People of Hamilton, Canada West. Harvard University Press, Cambridge, MA, 1975.

Katz, Michael, Michael Doucet, and Mark Stern
The Social Organization of Early Industrial Capitalism. Harvard University Press, Cambridge, MA, and London, U.K., 1982.

Kennedy, Jacqueline J.
"Qu'Appelle Industrial School: White Rites for the Indians of the Old North-West." M.A. thesis, Carleton University, Ottawa, 1970.

Kirk, D.W.
The Motherwell Story. Canada. Department of Agriculture, Regina, 1956.

LaDow, Beth
The Medicine Line: Life and Death on a North American Borderland. Routledge, New York and London, 2001.

Lalonde, Andre N.
"The North-West Rebellion and Its Effects on Settlers and Settlement in the Canadian West." *Saskatchewan History*, Vol. 27, No. 3 (Autumn 1974), pp. 95–102. Saskatoon.

Landis, Paul H.
Rural Life in Process. McGraw-Hill, New York, 1940.

Ledohowski, Edward M. and David K. Butterfield
Architectural Heritage: The Eastern Interlake Planning District. Winnipeg: Manitoba Department of Cultural Affairs and Historic Resources, 1983.
—. *Architectural Heritage: The MSTW Planning District*. Winnipeg: Manitoba Department of Cultural Affairs and Historic Resources, 1984.

Lehr, John C.
"Mormon Settlements in Southern Alberta." M.A. thesis, University of Alberta, Edmonton, 1971.
—. *Ukrainian Vernacular Architecture in Alberta*. Occasional Paper No. 1. Alberta Culture, Historic Sites Service, Edmonton, Alberta, 1976.

Le Roy Ladurie, Emmanuel
Montaillou: The Promised Land of Error (trans. Barbara Bray). Vintage Books, New York, 1979.

Leslie, Genevieve
"Domestic Service in Canada, 1880–1920," in Janice Acton, Pennie Goldsmith, and Bonnie Shepard (eds.), *Women at Work, 1850–1930*. Canadian Women's Educational Press, Toronto, 1974.

Levi, Giovanni
"On Microhistory," in Peter Burke, ed., *New Perspectives on Historical Writing*, 2nd Edition. Penn State University Press, University Park, Pennsylvania, 1991, pp. 99–100.

Lewis, Frank
"Farm Settlement on the Canadian Prairies, 1898 to 1911." *Journal of Economic History*, Vol. 41 (September 1981), pp. 517-35.

Lipset, Seymour M.
Agrarian Socialism: The Co-operative Commonwealth Federation in Saskatchewan. University of California Press, Berkeley, CA., 1971.

Loewen, Royden
Family, Church, and Market: A Mennonite Community in the Old and New Worlds. University of Toronto Press, Toronto, 1993.

Loveridge, Donald M.
"The Settlement of the Rural Municipality of Sifton, 1881–1920." M.A. thesis, University of Manitoba, Winnipeg, 1977.

Lüdtke, Alf, ed.
The History of Everyday Life. Reconstructing Historical Experiences and Ways of Life (trans. William Templer). Princeton University Press, Princeton, N.J., 1995.

Ludmerer, Kenneth M.
Genetics and American Society: A Historical Appraisal. John Hopkins University Press, Baltimore, MD., 1972.

Macdonald, N.A.
Canada: Immigration and Colonization, 1841–1903. Macmillan, Toronto, 1966.

Macdonald, Graham
"Mississippi River Valley: Historical Systems Plan Study." Ontario Ministry of Natural Resources, Eastern Regional Office, Kemptville, Ont., 1975.

MacIntosh, W.A.
Prairie Settlement: The Geographical Setting. Canadian Frontiers of Settlement Series, Vol. 1. Macmillan, Toronto, 1934.
———. *Economic Problems of the Prairie Provinces*. Canadian Frontiers of Settlement Series, Vol. 4. Macmillan, Toronto, 1935.

Macoun, John
Manitoba and the Great North-West. World Publishing Company, Guelph, Ontario, 1882.

Macpherson, C.B.
Democracy in Alberta: Social Credit and the Party System. University of Toronto Press, Toronto, 1953.

Malin, James C.
The Grassland of North America: Prolegomena to its History. P. Smith, Gloucester, Massachusetts, 1967.

Manitoba. University College of Agriculture.
Prairie Agriculture. W.J. Gage, Toronto, 1896.

Marr, William and Michael Percy
"The Government and the Rate of Canadian Prairie Settlement." *Canadian Journal of Economics / Revue canadienne d'Economique*, Vol. 11, No. 4 (November 1978), pp. 757–67.

Martin, Chester
"*Dominion Lands*" *Policy*. Canadian Frontiers of Settlement Series, Vol. 2. Macmillan, Toronto, 1938.

Maryn, Sonia
The Chernochan Machine Shed: A Land Use and Structural History. Occasional Paper No. 12, Alberta Culture, Historic Sites Service, Edmonton, 1985.

Mayor, James
Report to the Board of Trade on the North West of Canada, With Special Reference to Wheat Production for Export. King's Printer, London, 1904.

McClung, Nellie
Clearing in the West: My Own Story. Thomas Allen Ltd., Toronto, 1935.
———. *The Stream Runs Fast.* Thomas Allen Ltd., Toronto, 1945.

McCutcheon, Brian R.
"The Birth of Agrarianism in the Prairie West." *Prairie Forum*, Vol. 1, No. 2 (November 1976), pp. 79–94. Regina.
———. "The Patrons of Industry in Manitoba, 1890–1898." *Transactions of the Historical and Scientific Society of Manitoba*, Series 3, No. 22 (1965–66), pp. 7–26. Winnipeg.

McGowan, Don. C.
Grassland Settlers: The Swift Current Region During the Era of the Ranching Frontier. Canadian Plains Research Centre, Regina, 1975.

McKillican, W.C.
"Experiments with Wheat at the Dominion Experimental Farm, Brandon, Manitoba 1889–1923: A Summary." Issued by Canada, Department of Agriculture. Bulletin No. 42, New Series, Ottawa, 1924.

McLaren, Angus
Our Own Master Race: Eugenics in Canada, 1885–1945. McClelland and Stewart, Toronto, 1990.

McLeod, K.A.
"Education and the Assimilation of New Canadians in the North-West Territories and Saskatchewan, 1885–1934." Ph.D. thesis, University of Toronto, Toronto, 1975.

McManus, Sheila
The Line Which Separates: Race, Gender, and the Making of the Alberta-Montana Borderlands. University of Nebraska Press, Lincoln, 2005.
———. "'Their Own Country': Race, Gender, Landscape, and Colonization around the Forty-ninth Parallel, 1862–1900," in Stirling Evans, ed., *The Borderlands of the American and Canadian Wests: Essays on Regional History of the Forty-ninth Parallel.* University of Nebraska Press, Lincoln, Nebraska, 2006), pp. 117–30.

Meek, Lottie
"Early History of Blackwood District." Unpublished manuscript, Abernethy, Saskatchewan, 1955.

Mills, Ivor J.
Stout Hearts Stand Tall. Published by author, Vancouver, 1971.

Minifie, James M.
Homesteader. Macmillan, Toronto, 1973.

Moorehouse, Hopkins
Deep Furrows. George J. McLeod Ltd., Toronto and Winnipeg, 1918.
———. "Story of the Early Days - Hon. W.R. Motherwell, Regina" (excerpted from an interview by Moorehouse with Motherwell, 16 April 1916). *Saskatchewan History*, Vol. 8, No. 3 (Autumn 1955), pp. 108–12. Saskatoon.

Morris, Alexander
The Treaties of Canada with the Indians. Belfords, Clarke & Co., Toronto, 1880.

Morrison, W.R.
"'Their Proper Sphere': Feminism, the Family and Child Centered Social Reform in Ontario, 1875–1900," Parts I and II. *Ontario History*, Vol. 68, No. 1 (March 1976), pp. 45–64, and No. 2 (June 1976), pp. 65–74. Toronto.

Morton, A.S.
History of Prairie Settlement. Canadian Frontiers of Settlement Series, Vol. 2. Macmillan, Toronto, 1938.

Morton, W.L.
Manitoba: History. University of Toronto Press, Toronto, 1957.
———. "The Significance of Site in the Settlement of the American and Canadian Wests." *Agricultural History*, Vol. 25 (1951), pp. 97–104. Berkeley, Ca.

Morton, W.L.
"Victorian Canada," in W.L. Morton, ed., *The Shield of Achilles: Images of Canada in the Victorian Age.* McClelland and Stewart, Toronto, 1968.

Motherwell, W.R.
"Methods of Soil Cultivation Underlying Successful Grain Growing ...." Issued by Saskatchewan, Department of Agriculture, *Bulletin* No. 21, Regina, 1910.
———. "Oat Growing for the Qu'Appelle Valley." *Nor' West Farmer*, 20 April 1901, p. 239.

Murchie, R.W.
Agricultural Progress on the Prairie Frontier. Canadian Frontiers of Settlement Series, Vol. 5. Macmillan, Toronto, 1936.

Murchie, R.W., and H.C. Grant
Unused Lands of Manitoba. Manitoba Department of Agriculture and Immigration, Winnipeg, 1926.

Myers, Gustavus
A History of Canadian Wealth. Vol. 1. James Lewis and Samuel, Toronto, 1972.

Norrie, Kenneth
"Dry Farming and the Economics of Risk Bearing: The Canadian Prairies, 1870–1930," in Thomas R. Wessell, ed., *Agriculture in the Great Plains, 1876–1936.* The Agricultural History Society, Washington, D.C., 1977.
———. "The National Policy and the Rate of Prairie Settlement: A Review," in R. Douglas Francis and Howard Palmer, eds., *The Prairie West: Historical Readings.* Pica Pica Press, Edmonton, 1995, pp. 243-263.

Oakley, Ann
Housewife. Allen Lane Publishers, London, 1974.

Ostergren, Robert Clifford
A Community Transplanted: the Trans-Atlantic Experience of a Swedish Immigrant Settlement in the Upper Middle West, 1835–1915. University of Wisconsin Press, Madison, Wisconsin, 1980.

Owram, Douglas
The Promise of Eden: The Canadian Expansionist Movement and The Idea of the West, 1856–1900. University of Toronto Press, Toronto, 1980.

Palmer, Howard
Land of No Second Chance: A History of Ethnic Groups in Southern Alberta. The Lethbridge Herald, Lethbridge, AB, 1972.

Partridge, Michael
Farm Tools Through the Ages. New York Graphic Society, Boston, 1973.

Patton, Harald S.
Grain Growers' Cooperation in Western Canada. Harvard University Press, Cambridge, MA, 1928.

Peel, Bruce B.
"R.M. 45: The Social History of a Rural Municipality." M.A. thesis, University of Saskatchewan, Saskatoon, 1946.

Pinno, Erhard
"Temperance and Prohibition in Saskatchewan." M.A. thesis, University of Saskatchewan, Regina, 1969.

Pomfret, Richard
"The Mechanization of Reaping in Nineteenth Century Ontario: A Case Study of the Pace and Causes of the Diffusion of Embodied Technical Change." *Journal of Economic History,* Vol. 36, No. 2 (June 1976), pp. 399–415. Atlanta, GA.

Potyondi, Barry
In Palliser's Triangle: Living in the Grasslands, 1850–1930. Purich Publishing, Saskatoon, Saskatchewan, 1995.

Powell, Aaron M.
The National Purity Congress: Its Papers, Addresses, Portraits. First Published, 1896; Arno Press Reprint, New York: New York, 1976.

Rasmussen, Linda, Lorna Rasmussen, Candace Savage, and Anne Wheeler.
A Harvest Yet to Reap: A History of Prairie Women. The Women's Press, Toronto, 1976.

Ray, Arthur J.
Indians in the Fur Trade: Their Role as Trappers, Hunters, and Middlemen in the Lands Southeast of Hudson Bay, 1660–1870. University of Toronto Press, Toronto, 1974.

Ray, Arthur J., Jim Miller, and Frank Tough
Bounty and Benevolence: A History of Saskatchewan Treaties. McGill-Queen's University Press, Montreal and Kingston, 2000.

Rea, J.E.
"The Roots of Prairie Society," in David Gagan, ed., *Prairie Perspectives.* Vol. 1. Holt, Rinehart and Winston, Toronto, 1970.

Reynolds, Thomas M.
"Justices of the Peace in the North-West Territories, 1870–1905." M.A. thesis, University of Regina, 1978.

Richards, John, and Larry Pratt
Prairie Capitalism: Power an Influence in the New West. McClelland and Stewart, Toronto, 1979.

Richtik, James M.
"Manitoba Settlement: 1870–1886." Ph.D. dissertation, University of Minnesota, 1971.
——. "The Policy Framework for Settling the Canadian West, 1870–1880." *Agricultural History,* Vol. 49, No. 4 (October 1975), pp. 613–28. Berkeley, CA.

Ricou, Lawrence
Vertical Man/Horizontal World: Man and Landscape in Canadian Prairie Fiction. University of British Columbia Press, Vancouver, 1973.

Rotundo, E. Anthony
"Romantic Friendship: Male Intimacy and Middle-Class Youth in the Northern United States, 1800–1900." *Journal of Social History*, Vol. 23, No. 1 (Autumn 1989), pp. 1–25.

Rutherford, Paul
"The New Nationality, 1864–1897: A Study of the National Aims and Ideas of English Canada in the Late Nineteenth Century." Ph.D. thesis, University of Toronto, 1973.

Sandwell, R.W.
Contesting Rural Space: Land Policy and Practices of Resettlement on Saltspring Island, 1859–1891. McGill-Queen's University Press, Montreal and Kingston, 2005.

Saskatchewan, Department of Agriculture
Report of the First Annual Convention of the Homemakers' Clubs of Saskatchewan. Regina, 1911.

Schrodt, Barbara
"Sabbatarianism and Sport." *Journal of Sport History*, Vol. 4, No. 1 (Spring 1977), pp. 22–33. Radford, VA.

Smith, Allen
"The Myth of the Self-Made Man in English Canada, 1850–1914." *Canadian Historical Review*, Vol. 59, No. 2 (1978), pp. 189–219. Toronto.

Smith, David
Prairie Liberalism: The Liberal Party in Saskatchewan, 1905–71. University of Toronto Press, Toronto, 1975.

Snell, J.F.
"The Cost of Living in Canada in 1870." *Social History/Histoire sociale*, Vol. 12, No. 23 (May 1979), pp. 189–93. Ottawa.

Spafford, D.S.
"The Elevator Issue, The Organized Farmers and The Government, 1908–11." *Saskatchewan History*, Vol. 15, No. 3 (Autumn 1962), pp. 81–92.

Spector, David
"Agriculture on the Prairies, 1870–1940." *History and Archaeology/Histoire et archéologie*, No. 65, Parks Canada, Ottawa, 1983, pp. 215–60.

Sprague, D.N.
"Government Lawlessness in the Administration of Manitoba Land Claims, 1870–1887." *Manitoba Law Journal*, Volume 10, Number 4 (1980), pp. 415–41.

Stavenhagen, Rodolfo
Social Classes in Agrarian Societies. Anchor Books, New York, 1975.

Street, A.G.
Farmer's Glory. Macmillan, Toronto, 1934.

Stueck, Nelson
North of the Qu'Appelle. United Church Women's Committee, Abernethy, Saskatchewan, 1967.

Swierenga, Robert
"Towards the 'New Rural History': A Review Essay." *History Methods Newsletter*, Vol. 6, No. 3 (June 1973), pp. 111–21.

Sykes, Ella C.
A Home Help in Canada. G. Bell & Sons, London, 1912.

Sylvester, Ken
The Limits of Rural Capitalism: Family, Culture, and Markets in Montcalm, Manitoba, 1870–1940. University of Toronto Press, Toronto, 2001.

Szijártó, István
"Four Arguments for Microhistory." *Rethinking History*, Vol. 6, issue 2 (June 2002), pp. 209–15.

Talmage, T. De Witt
The Night Sides of City Life. J. & A. McMillan, Saint John, New Brunswick, 1878.

Thomas, Lewis G.
"Alberta Perspectives." *Alberta History.* Vol. 28, No. 1 (Winter 1980), pp. 1–5.

Thomas, Lewis H.
The Struggle for Responsible Government in the North-West Territories, 1870–97. University of Toronto Press, Toronto, 1978.

Thompson, John H.
"Bringing in the Sheaves: The Harvest Excursionists, 1890–1929." *Canadian Historical Review*, Vol. 59, No. 4 (Summer 1978), pp. 467–89.
——. *The Harvests of War: The Prairie West, 1914–1918.* McClelland and Stewart, Toronto, 1978.
——. "'Permanently Wasteful but Immediately Profitable': Prairie Agriculture and The Great War." Canadian Historical Association, *Historical Papers 1976* (1976), pp. 193–206. Montreal.
——. "The Prohibition Question in Manitoba, 1892–1928." M.A. thesis, University of Manitoba, 1969.

Thwaite, Leo
Alberta: An Account of Its Wealth and Progress. George Routledge & Sons, London, 1912.

Tobias, J.L.
"Canada's Subjugation of the Plains Cree, 1879–1885." *Canadian Historical Review*, Vol. 50, No. 4 (December 1983), pp. 519–48.

Tracie, Carl
"Toil and Peaceful Life": Doukhobor Village Settlement in Saskatchewan, 1899–1918. Canadian Plains Research Centre, Regina, 1996.

Treiman, Donald
"A Standard Occupational Classification in History." *Journal of Interdisciplinary History.* Vol. 3, No. 1 (Summer 1972), pp. 63–88.

Troeltsch, Ernst
The Social Teachings of the Christian Churches, Vol. 2. Harper Torchbooks, New York, 1960.

Turner, Allan R.
"W.R. Motherwell and Agricultural Development in Saskatchewan, 1905–1918." M.A. thesis, University of Saskatchewan, Saskatoon, 1958.
——. "W.R. Motherwell and Agricultural Education, 1905–1918." *Saskatchewan History*, Vol. 12, No. 1 (Winter 1959), pp. 81–96. Saskatoon.
——. "W.R. Motherwell: The Emergence of a Farm Leader." *Saskatchewan History*, Vol. 11, No. 3 (Autumn 1958), pp. 94–103. Saskatoon.

Vanek, Joann
"Work, Leisure and Family Roles: Farm Households in the United States, 1920–1955." *Journal of Family History*, Vol. 5, No. 4 (Winter 1980), pp. 422–31. Minneapolis.

Veblen, Thorstein
The Theory of the Leisure Class. Modern Library, New York, 1934.

Voisey, Paul
Vulcan: The Making of a Prairie Community. University of Toronto Press, Toronto: 1988.

Vogt, Evon Z.
Modern Homesteaders: The Life of A Twentieth Century Frontier Community. Harvard University Press, Cambridge, MA, 1955.

Warkentin, John
"The Mennonite Settlements of Southern Manitoba," Ph.D. dissertation, University of Toronto, 1960.

Webb, Walter Prescott
The Great Plains: A Study in Institution and Environment. Ginn and Co., Boston, 1931.

Widdis, Randy William
With Scarcely a Ripple: Anglo-Canadian Migration into the United States and Western Canada, 1850–1920. McGill-Queen's University Press, Montreal and Kingston, 1998.

Wilson, M.C. and I.J. Dijks
"Introductory Essay – Land of No Quarter, the Palliser Triangle as an Environmental-Cultural Pump," in R.W. Barendregt, M.C. Wilson, and F.J. Jankunis, eds., *The Palliser Triangle, A Region in Space and Time.* University of Lethbridge, Lethbridge, Alberta, 1993.

Wilson, W.A.
A Century of Canadian Grain: Government Policy to 1951. Western Producer Prairie Books, Saskatoon, 1978.

Wood, Lewis Aubrey
A History of Farmers' Movements in Canada: The Origins and Development of Agrarian Protest, 1872–1924. University of Toronto Press, Toronto, 1975.

Wood, Susan
"God's Doormats: Women in Canadian Prairie Fiction." *Journal of Popular Culture*, Vol. 14, No. 2 (Fall 1980), pp. 350–59. Bowling Green, OH.

Yule, Annie I.
Grit and Growth: The Story of Grenfell. United Church Women's Committee, Wolseley, Sask., 1967.

Zimmerman, C.C., and Seth Russell
Symposium on the Great Plains of North America. Institute for Regional Studies, Fargo, ND, 1967.

INDEX

A

abandonments. *See* cancellations
Abernethy *Abernethan*, 29–30, 163, 169
Abernethy Agricultural Society, 145
Abernethy district, 43, 86, 218
   age of settlers, 32, 48
   breaking or land clearing, 61, 66
   climate, 1–2
   cost of building shelter (*See* homesteading costs)
   crop yields, 83
   cultivation (first year), 59–60
   economic development, 67–69
   farm size, 208
   farmers (*See* Abernethy farmers; Anglo-Canadian settlement group)
   geographical setting, 1–3
   land rush, 20 (*See also* land selection)
   precipitation, 3
   Protestant Anglo-Canadian community, 140–41, 179 (*See also* religion)
   rail linkage, 29, 44, 68, 82, 84
   social and economic structure (*See* Abernethy's social and economic structure)
   social creed (*See* Abernethy's social creed)
   soil characteristics, 1–3, 111, 113, 208
   stone houses (*See* houses)
   surveyors' comments, 5–6
   topography, 2
   vegetation, 1
   wheat production and export, 69, 82, 96, 98
Abernethy farmers. *See also* Anglo-Canadian settlement group
   "conspicuous consumption," 227
   identification with business class, 227
   joint stock company, 169
   mortgage money, 86
   optimistic orientation, 218
   prosperity, 81–84, 227
   starting capital, 48–49
Abernethy, township around, 34
Abernethy village, 84, 137
Abernethy's social and economic structure, 133–34, 136, 226
   formative period, 144, 167
   hierarchy, 167
   leadership, 144

Abernethy's social creed, 171
   individual enterprise, 172–73
   materialistic orientation, 173
   Protestant creed and, 179
   religious orientation, 175, 179
   respectability, 174
   "social purity" movement, 180, 182
   work ethic, 173–74
Abernethy's socio-economic groupings (pre-1914), 135
Aboriginal peoples. *See* First Nations peoples
absentee business class, xxvi
absentee landlords, 137
Acton, Richard, 37
Adelman, Jeremy, *Frontier Development*, xxvii
Adolf, Jacob, 43
adultery. *See* sexuality
age of population, 157
   Abernethy settlers, 48
   German settlers, 32, 157
agrarian movement, xiv, xv, xxiii, 44, 134, 191, 225
   anti-corporate sentiments, 217
   Central Canadian dominance and, 135, 217, 221, 226
   contradictions in, xxiv
   divisions in, 211, 214, 218, 221
   farmers' "revolt," 200
   Grain Growers' Grain Company, 211
   "indignation meeting," 200, 216
   Manitoba and North West Farmers' Union, 192–93
   and monopolistic transportation and grain handling, 73, 194–95, 198–99, 203–4
   National Policy and, 192
   Partridge, E.A., 211
   Patrons of Industry, 196–98, 216
   Protective Union, 193–96
   radical/conservative split, 211, 215
   repeated hope and disappointment, 217–18
   significance, 192
   socio-economic differences within, xxv, 218–19
   temporary unified front, 216
   Territorial Grain Growers' Association, 191, 202–10, 216

wheat blockade and, 201–2
agrarian radicalism
    ephemeral nature of, 215
    precipitating issues, 215
Agricultural Credit Commission, 88
Alberta, 142, 188
alcohol consumption, 154, 179, 181. *See also*
        Prohibition; temperance
    moral battle against, 183
"aliens," 141–42
Allan, Hugh, 7
Allen, John, 157
Allen, Richard, 176
American agricultural historians, xix
American settlement studies, xxx
Anglican Church, 141
Anglo-Canadian dominance, 188, 225, 230
    economic and political power, 138–39, 184
    use of Protestant creed, 179
Anglo-Canadian settlement group, xix, 19.
        See also Abernethy farmers
    advantage over subsequent settlers, 44
    age, 32, 48
    assimilationist drive (*See* assimilation)
    cost of stables, 54
    as ethnic group, xxvi
    ethnocentrism, 141, 188–89
    marital status, 32
    materialistic values, 173
    persistence rates, xxiv, 34
    petitions for extensions, 39
    political dominance, 39, 44
    pooled labour and sharing of resources, 62
    Protestant creed, 174–75, 179 (*See also* religion)
    pursuit of respectability, 174–75
    stratified social structure, 133
    symbols and rituals of parent society, 159, 175, 223
    temperance movement, 142, 177, 183, 225
"Anglo-Saxon predominance," 190
Ankli, Robert, 50, 54, 57–58, 64, 97
    "The Growth of Prairie Agriculture," 244n1
Annis, A.W., 204
anti-corporate sentiments, 217
anti-immigrant sentiment, 190
application for patent, 49. *See under* patent
Arcola, 160
Arsenault (homestead inspector), 44

assimilation
    Aboriginal peoples, 183–84
    of immigrants, 183–84, 189
    language issues, 184, 188–89
    of non-Christian cultures, 177
Assiniboine River, 2

B

"Bachelor Billy," 153–54
bachelors, 147, 156
    bachelor homesteads, 52, 148–49, 154, 255n26
    cohabiting, 149–50
    competition for women, 153
    isolation, 151
    living alone, 150
    loneliness, 148
    sexual behaviour, 154
    social taboos and, 179, 181
"Bachelors, beware...", 180
bachelors' balls, 153
Balcarres, 29, 43, 49, 84, 217
Balfour, George, 43
Banish-the-Bar Crusade, 183
Bantjes, Rod, xxxi
    Improved Earth, xxx
Baptist Church, 141
barns and outbuildings, 54, 85, 97, 103, 129
    farm capital in, 96
    Motherwell barn, 168
barter, 69, 78
Batoche, 49
Battleford, 25–26
Bearden, Arthur, 125
Beaty, James, 28, 30
Beaver Hills, 69
Belfast Newsletter, 79
Bell, Charles N., 200
Bennett, John W., xx, 139
    Northern Plainsmen, xxvi, 138
    Settling the Canadian-American West, xxx
    "social credit," 142
Benson, Nathaniel, *None of It Came Easy*, 237n1
Beynon, Lillian, 144
Bidney, David, 171
binders, 81, 114
Binnie-Clark, Georgina, 54, 56, 158
    crop yields, 89
    excessive reliance on wheat, 91
    financial over-extension, 91
    grain-pickling, 108
    mortgages, 89, 91
    revenues and expenses, 90

Bittner, Jack, 104
Blackwood, J.A.R., 19
Blackwood district, 77
Blair, Andy, 72
Blakesley, E.A., 166
Bloch, Marc, *French Rural History*, xix, 101
"bluestone," 108
Boam, Henry J., *Prairie Provinces of Canada, The*, 57
Bogue, Allan G., 144, 172
    "Farming in the North American Grasslands," 238n5
    "Social Theory and the Pioneer," xix
bona fide settlers, 7, 14, 16–17, 23–24, 42
borderlands studies, xxvii, xxviii
Bourgeois, J., 5
"Box Social," 153
"Boys' Brigade" of the Methodist Church, 181
Brandon, 19
Brandon Sun, 193
Braudel, Fernand, *The Mediterranean and the Mediterranean World in the Age of Philip II*, xx
breaking (land clearing), 61–62, 66, 248n7
    cost of, 60
    slow rate of, 208
brome grass and seed, 81, 110, 112
brome hay, 104
buffalo, 11, 45
Burnet, Jean, *Next Year Country*, xviii
Burton, John, 49

C

"cabooses," 163
Caffrey, Kate, 129
Calvinism, xiii, 175–76, 178
Campbell, Dave, 152
Canadian-American Joint High Commission (1898-99), 216
Canadian Frontiers of Settlement series, xviii, xix
Canadian National Committee on Mental Hygiene, 188
Canadian Order of Foresters, 158
Canadian Pacific Railroad. *See* CPR
Canadian Pacific Railway Act (1872), 7
cancellations, 22–23, 25, 41–42, 77, 82, 157
    German *vs.* Anglo-Canadian rates, 32
    owing to lack of water, 57
    Neudorf District, 32
    rail service and, 29
capital, 134
    farm buildings, 96
    farm capital expenditures (or cost of production), 92
    shortage of, 69, 78, 86, 88
capital accumulation, 84, 94, 136
    Abernethy farmers, 94, 97, 219
    from increase in land value, 84
    Neudorf, 94, 97
capital-intensive agriculture, 68. *See also* commercial agriculture
car distribution. *See* grain cars
card playing, 154, 179, 181
Carter, Sarah, xxii, xxviii, xxix, 13
    "Materials History of the Motherwell Home, A," xxxi
cash. *See* capital
Castle, C.C., 220
Caswell, Stephen H., 17
cattle losses, 77–78
C.C.F., xxiv, 133
Census of Canada (1891), 147
Census of Canada (1901), 149
Central Canada
    business interests, 135
    exploitation, 217
    western economic dependence, 214, 221, 226
Charlebois, A., 27
Chipperfield, Samuel, 76–77, 116, 128–29, 209
Chipperfield family, 145
Chipperfield farm, 113
Christian church, 225. *See also* religion
    Anglican Church, 141
    Baptist Church, 141
    evangelical churches, 181
    on issues of sexuality, 181
    Methodist Church, 141, 175, 180–81
    Presbyterian Church, xiii, 141, 144, 175, 180
    role in Victorian Canada, 175
    Roman Catholic clergy, 15
    social and moral authority, 175
Christianizing the community, 105, 175–77
church. *See* Christian church
church union, 144
"claim jumpers," 43
Clarke, Ian, xxxi
Clarke, W.A., 18
classical political economy, xxvi
Clemow, Francis, 26, 30
cliometrics, xxi
co-operative activities, 66
    beef ring concept, 169–70
    cultivating crops "on shares," 61–62
    effect on homesteading costs, 61–62
    joint stock companies, 169

myth of co-operative society, xxiv
pooled labour and resources, 31, 53, 61, 66, 168
shared ownership of implements, 119, 168–69
through necessity, 224
Co-operative Commonwealth Federation, xxiv, 133
co-operative principle
Partridge's belief in, 213–14
Codd, J.A., 26–27
cohabiting males, 149–50
farmer with hired man, 149–50
Colonization Companies, 78
commercial agriculture, xxvii, 63, 69, 229
First Nations' interest in, 12–13, 242n19
land requirements, 35
transition to, 67
community prestige, 138
community service, 131, 139, 142. See also social roles and responsibilities
women, 142, 186
conservatism, 215–16, 239n25
Women Grain Growers Association, 186
Conservative government, 9. See also Dominion government
continued extensions to railway, 45
corruption charges, 44–45
CPR concept, 7
National Policy, 3–4, 7, 192
Conservative Party, 201, 216
consolidation of farmlands, 134, 228
conspicuous consumption, 145, 173, 227, 230
Conzen, Michael, xix
Frontier Farming in an Urban Shadow, 20
Copithorn, Samuel, 49
corruption charges, 30, 44–45
Great North West Central Railway, 27–28
Cowie, Isaac, *Edmonton Country, The*, 54
CPR, 21, 23, 31, 68–69, 126–29, 217–18
assumed Great North West Central Railway, 28
building of, xvii
employment on, 62
Kirkella branch, 84
land and land prices, 7, 15–16, 36, 50, 64, 85, 198
monopoly, 9, 192, 194–95
priority to elevator companies, 198, 203–4
terminus at Brandon, 19
wheat blockade (1901), 88, 200–202, 215, 217, 261n16

Cree agriculture, 10–11, 13, 142, 225, 242n19
implements of cultivation, 11
Cree entrepreneurship
discouraged by Dominion government, 12
crop failures, 22, 39, 67, 75–76, 78
drought, xxiv, xxx, 67, 76, 78
early frost, 67, 70, 76–78
gophers, 78, 96, 248n17
crop rotation, 81
crop yields, 67–69, 75–82. See also wheat yields
soil packing and, 112
summer fallowing and, 80
variability, 75, 77, 83
Crossley, H. T., 182
crown lands. *See* land
Crow's Nest rates, 73
cultural dominance. *See* Anglo-Canadian dominance
cultural superiority of Anglo-Saxon "races" asserted, 188
cultural transplantation, xix, xxii, xxvi, 175, 224–25
Cummins Rural Directories, 37
Curti, Merle, xix

D

daily life (Motherwell farm), 101–31
dances, 153–54
David, Ferdinand, 151
Davin, N.F., 39
Davis, T.O., 205–6
Dayman, Peter, xxv, 191, 200–203, 218
Conservative supporter, 216
"Decline of Marriage, The," 180
Deker, Herbert, 150
democracy, 147
demographics, xxii, xxxii, 32
age of settlers, 48
farm population, 68
male/female population imbalance, 147, 255n1
demurrage charges, 199
Department of the Interior, 31, 37–38, 43–44, 77
Dewdney, Edgar, 15, 18, 39
Dick, Lyle
Farmers "Making Good" (1st ed.), xx, xxi, xxii, xxiii, xxvi, xxvii, xxiv, xxv, xxvi
challenge to prevailing notions, xxiii
socio-economic analysis in, xxvii
"History of Prairie Settlement Patterns, A," xxxi, 240n33

Dickenson, C.S., 76
Diehl, Marian, 229
disallowance legislation, 192
District of Assiniboia, 39, 49, 192, 200
  crop yields, 83
  wheat production for export, 69–70
diversification, 80, 94, 98
Dixon Brothers, 76
Dodd, R.J., 17
Domestic Engineering, 164
domestic servants, 160, 185. *See also* "hired girls"
  low status, 135, 164
Dominion Experimental Farm, 80
Dominion government, 9, 199
  accession to TGGA demands, 220
  corruption charges, 44
  discouraging Cree entrepreneurship, 12
  extensions to railways, 44–45
  National Policy, 3–4, 7, 192
  patriarchal policies, 180
  railway policy, 221
  responsibility for railway failures, 30
  responsiveness to farmers, 217, 221
  seed grain advances, 58
  transportation and tariff policies, 7, 196, 216
Dominion Land Office at Fort Qu'Appelle, 19
Dominion land surveyors, 5
Dominion Lands Act and Regulations, xx, 4–5, 7, 31, 38, 40–41, 50, 225. *See also* National Policy
  administration of, 41–43
  amendments (1884), 9–10, 39
  American expansionism and, 4
  free homestead policy (*See* free-grant land)
  frequent changes, 9, 38
  grid survey, xxxi, 4, 17–18, 38
  human toll, 38, 40 (*See also* hardship)
  inequities in legislation, 42
  pre-emptions, xxiv, 9–10, 35, 39–40, 44–45, 76–77, 91
  quarter-section as homestead unit, 38, 97
  residence and cultivation requirements, 10, 33, 48, 53
  second homestead entry, 9–10, 208
  success of, 37–38
  unsuitable land, 40
Doucet, Michael, 134
Douglas, Dr., 182
Douglas, J.M., 198–99
Douglas, James, 205
Doukhobor settlements, xxxi

dress, 130, 145, 173, 224
drill-seeders, 110
drought, xxiv, xxx, 67, 76, 78
dry farming, 80, 91, 99, 109
duck-foot cultivator, 112
Duck Lake, 15
"dude" ranches, 154
dugout dwellings, 51

E
early frost, 67, 70, 76–78
early settlers (pre-1900), 68–69. *See also* timing of settlement
  capital gain from increase in land value, 84, 99
  economic advantage, 97
  egalitarianism, 223
  free grant land, 97
  leading positions in Abernethy society, 144–45
Eastern European immigrants. *See also* German settlers
  identified with enemy (First World War), 183
  prohibition and, 259n46
Eastern European settlements
  basic problems of subsistence, 139
  disadvantages, 225
economic and political power, 138–39
economic change, xxxii
economic depression (1870s), 3
economic development (Abernethy district)
  crop yields (1880–1900), 67–69
  developing prosperity (1900–1910), 82–85
  farm debt, 86–91
  initial settlement period (1880–1900), 67–82
  wheat prices (1880–1900), 67–75
  wheat prices (First World War), 92–99
egalitarian cooperative society, myth of, xxiv
egalitarianism, xxiv, 147, 156, 224
  early settlement era, 223
elevator companies, 73, 128, 209
  discriminatory practices, 89
  monopoly, 198–99, 203–4
Elkington, Walter, 57, 148, 151, 154
Ellisboro, 151
Elofson, Warren, xxviii
Epworth League of Christian Endeavour, 157, 181
Equal Rights League, 144
ethnic prestige, 138

ethnicity, xviii, xxvi. *See also* Anglo-Canadian settlement group; First Nations; German settlers; Métis
  ethnic prestige, 138
  language barriers, 44
  status and, 138, 140–41, 225
ethnocentrism, 141, 188–90
Eugenics movement, 187–88
Euro-Canadian and British settlers. *See* Anglo-Canadian settlement group
European immigrants. *See* Eastern European immigrants; German settlers
evangelical churches, 181
evangelical Presbyterianism. *See* Presbyterian evangelism
Evans, F.S., 76
Evans, Stirling, *Borderlands of the American and Canadian West, The*, xxvii–xxviii

F

failure. *See also* cancellations
  domestic, 230
  post-1900 settlers, 89–91
fall harvest. *See* harvesting
fallow. *See* summer fallowing
family, xxi
  and kinship relationships, xxi, xxii, 31
  nuclear family, xxix, 181
family size, 32, 52
  homesteading costs and, 47, 64
fanning mills, 81, 107, 126
farm consolidation, 224
farm debt, 86–92, 94, 96, 99, 218. *See also* mortgages
  to Colonization Companies, 78
  Farm Indebtedness study, 94, 96–97, 137
  to general stores, 87–88
  to implement companies, 86–87, 217
  smallest for early settlers, 97
  soil characteristics and, 97
farm failure. *See* failure
farm implements, 101. *See also* mechanization
  cost of, 58, 81, 93, 246n52, 247n53, 247nn55–57
  cost of (table), 60
  Cree, 11
  financing, 85, 217
  implement companies, 86–87, 217
  tariff on, 81
farm incomes, 97–98. *See also* prosperity
Farm Indebtedness study, 94, 96–97, 137
farm labourers. *See also* hired men
  booklet for, 174, 224
  decreased need for, 133, 228
  harvest gangs, 62, 119, 122, 125
  not a permanent class, 135
  status, 140, 224
farm machinery. *See* farm implements; mechanization
farm-making costs. *See* homesteading costs
farm population, 68
farm proprietor class, 134–35. *See also* property ownership
  economic differences within, 136
  "quasi-colonial" status, 214
  rentier component, 136
  separate interests from labourers, 197, 224
  status, 140
farm protest movements. *See* agrarian movement
farm size, 38, 68, 97–98, 136, 207
  farm consolidation, 134, 224, 228
  income and, 98
  increase after First World War, 228
  small, 97, 228
  trend toward larger, 97
farm tenancy. *See* tenant-operated farms
farm women. *See* women
farmer/employee relations, 151, 164–67
  booklet for farm labourers, 174, 224
  close and unstructured, 151
  cohabiting arrangements, 149–50
farmers. *See also* Abernethy farmers; German settlers; homesteading
  central Canadian business domination, 214, 221, 226
  as competitors in free market, 213–16, 221
  farmers' era in Canadian politics, xv
  First Nations, 13
  illusion of independent entrepreneurship, 135, 213–16, 221
  Métis, 14, 19
Farmers' Alliance, 196
farmers' movement. *See* agrarian movement
Farmers' Union convention (1925), 213
farmers' wives
  expanded role of, 185–86
  farm labour, 32
  social responsibilities, 158
  status and influence, 145
farmhouses. *See* houses
"feeblemindedness." *See* Eugenics movement
female domestics. *See* domestic servants; "hired girls"
female suffrage, 186, 198

feminism, 186
fencing, 114, 129
  costs, 56, 246n36
fictionalized accounts of settlement experience, 229
  failure of domestic life in, 230
  Grove, F.P., 139, 157, 164, 229
  McClung, N. 106, 144, 166, 169–70, 172, 179
File Hills area, 18, 142, 225
  soil characteristics, 2
File Hills Cree. *See* Cree
File Hills Métis. *See* Métis
financial over-extension. *See* farm debt
fireguards. *See under* prairie fires
First Nations peoples, xxix
  assimilation, 13, 183–84
  cross-border movements of, xxviii
  farmers, 13
  "hired girls," 224
  impact of Euro-Canadian settlement, 13–14, 45, 225
  Indian Treaty No. 4, 10–13
  North-West Rebellion and, 22
First World War, 228
  demand for exported farm products, 92
  ideology of womanhood, 185
  impact on prairie economy, 92–93
  inflated prices, 93–94
  temperance movement during, 183
Fisher, Albert, 17
"formalin," 108
Fort Ellice, 11, 25–26
Fort Garry, 4
Fort Qu'Appelle, 12, 15, 19
  crop yields, 76–77
  treaty signing, 11
Fort Qu'Appelle Tennis Club, 158
Fort William, 70, 206, 212
Fort William price, 82
Fowke, Vernon, xviii, xx, 38, 69, 135, 208, 211, 213, 220–21
  National Policy and the Wheat Economy, The, 207, 237n3, 260n3
Franco-Manitobans, xxvi. *See also* Métis
fraternal orders, 225
fraud, 17, 205–6, 225
free-grant land, xxiv, 7, 19, 37, 64, 97
  eligibility for, 91, 180
  exclusion of women, 180
  land speculation and, 23
  "proving-up" period (*See* "proving-up")
  unequal access to, 38, 40, 98, 180, 208
free homestead system. *See* free-grant land

free trade, 195
freight rates, 73, 82
French *Annales* tradition, xix
French language rights, 188–89
Fried, Morton, 133
Frontiers of Settlement series. *See*Canadian Frontiers of Settlement series
frosted wheat. *See* early frost
Fruits of the Earth (Grove), 139, 150, 164
funerals, 152–53

G

Gaddis, James, 43
Gagan, David, 48
Gallant, Dan, 104
gambling, 154, 181. *See also* card playing
gardens, 62, 80, 102, 105, 114
Gardiner, James, xiii
Garratt, W.H., 76
Garratt, Wallace, 76
Gelder, Willem de, 52, 157
gender, xxviii, xxxii
gender relations
  prevailing views on, 181 (*See also* patriarchy)
general stores, 87–88
German settlers, xxiii, 30. *See also* Neudorf District
  age and marital status, 32
  breaking (or land clearing), 66
  cancellation rates, 32
  capital accumulation, 97
  cultivation (first year), 59–60
  disadvantage in grain marketing, 208
  Dominion Lands Act and, 40–41, 44
  early success (seeming), 32, 36
  emigration to US, 37
  family size, 32–33
  hardship, 37
  land selection, 30–31
  language barriers, 31, 41, 44
  New Toulcha colony, 49
  persistence rates, xxiv, 32, 34, 36
  pooled labour and shared resources, 62
  rapid disappearance after issuance of patent, 36–37
  time of settlement (after pre-emption), 35, 40, 44–45
GGGC. *See* Grain Growers' Grain Company
Gibbons, David, 150
Gillespie, Catherine. *See* Motherwell, Catherine
Gillespie, J.B., 204
Gillespie, J.R., 208

Gillespie, Janet, 129, 186
Ginzburg, Carlo, xx
Gjerde, John, xix
Glassie, Henry, xxxi
gophers, 67, 78, 96, 248n17
Govan, Walter, 218
government ownership, 213, 220–21. *See also* publicly owned elevators
grain "blockade," 200–202, 215, 217, 261n16
grain cars, 198–99
    distribution inequities, 205–6, 208–9
Grain Commission. *See* Royal Commission on the Shipping and Transportation of Grain
grain cultivation, 1, 21. *See also* wheat
Grain Growers' Grain Company, xxv, 211
grain handling reforms, 91, 198, 220–21, 226. *See also* Manitoba Grain Act
    reinforced social and economic differences, 205–7, 226
grain harvest. *See* harvesting
grain hauling, 73–74, 126–29
grain inspection system, 198
grain prices. *See* wheat prices
grain separators, 119, 122
granaries. *See* barns and outbuildings
grasshopper infestations, 76
Gray, James, 154
Great North West Central Railway, 25–30, 217. *See also* North West Central Railway Company
    corruption charges, 27–29, 44–45
Greenway, Thomas, 198
Greenwood, David, 233
Grenfell, 69, 84
Grenfell-Wolseley area
    farm debt, 94
grid survey, xxxi, 4
    problems with, 17–18, 38
Grove, Frederick Philip, 157, 229
    Fruits of the Earth, 139, 150, 164

H

hailstorms. *See* crop failures
Hall, D.J., 191, 198, 221
Hall, John R., 16–17
Hamilton, Ont., 134
Hansen, Per (fictional character), 230
hardship, 37, 78–79
    distance from railway and, 44
    harvest gangs, 125
    Métis, 18
hardship prestige, 138–39
Harrison, Dick, 229

harrowing, 109, 112
Hart-Parr tractor, 113
Hartwell, George, 168
harvesting, 114–25
    farm women, 106
    harvest gangs, 62, 119, 122, 125
    stacking, 115–17
    stooking, 114–15
    threshing, 62, 119, 122–24
Haultain, F.W.G., 201
hay and haying, 51, 58–59, 62, 103, 110, 114, 168
    brome, 104
    hay scarcity, 78
    sale of hay, 41, 69, 98
Herd and Estray Animals Ordinances, 169
heterosexuality within marriage, 181–82
hierarchical organization of homes, xxxi, 160–61, 163. *See also* Motherwell house
hierarchical social conventions, xxv, xxxi, 133–34, 147, 158, 223–24
"hired girls," 102, 106, 159. *See also* domestic servants
    duties, 104–6
    European or First Nations, 164, 224
    hired girls' room, 161
    isolation, 164
    status, 164, 166, 224
    wages, 164
hired men, 128–29. *See also* farm labourers
    care of livestock, 103–4
    difficulty collecting wages, 138
    duties, 102
    fall harvest, 114–25
    hired men's room, 161
    insecure and transient, 224
    lack of status or power, 138
    social status, 137, 165–67
    work day, 103–4
hired "teamsters," 73
Historic Sites and Monuments Board of Canada, 231
Hofstadter, Richard, 188, 229
Homestead Act of the United States (1862), 7
Homesteader (Minifie), 50
homesteading
    breaking (land clearing), 60–62, 66, 208, 248n7
    cancellations, 22–23, 25, 29, 32, 41–42, 57, 77, 82, 157
    economic advantage over later purchasers, 96

employment during, 62
gradual process, 62
isolation, 23
land selection, xix, xxiii, 19–20, 22, 25, 30–31
open to most settlers in earliest period, 64
persistence or success, xxiii, xxiv, xxvii, 33–34, 68, 79, 98–99, 144, 156–57
subsistence farming, 63
homesteading costs, 63–64
   Abernethy district, 47, 50, 66
   breaking, 60
   chattels and furnishings, 54
   co-operative activities and, 61
   contracting out, 62
   farm implements, 58, 60, 81, 93, 246n52, 247n53, 247n55, 247n57
   fencing, 56, 246n36
   forms of financing, 50
   horses, 56–57
   livestock, 57–58
   oxen, 56–58
   provisions (groceries), 56
   seed, 58
   shared ownership and, 49
   shelter costs, 50–53
   shelter for livestock, 54
   sleigh runners, 58
   starting capital, 48–50, 64, 68, 247n70
   tools, 54–55
   variables or factors affecting, 47, 64
   wagons, 58
   wells, 56
homesteading costs (table), 55, 65
Hopp, Karl, 40–42
horses, cost of, 57–58
household as centre of social life, 156
houses, 13, 50–53, 82, 173, 175. *See also* shelter
   cost of building, 50–53
   fieldstone, frame and brick residences, xxv, 13, 82, 156, 160, 164, 173, 175, 219
   hierarchichal arrangement of space, xxxi, 156, 160, 163
   Motherwell house, 53, 160–63
   representative prairie dwellings, xxxi, 173
   Victorian layout, 156
Hudson's Bay Company lands, 36
Hudson's Bay Company Territory, xvii, 3–4, 7–10
Hudson's Bay Railway, 193
human interaction. *See* social intercourse
"human wastage" in settlement, 38, 40

I

immigration. *See also* settlement
   anti-immigrant sentiment, 190
   Colonization Companies, 78
   National Policy of, xviii, xxi, xxvii, 3–4, 7, 38, 180, 188, 220–21, 226
   out-migration to US, 3, 9, 23
immigration literature, 54, 56, 75, 78–79
immigration policy, 38, 188
implement companies, 86–87, 217
implements. *See* farm implements
improved land, 9, 23–24, 47, 58–59, 62, 78–80, 82–83, 136, 218, 223
Independent Order of Foresters, 172
Indian Head, 69, 88, 200–201, 218
   cohabiting males, 150
   crop yields, 76, 80, 84
   prosperity, 82
   rail linkage, 68
Indian Head–Balcarres area
   capital accumulation, 94
   farm debt, 94
   owner-operated farms, 136
   part-owner operators, 136
   tenant-operated farms, 136–37
Indian Head Dominion Experimental Farm, 110, 202
Indian Treaty No. 4, 10–13
individual enterprise, xxviii, 50, 169–70, 172–73
   illusion of independent entrepreneurship, 135, 213–16, 221
   myth of independent yeomanry, 226, 228
   myth of the self-made man, xxiii, 172, 226
   self-made men, 139
Innis, Harold, 135
Grain and Grain Inspection acts (Manitoba), 212
international boundary, xxvii, xxviii, 4
Interprovincial Council of Grain Growers, 212
Irving, John A., xviii
   Social Credit Movement in Canada, 237n3
Ismond, W.H., xxv, 219
isolation, 148, 151, 156
   "hired girls," 164
Ituna, 141

J

Jackson, T.W., 15–17
joint stock elevators, 200
Jones, David, xxx, 150
  Empire of Dust, xxiv
Jones, Ernest, 150

K

Katepwa, Lake, 2
Katz, Michael, 134, 140
Kenlis, 150, 157, 181
King Edward Hotel, 163
kinship ties, xxi, xxii, 31
Kirkella branch of the CPR, 30, 73, 84, 217
Knox Presbyterian Church, 141, 144, 175
Kohl, Seena, xxx

L

labour, 32–33, 73, 90, 134, 137, 228
  harvest gangs, 62, 119, 122, 125
  hired girls, 102, 104–6, 159, 161, 164, 166, 224
  hired men, 102–3, 114–25, 128–29, 133, 137–38, 165–67
  initial settlement period, 67–68
  lodgings, 149–50, 161, 163
  pooled labour and sharing of resources, 31, 53, 61, 66, 168
  seasonal, 141
  shortage, 23
LaDow, Beth, xxviii
  Medicine Line, The, xxvii
Lalonde, André, 22
Lambton County, Ont., 196
Lanark County, Ont., 49
Lanark Place, 104, 159, 164, 230–31. *See also* Motherwell house
land, xiv, 44. *See also* Dominion Lands Act and Regulations; soil characteristics
  additional farm lands, xxiii, xxiv, 7, 9–10, 36, 39–40, 85
  free-grant land, xxiv, 38, 40, 91, 97, 180, 208
  improved, 9, 23–24, 47, 58–59, 62, 78–80, 82–83, 136, 218, 223
  land suited to grazing animals, 2, 64, 94
  marginal land, 40, 42, 44, 94
  marshy areas, 5–6, 30–31, 42, 45, 69, 96
  prices, 50, 68, 84–85
  provincial control of public land, 195, 197
  second homestead entries, xxiv, 9–10, 39–40, 85
  wooded areas, 20–21, 30, 35, 52, 69, 94, 96, 208

land selection, xxiii, 22
  Pheasant Plains district, 19
  proximity to friends and relatives, 31
  proximity to railway, 20, 25, 31
  statistical study of, xix
land selection (en bloc) German settlers, 30
land speculation, 23–24, 38–39, 136
language
  assimilationist drive, 184
  compulsory English language, 188
  French language rights, 188–89
  unilingual English language, 189
language barriers, 31, 41–42, 44
Laurie, P.G., 189
Laurier, Wilfrid, 28, 217
  promotion of western settlement, 220–21
Le Roy Ladurie, Emmanuel, *Montaillou*, xix
Leader, 180
leadership, 144
"Leadership Values" (public lecture), 157
LeBoeuf, Jean-Claude, xxxi
Lehr, John, xxxi
  "Mormon Settlements in Southern Alberta," 240n34
Lemberg, 37, 98, 157
Lemberg-Neudorf area
  farm debt, 94
  wooded and marshy areas, 96
Leslie, Geneviève, 164
Levi, Carlo, xx
Lewis, Frank, xxi
Liberal government, 199. *See also* Dominion government
  railway policy, 7, 217, 221
  responsiveness to farmers, 221
  tariff policy, 7, 216
Liberal Party, 189, 198, 201, 262n61
  Motherwell's involvement in, 188, 216
  Protective Union and, 195–96
Lipset, Seymour Martin
  Agrarian Socialism, xxvi
  study of C.C.F., 133
Litt, Robert M., 50, 54, 64, 244n1
Little Black Bear's Reserve, 12
livestock, 31, 56
  cost of, 57–58
  feeding and care of, 103–4, 129
  impounding of, 169
  Métis farming operations, 14
  sharing, 62
  slaughter and butchering, 117
Loewen, Royden, xxvi
  Family, Church, and Market, xx
logs, 51–52

loneliness. *See* isolation
Lord's Day Act, 177–78
Lorlie, 79
Loyal Orange Lodge, 158
luck, xxiv, 172
Ludmerer, Kenneth, 188
Lutz, "Lizzie," 160

M

Macdonald, John A., 3, 7, 16, 192–93
MacKay, Angus, 80, 202
Mackenzie, Alexander, 7
Mackintosh, W.A., 97
Macoun, John, 56
Macpherson, C.B., xviii, xx, xxiv
    Democracy in Alberta, 133, 214, 237n3
"making good" theme, xiii, 173, 227
male/female population imbalance, 147,
    255n1. *See also* bachelors
male homosexuality, 182–83, 240n27,
    259n41
Malin, James C., xix, xxx
Manifest Destiny, 4
Manitoba, 3, 21, 49, 62
    compulsory English language instruction,
        188
    Eugenics movement, 188
Manitoba and North West Farmers'
    Cooperative and Protective Union.
    See Protective Union
Manitoba and North West Farmers' Union,
    192–93
    failure, 215
Manitoba boundaries case, 197
Manitoba Farmers' Mutual Hail Insurance
    Company, 217
"Manitoba for Manitobans," 197
Manitoba Free Press, 183
Manitoba Grain Act and amendments, xxv,
    89, 202–4, 208–9, 212
    farmers' subordinate status, 221
    favouring large-scale farmers, 205–6
    unequal benefits under, 205–7, 221, 226
Manitoba Grain Growers' Association, 211
Maple Creek, 138
marginal land, 40, 44, 94
market-oriented agriculture. *See* commercial
    agriculture
markets, xx, xxii, xxvii. *See also* wheat prices
    farmers' economic position defined by,
        215
    impact on wages, 137
    instability, 67
    local market centres, 84
    for wood and hay, 69

marriage, 52
    heterosexuality within, 181–82
    promotion of, 180
    tax on unmarried men, 180
"Marriage an Aid to Longevity," 180
marshy areas, 5–6, 30–31, 42, 45, 69, 96
Martin, Chester, 38
Masons, 158
material culture, xxiii, xxxi, 172
materialistic values, 173–74
Mavor, James, 63
McCarthy, D'Alton, 199
McClintock, Alexander, 150
McClung, Nellie, 106, 144, 170, 172
    on "pioneer co-operation," 169
    Sowing Seeds in Danny, 164
    Stream Runs Fast, The, 179
McCuaig, D.W., 211
McCutcheon, Brian, 193, 215
McKay, Angus, 110
McKay, William, 43–44
McManus, Sheila, xxviii
    Line Which Separates, The, xxvii
mechanization, 92–93, 228
    tractors, 94, 97, 228
Mennonite settlements, xxvi, xxxi
mentally deficient. *See* Eugenics movement
Methodist Church, 141, 175, 180–81
methodology, xix, xxii, xxviii, 20
    multiple regression model, 20
Métis, 13–19, 45
    closely knit community, 18
    Dominion Lands Act and, 44
    economic hardship, 18
    employment, 18–19
    land claims, 14
    language barriers, 44
    left land after Euro-Canadian settlement,
        14
    loss of land, 14, 17, 45
    marginal position, 14, 19, 45, 225
    Ontario and Qu'Appelle Land Company
        and, 15–17
    participation in agriculture, 14, 19
    petition to Edgar Dewdney, 15
    problems with grid survey, 17–18
    prosecution under Lord's Day Act, 178
    scrip, 14–15
Métis resistances of 1870 and 1885, xvii, 15,
    22, 76, 195–96
MGGA. *See* Manitoba Grain Growers'
    Association
microhistory, xvii, xx, xxv, xxix, xxxii,
    239n24

middle-class feminists
    blaming the victim, 186–87
middle-class lifestyle, 129, 157–58, 163
    attempts to entrench authority, 179
    establishment of acceptable social
        relations and behaviour, 170–71
    hired labour, 101
    "invited parties," 159
Miles, C.F., 5
military scrip, 19
Millar, John, 200, 202, 209, 211, 216, 218, 220–21
Miller, Jim, 12
Mills, Ivor J., *Stout Hearts Stand Tall*, 237n3
Minifie, James M., 117, 125, 127, 168
    Homesteader, 50
mixed farming, 69, 77
money. *See* capital
monoculture, 94, 98
monopolistic practices
    CPR monopoly, 9, 192, 194–95
    elevator companies, 198–99, 203–4
    grain marketing, 194–95
    wheat prices, 73
Moorehouse, Hopkins, 221
    Deep Furrows, 191
Moose Jaw, 154
Moose Jaw district
    crop yields, 77
Moose Mountain, 69
"moral statutes," 171
moral virtue. *See* religious creed and accompanying code of morality
Mormon settlements, xxxi
Morris, Alexander, 11–12
Morrison, James, 219
Morrison, W.R., 186
Morrison family, 145
Morse, Saskatchewan, 57, 157
mortgages, 50, 88–92, 200
    for barns and farm implements, 78, 85, 87
    cash for operating expenses, 85–86
    defaults on, 86, 228
Morton, W.L., 174
Motherwell, Catherine, 165, 187, 230
    on church union, 144
    community involvement, 142
    domestic accounts, 129
    entertaining, 129
    image of leisured matriarch, 131
    Presbyterian missionary (File Hills), 142, 177
    role in Presbyterian Church, xiii
    on role of women, 186
    speaking engagements, 142, 144, 253n103
    surrogate farm manager, 129
    teacher, 142
    Women Grain Growers Association, 144
    Women's Missionary Society of Knox Presbyterian Church, 144
Motherwell, Talmage, 102–3
    marriage to non-Anglo-Canadian, 230
    shift in values, 229
Motherwell, W. R., xv, xxv, 3, 76–77, 80, 169, 200–203, 208–9, 217–18, 221
    agrarian activism and, 28, 205, 211–13 (*See also* Territorial Grain Growers' Association)
    on assimilation, 189
    barn (*See* barns and outbuildings)
    community affairs, 131, 139, 142
    defence of French language instruction, 189
    on dry farming, 109
    elegant house (*See* Motherwell house)
    figure of national historic significance, xiv, 231
    first log house, 5, 52
    interest in new techniques, 81
    land holdings, 85
    land speculation, 136
    Liberal Party position, 188, 216
    marriages, 230 (*See also* Motherwell, Catherine; Motherwell family)
    Moorehouse's description, 191
    not "typical prairie settler," 231
    on overcropping, 228
    political career, xiv, 129, 165
    role in Presbyterian Church, xiii, 105, 177
    scientific agriculture, xiv
    on soil packing, 112
    starting capital, 49
    status of hired men, 165–67
    support of Prohibition, 188
    temperance movement and, 142, 177, 183
    Victorian Ontario mindset, xv, 231
Motherwell family, 145, 228
    Edwardian dress codes, 130
    entertaining, 130, 159
    role as leading middle-class family, 101, 129–30
Motherwell farm
    crop yields, 80
    dry farming, 109
    emulation of Anglo-Canadian model, 175
    heavy clay soil, 113
    hired labour, 102, 104–5, 161 (*See also* hired men)

livestock, care of, 103–4
meals and meal preparation, 104–5
new techniques, 81
tradition and ceremony, 105
work and daily life, 101–31
yearly cultivation cycle, 107–29
Motherwell homestead, 5
Motherwell Homestead National Historic Site, xvii, xxxi
themes, xiv
Motherwell house, 13
family's "chambers," 161, 163
floor plan, 162
hierarchichal arrangement of space, 160
hired girls' room, 161
hired men's room, 161
mouldboard ploughs, 37, 61, 207
Mounted Police, 19, 154
mowers, 81
Murray, A.T., 177
myth of egalitarian co-operative society, xxiv
myth of independent yeomanry, 223, 226, 228
myth of the self-made man, xxiii, 172, 226

N

National Policy, xviii, xxi, xxvii, 3–4, 7, 38, 180, 226, 260n3
promotion of western settlement, 220
and reforms in grain handling and, 220–21
natural environment, xxii, xxvi, xxx, xxxii
Neudorf District, xxiii, 30, 34, 40, 42–43, 96, 140–41. *See also* German settlers
basic problems of subsistence, 139
breaking (or land clearing), 66
commercial agriculture and, 35–36
cultivation (first year), 59–60
farm incomes, 98
fencing, 56
homesteading costs, 47, 51–52, 54
livestock, 62
mixed farming, 69
poor land, 42, 44
size of farms, 36, 208
time of settlement, 35
water, 36
wooded areas, 35
"new rural history," xix, xx
New Toulcha colony, 49
Nichols, John, 74–75
Norquay, John, 195
draft "Bill of Rights," 193

Norrie, Kenneth, xxi
North West Central Railway Company, 25–26. *See also* Great North West Central Railway
North-West Halfbreed Commission, 15
North West Mounted Police, 19, 154
North-West Rebellion, xvii, 15, 22, 76, 195–96
impact on settlement, 22
North-West Territorial Council, 39
Lord's Day Act, 177–78
position on scrip, 14
North-West Territories, 3. *See also* Rupert's Land
demographic imbalance, 147
Patrons of Industry, 197
prostitution, 154
nuclear family, 180

O

occupational prestige, 138, 140
Oddfellows, 158
Okanese Reserve, 12
Ontarian Agricultural College, 49
Ontarian inheritance system, 48, 244n2
Ontarian migration to western Canada, 48. *See also* Anglo-Canadian settlement group
Ontarian political institutions, 188
Ontarian settler theme, xiv, 139
Ontario and Qu'Appelle Land Company
land dispute with Métis settlers, 15–17
Osler, Hammond and Nanton, 81
Ostergren Robert, xix
over-cropping, 93, 98
ownership of land. *See* property ownership
oxen, 56, 207
cost, 57–58

P

Pacific Scandal, 7
packing the soil, 110, 112
Palliser Triangle, xxiv, 1–2, 241n1
cost of building shelter, 52
Palmer, Howard, 142
Parks Canada, 231
Partridge, E.A., xxv, 209, 211–12, 219
on role of government in economy, 213
Partridge, Henry O., xxv, 204
Partridge Plan. *See* publicly owned elevators
pasture lands, 64, 94
patent, 7, 10, 16, 29, 32–34, 36, 39, 41, 43, 66
Application for Patent form, 47–48, 50
paternalism, 166

patriarchy, 180
patriotism, 183, 187
Patrons of Industry, 196, 262n61
   divisions among ranks, 216
   on enfranchisement of women, 198
   Manitoba, 197
   North-West Territories, 197
   political activity, 198
   provincial rights plank, 198
   on tariff barriers, 198
Patton, Harald S., 206
Pearce, William, 24
Peel, Bruce, 127
Peepeekisis's Reserve, 12
pemmican, 11
Percy, Michael, xxi
Perley, W.D., 36, 142
Perry, A.B., 81
perseverence, 172
persistence. *See* success or persistence of farmers
Pheasant Creek, 2, 6, 19
Pheasant Forks, 26, 157, 159, 168
Pheasant Hills, 2, 31, 40, 44, 69
   physical characteristics, 30
Pheasant Plains (around Abernethy), 3, 49
   high-quality clay soils, 2
   prosperous, settled community, 223
   settlement, 19–23 (*See also* Anglo-Canadian settlement group)
   settlers' expectation of rail service, 21
Piapot, Chief, 12, 22
"pickling" of grain, 108–9
pioneer hospitality, 163
pioneering process. *See* homesteading
Plains Cree. *See* Cree
platform loading, 195, 209
ploughing, 110
ploughs, 37, 61, 81, 207
political behaviour (of prairie farmers), 214. *See also* agrarian movement
   ephemeral radicalism, 214
   Macpherson's account of, 214–15
pooled labour and sharing of resources, 31, 53, 61, 66, 168
Popp, Jacob, 42
Port Arthur, 68, 212
Potyondi, Barry, *In Palliser's Triangle*, xxx
Practical Pointers for Farm Hands, 166–67
prairie farmers as failures stereotype, xviii
prairie fires, 67, 76
   fireguards, 114
prairie historiography, xviii
   prevailing notions, xxiii, xxiv

pre-emptions, xxiv, 9, 35
   abolished, 39, 44–45
   cancellation, 40
   difficulty paying, 76–77
   extensions, 39
   revival (1908), 91
predestination, doctrine of, 175
Presbyterian Church, xiii, 141, 144, 175, 180
Presbyterian evangelism, 142, 176–77
price fixing, 70, 72. *See also* monopolistic practices
Primitive Methodist Colonization Company, 26, 51
Prince Albert Settler's Union, 195
Prohibition, 188, 259n46
   equated with patriotism, 183
Prohibition laws, 154
property ownership, xxvii, 137, 197
   conservatism and, 216
   part-owner operators, 136
   social and political power, 135, 140
prosperity, 82–86, 94, 218
   Abernethy settlers, 81–82, 223
   community service and, 139
   expensive farm buildings, 97
   fieldstone, frame and brick residences, 82, 164 (*See also* houses)
   First World War, 92
prostitution, 154
Protective Union, 193–96
   appeal to Queen Victoria, 195
   CPR monopoly and, 194
   failure, 215
   free trade, 195
   on the grain marketing system, 194–95
   North-West Rebellion connection, 195–96
   political orientation, 195
Protestant creed, 179
Protestant ethic, 174
provincial control of public land, 195, 197
Provincial Orange Lodge of Saskatchewan, 190
provincial power to charter railways, 193
"proving-up," 32, 39
"proving-up" costs. *See* homesteading costs
"proving-up" period, 7, 50, 53, 245n14
provisions (groceries), 56
public lectures (speaking engagements), 157, 253n103
publicly owned elevators, 211–12

Q

Qu'Appelle, 76–77, 201
Qu'Appelle district, 22, 201
   cancellations, 23
   co-habiting males, 150
   crop yields, 76–77, 84
   land speculation, 24
   prostitution, 154
   wary of "alien" immigrants, 142
Qu'Appelle Métis. *See* Métis
Qu'Appelle Progress, 79–80, 153, 157, 180, 183, 209
   serialized sermons, 182
Qu'Appelle River, 2
Qu'Appelle River Valley, 2, 126
Qu'Appelle Valley farming region, 68, 80
Qu'Appelle Vidette, 25, 54, 70, 72, 76, 142, 154–55, 159, 180, 189
quarter-section
   as basic homestead unit, 38
quarter-section farm
   unprofitability, 97

R

race, xxviii
Railton, David, 219
Railway Committee of the Commons, 199
railway lands, 7, 198
   Métis settlers on, 15–16
railways, 3, 44, 82, 84
   agrarian unrest and, 193
   Canadian Pacific Railway (*See* CPR)
   conflicts of interest, 28
   construction delays, 22–23, 25–29
   corruption charges, 27–30, 44–45
   government policies, 30, 44–45, 221
   Great North West Central Railway, 25–30, 217
   Hudson's Bay Railway, 193
   impact on settlement, 20–21, 29
   Indian Head, 68
   North West Central Railway Company, 25–26
   petitions for, 216
   provincial power and, 193
   Souris and Rocky Mountain Railway Company, 25
rakes, 81
Ray, Arthur R., *Bounty and Benevolence*, 12
Rea, J.E., 188
red light districts, 154
Red River carts, 58

Regina, 2, 82
   demographic imbalance, 148
   prostitution, 154
   rail linkage, 68
Regina Leader, 36, 49, 68, 217
Regina Standard, 182–83, 209
regional studies, xxviii, xxix, xxxii
religion, 157, 181, 258n15. *See also* Christian church
   Calvinism, xiii, 175–76, 178
   Christian religion and the family, 259n36
   justification of territorial control, 177
religious creed and accompanying code of morality, 174–75, 179
remittance men, 154, 156
research design, 233–36
research methodology, xx, xxi, xxii
respectability, 174–75
Richards, A.E., 220
Richardson, Hugh, 204
Richardson, R.L., 198
Riel, Louis, 15
Roblin, Rodmond P., 201, 211
role of women, xxviii, 184, 258n32
   expansion of, 186
rolling the soil, 110–12
Rolvaag, Ole, 230
Roman Catholic clergy, 15
Rosewood, 19
Royal Commission on Agricultural Credit, 86
Royal Commission on Immigration and Settlement, 190
Royal Commission on public elevators, 212
Royal Commission on the Shipping and Transportation of Grain, 199–200, 220
Royal Templars of Temperance, 157
Rupert's Land, xvii, 3–4, 10
rural historiography (or prairie agricultural historiography), xix
rural modernization, xviii
Rural Municipality of Abernethy, 2, 5
Rutherford, Paul, 175

S

Sabbatarian movement, 177–78
Saltcoats settlers, 77
Saskatchewan, 3, 133, 180
   crop yields, 83
   Eugenics movement, 188
   political divisions, xxiv
Saskatchewan Archives Board, 50
"Saskatchewan Co-operative Elevator Company," 212

Saskatchewan College of Agriculture, 97, 103
  Farm Indebtedness study, 94, 96–97, 137
Saskatchewan Department of Labour, 86, 92
Saskatchewan Grain Growers' Association, 211–12, 219
Saskatchewan Herald, 182, 188
Saskatchewan Homemakers, 142, 144
school lands, 36
schools, 11, 23, 32
schoolteachers, 153, 163
scientific agriculture, xiv
Scott, Walter, 48, 205, 212–13
scrip, 14–15, 225
  military, 19
second-generation farmers, 229
second homestead entries, xxiv, 9–10, 39, 85
  cancellation, 40
seed grain assistance, 58–59, 76
seeders and binders, 81
  shared ownership, 168
seeding, 110, 113, 224, 251n42
  "pickling" of grain, 108–9
  preparation for, 107–9
Seibold, Frederick, 36–37
Senklar, Justice, 220
settlement, xvi. *See also* German settlers; homesteading; Neudorf District
  American expansionism and, 4
  Anglo-Canadian (*See* Anglo-Canadian settlement group)
  bona fide settlers, 7, 14–17, 23–24, 42
  Canadian outflow to US, 3, 9, 23
  Laurier's promotion of, 220–21
  National Policy of, 3, 7
  timing of, xiv, xxiii–xxiv, 20–21, 30, 35, 39, 44–45, 189, 225, 247n70
settlement, second phase (post-1900)
  capital-intensive agriculture, 68
  decrease in farm population, 68
settlement period (initial) (1880-1900), 68–69
  labour intensive, 67–68
settlement prestige, 138–39
settlement process, xxiii, 20
  selective nature of, 172
sexuality, xxix, xxx, xxxii
  campaign against non-procreative forms, 181–82
  heterosexuality within marriage, 181–83
  legislation, 181
  male same-sex sexuality, 182, 240n27, 259n41
  prostitution, 154
  sexual laxity, 155, 179
  sexual licence, 154, 181
  of teenaged boys, 181

SGGA. *See* Saskatchewan Grain Growers' Association
shared living arrangements, 53, 149–50. *See also* cohabiting males
shared ownership, 49, 61–62, 64
Shaw, Elmer, xxv, 218–19
shelter. *See also* barns and outbuildings; houses
  cost of building, 50–53
  dugout dwellings, 51
  materials, 51–52
  Métis dwellings and outbuildings, 14
  for seasonal or transient staff, 163
Sheppard, Walter F., 150, 213
Sibbold, John, 200, 216
Sifton, Clifford, 199–200, 204
silver tea service, 159, 175, 224
single-crop economy, 94, 98. *See also* wheat
Sintaluta, 69, 81, 168, 201, 203–4, 211
  stone houses, 160
Sintulata (CPR line), 129
slaughter and butchering of livestock, 117, 119
sleigh runners, 58
Smith, Allan, 172–73
Smith, H.H., 43–44
Snow, Matthew, 202
social amenities, 23
social class, xviii
  definition, 134
  historical category, 226
"social credit," 138, 142
Social Credit in Canada series, xviii, xix
social creed, 171, 173–75, 179–80, 182
  individual enterprise, 172–73
  shift in values, 229
social gospel movement, 176
social intercourse
  bachelors' balls, 153
  "Box Social," 153
  dances, 153–54
  Friday evening at the post office, 152
  funerals, 152–53
  informal, 150–51
  institutionalizing and regulating social behaviour, 158
  "invited parties," 164
  loss in, 164
  new middle-class approach to, 158, 160, 164
  pioneer hospitality, 163
  public lectures, 157, 253n103
  spontaneous, 158
  surprise parties, 151

social mobility, 134, 174, 226
"social purity" movement, 180, 182
social relationships, xxi, xxii
    early settlement era, 147–56
    exclusiveness, 158
    formalized between farmers and employees, 164
    mature society (1900–1920), 156–67
social roles and responsibilities, 101, 157
    leading middle-class families, 129–30
social status, 138, 158, 160, 179, 223–24, 231
    community prestige and, 138–40, 142, 144
    leisure and, 129–30
    low, 138, 142, 164–65, 184, 224
    occupational, 140
    privileged, xxiv
    "social credit," 142
    "status revolution," 188
    status symbols, 159, 175, 224
    wealth or prosperity, 173
social structures, xxiii, 133–34, 136, 144, 226
    stratification, 133–34, 147, 158, 167, 223–24
    two-class model, 134
sod, 51–52
soil characteristics, 2, 6, 20, 98–99, 208, 218
    Abernethy district, 1–2, 111, 113
    farm debt and, 97
    marshy areas, 5–6, 30–31, 42, 45, 69, 96
    Motherwell farm, 113
    pasture rather than crop, 69
    profitability of clay soils around Indian Head and Balcarres, 96
soil exhaustion, 93, 98, 228
Souris and Rocky Mountain Railway Company, 25
Souris and Rocky Mountain stock, 28
Spafford, D.S., 211
Spalding, Abe (fictional character), 139, 157, 164, 230. *See also* cohabiting males
Spector, David, "Agriculture on the Prairies," xiv
Spring Bank, 82
squatters or pre-entrants, 20
stables. *See* barns and outbuildings
Star Blanket, Chief, 22
Star Blanket's Reserve, 12
starting capital. *See under* homesteading costs
Stavenhagen, Rodolfo, 226
stem rust, 93
Stern, Mark, 134
Stevenson, Rufus, 24

Stewart, Charles, 195
Stilborne, Edith, 159
stock raising. *See* livestock
Stonechild, Nina, 164–65
stooking, 114–15
stoves, 54
Street, A.G., 115
"street" wheat, 89, 206, 208
Stueck, Englehart, 144
subsistence farming, 63, 98
    barter, 69
success or persistence of farmers, xxiii, xxiv, xxvii, 33, 68, 79, 82, 144, 156–57, 172
    accessible rail service and, 20, 29
    adjustment to dry farming, 80
    Anglo-Canadian settlers (or Ontarians), 34
    Calvinist interpretation of, xiii
    capital gain on the land, 98–99
    diversification, 80
    German settlers, 34
    soil quality and, 99
    timing of settlement and, xiv
summer fallowing, 80, 83, 110
    effect on crop yields, 80
    purpose, 113
summer frosts. *See* early frost
Sun Dance, 13
Supreme Court of the North-West Territories, 204
swearing, 179
Swierenga, Robert, xix
Sylvester, Ken, xxvi
    Limits of Rural Capitalism, The, xx

T

taboos, 179
Talmage, Thomas De Witt, 178
    Night Sides of City Life, The, 182
tariffs, 3, 7, 192, 196, 198, 216
    on American machinery, 81
"teaming" supplies to government troops, 76
technology. *See also* mechanization
    daily life and, 101
Teece, John, xiii, 61, 76, 168–69, 217
    "made good," 173
temperance, 142, 177, 183, 225
tenant-operated farms, 136–37
Territorial Grain Growers' Association, xiv, xv, 89, 139, 226
    changes to Manitoba Grain Act, 202
    complaint before Warehouse Commissioner, 204

divisions in, 214
Dominion government's response to, 220
formation, 88, 202
hierarchical assumptions, xxv
minute book, xxv, 219
non-partisan leadership, 201, 216
privileged elite leadership, 219
significance of, 191–92
success of, 216
upper stratum of farming population, xxv, 218–19
in western Canadian folklore, 191
TGGA. *See* Territorial Grain Growers' Association
Thompson, John Herd, xxii, 48
threshing, 119
    accidents, 125
    grain separators, 122
    winter months, 123–24
threshing gangs, 62, 122
threshing machines, 119, 122
    shared ownership, 169
timber, 6. *See also* wood; wooded areas
timing of settlement, xiv, xxiii–xxiv, 20, 44–45, 225, 247n70. *See also* free-grant land
    Abernethy district, 21, 39
    early, 68–69
    effect on cost, 64
    effect on status, 189
    Neudorf, 30, 35
Tobias, J.L., 12
tools, 54–55
Touchwood, 69
Touchwood-Qu'Appelle Land and Colonization Company, 24
Tough, Frank, 12
Tracie, Carl, xxi, xxxi
    "Toil and Peaceful Life," 240n34
"track" wheat prices, 73, 195, 206
tractors, 94, 97, 113, 228
transportation, 68, 196. *See also* railways
    freight rates, 73
    grain hauling, 73–74, 126–29
    hauling brome, 81
    monopoly, 226
    "teaming" supplies, 76
Treaty No. 4 (Indian Treaty No. 4), 11–12
Troeltsch, Ernst, 175
typical prairie dwellings, xxxi, 173
"typical prairie settler," 231

U

Ukrainian Orthodox or Russian-German Lutheran seasonal labourers, 141
Ukrainian settlements, xxxi, 49
United States
    competition for settlers, 9
    Eugenics movement, 188
    land disposal policies, 9
    Manifest Destiny, 4
    Patrons of Industry, 196
unmarried male settlers. *See* bachelors
upward mobility, 134, 174, 226

V

Veblen, Thorstein, 130
    Theory of the Leisure Class, 129
Victorian Canadian identity, 172
Victorian houses, 156, 164, 175
    status symbols, 224
Victorian middle-class order, 157
Victorian Ontario mindset, 231
Victorian prudery, 179
Victorian sense of mission, 183, 249n43
Vogt, Evon, 218
Voisey, Paul, xxvi
    Vulcan, xx

W

wage labourers. *See* farm labourers
wages, 48–49, 137, 244n3
    for experienced farm labour, 64
    farm labourers, 64, 92
    "hired girls," 164
    non-payment, 254n19
wagons, 58, 81, 246n48
Walsh, James, 16–17
Warkentin, John, xxxi
    "Mennonite Settlements of Southern Manitoba, The," 240n34
water, 20, 31, 36, 42, 47, 56
    access to, 20, 36, 47
    cost of wells, 56–57
    homesteading costs and, 64
    lack of, 57, 78
Webb, Walter Prescott, xxx
Weber, Max, 138
Welsh, Norbert, 18
wheat. *See also* crop failures
    excessive reliance on, 91
    monoculture, 94, 98
    over-cropping, 93
    principal cash crop in Abernethy area, 69–70
    risky proposition, 99

wheat acreage, 94, 96
wheat blockade, 88, 200–202, 215, 216n16, 217
wheat prices, 66, 195, 248n4
    First World War, 92–99
    Fort William price, 82
    monopolistic practices and, 73
wheat prices (1880–1900), 67–75
    farm-gate price, 70
    inflated wartime prices, 93–94
    new European markets (1896), 81
    price fixing, 70, 72
wheat prices (1901–1910), 82–85
wheat production costs, 92, 249n52
wheat yields, 76, 83, 261n40. *See also* crop yields
    bumper year (1901), 73, 84, 88, 200, 218
    decline (1916), 93
    new European markets (1896), 81
whisky smugglers, 154
Whyte, William, 203
Widdis, Randy, xxvi
Wide Awake, 82, 201
Wilde, Oscar, 183
Winnipeg, 135
Winnipeg businessmen, 194
Winnipeg Grain Exchange, 211
winter activities
    building and repairing granaries, 129
    cutting and hauling wood, 62, 129
    fencing, 62, 129
    grain hauling, 126–29
    loneliness, 148
    yearly cultivation cycle, 123–24
winter surprise parties, 150
winter threshing, 123–24
Wolseley, 36, 69, 201–2
    crop yield, 84
women, 106
    community service, 186
    dependence on husbands, 129
    dual ideology of, 184
    farm labour, 32
    female suffrage, 144, 186, 198
    feminism, 186
    free grant lands and, 180
    "hired girls," 102, 104–6, 159, 164, 166, 224
    ineligible for free-grant lands, 91
    as inspirational force of the home, 184
    middle-class lifestyle, 129, 157
    role of women, xxviii, 145, 158, 184–86, 258n32
    teachers, 153, 163

Women Grain Growers Association, 144, 145, 186, 188
Women's Christian Temperance Union, 145, 157, 184
women's groups
    blaming the victim, 186–87
    maternalistic mission, 187
women's missionary societies, 145
Women's Missionary Society of Knox Presbyterian Church, 144
wood, 69
    access to, 20, 36, 47
    cutting and hauling, 62, 114, 129
    gathering, 63
    sale of, 69
Wood, Lewis Aubrey, 204
wooded areas, 20–21, 30, 35, 52, 69, 94, 96, 208
Woodsworth, D.W., 28
work and daily life, 101–31
work day, 106
    grain hauling, 126–29
    harvesting, 115
    "hired girls," 104–6
work ethic, xii, 173–74
work for work's sake, 174
working classes, 197. *See also* labour
world views, xxiii, xxix, 172, 188
World War I. *See* First World War
Wright, Archie, 125

Y

yearly cultivation cycle
    fall cultivation of stubble fields, 117
    grain hauling, 74, 126–29
    harrowing, 112–13
    haying, 114
    packing the soil, 110, 112
    ploughing, 112–13
    preparation for seeding, 107–10
    rolling the soil, 110–12
    seeding, 110, 113
    summer fallow, 113
    threshing, 119, 122–25
    winter months, 123–24, 126–29
Yule, Annie I.
    Grit and Growth, 237n2

www.ingramcontent.com/pod-product-compliance
Lightning Source LLC
Chambersburg PA
CBHW052057300426
44117CB00013B/2161